Smart Nanomaterials Technology

Series Editors

Azamal Husen, Department of Biotechnology, Smt. S. S. Patel Nootan Science and Commerce College, Sankalchand Patel University, Gujarat, India

Mohammad Jawaid, Department of Chemical and Petroleum Engineering, United Arab Emirates University, Al Ain, United Arab Emirates

Nanotechnology is a rapidly growing scientific field and has attracted a great interest over the last few years because of its abundant applications in different fields like biology, physics and chemistry. This science deals with the production of minute particles called nanomaterials having dimensions between 1 and 100 nm which may serve as building blocks for various physical and biological systems. On the other hand, there is the class of smart materials where the material that can stimuli by external factors and results a new kind of functional properties. The combination of these two classes forms a new class of smart nanomaterials, which produces unique functional material properties and a great opportunity to larger span of application. Smart nanomaterials have been employed by researchers to use it effectively in agricultural production, soil improvement, disease management, energy and environment, medical science, pharmaceuticals, engineering, food, animal husbandry and forestry sectors.

This book series in Smart Nanomaterials Technology aims to comprehensively cover topics in the fabrication, synthesis and application of these materials for applications in the following fields:

- Energy Systems—Renewable energy, energy storage (supercapacitors and electrochemical cells), hydrogen storage, photocatalytic water splitting for hydrogen production
- Biomedical—controlled release of drugs, treatment of various diseases, biosensors,
- Agricultural—agricultural production, soil improvement, disease management, animal feed, egg, milk and meat production/processing,
- Forestry—wood preservation, protection, disease management
- Environment—wastewater treatment, separation of hazardous contaminants from wastewater, indoor air filters.

Azamal Husen
Editor

Plant Response to Gold Nanoparticles

Plant Growth, Development, Production, and Protection

Springer

Editor
Azamal Husen
Wolaita Sodo University
Wolaita, Ethiopia

ISSN 3004-8273 ISSN 3004-8281 (electronic)
Smart Nanomaterials Technology
ISBN 978-981-97-7173-8 ISBN 978-981-97-7174-5 (eBook)
https://doi.org/10.1007/978-981-97-7174-5

This Springer imprint is published by the registered company Springer Nature Singapore Pte Ltd.
The registered company address is: 152 Beach Road, #21-01/04 Gateway East, Singapore 189721, Singapore

Preface

Desired gold nanoparticles (Au-NPs) have been fabricated using chemical, physical, and biological methods. Fabricated Au-NP shape and size of NPs depend on the concentration of the precursor and the reducing agents regardless of the in vitro or in vivo process of synthesis. These fabricated nanoparticles are utilized in almost all discipline of science and technology. In this book, overall plant and soil system are examined exclusively in terms of Au-NP exposure. Au-NPs influence plant growth and yield performance. They also affect plant and soil system interactions. These particles have shown both beneficial and harmful effects on plant and soil system. Au-NP responses primarily depend on the concentration, plant species or cultivars, exposure time, shape and size of NPs. Investigation has shown that the lower concentrations of Au-NPs increase seed germination, beneficial biochemical components, rate of photosynthesis, and overall growth, however, at higher concentrations, all these responses are declined in different plant species. At higher concentration, Au-NPs induce stress in plant, create phytotoxicity, and produce reactive oxygen species, which leads to the disruption of cellular metabolism. Further, at higher concentration, Au-NPs hampered the physico-chemical process of plant and soil system. Moreover, the molecular and physiological performance showed that the exposure of Au-NPs resulted in oxidative stress and induced antioxidative and defence responses. However, these responses may differ from species to species depending on the concentration and exposure time of Au-NPs. In a few studies, the energy production related to metabolic pathways was also influenced by Au-NP exposure.

Au-NP exposure accelerated the number of chromosomal aberrations, micronuclei, and decreased the mitotic index in plant root tip cells. Proteomic study has shown that the exposure of Au-NPs resulted in an accumulation of protein precursors, indicative of the dissipation of a proton motive force. Au ions were noticed in roots and shoots, while Au-NPs were absorbed only in the plant root tissues. Upregulation of genes involved in the plant stress response like glutathione transferases, cytochromes P450, glucosyl transferases, and peroxidases have been observed. Likewise, downregulation of gene-encoding proteins involved in the transport of copper, cadmium, iron, and nickel ions and aquaporins bonded to Au have been found. Au disrupts protein structure and may also displace some essential metal nutrients from proteins. Au-NP exposure also influences the flowering process and seed germination. Beside the terrestrial plants, these particles have also influenced the growth of some wetlands and aquatic plants. It was also noticed that the addition of Au-NPs into the culture medium had also influenced the in vitro development and multiplication of plants. Studies have begun to examine how arbuscular mycorrhizal fungi might affect the uptake of specific nanoparticles including Au and how these particles might influence the plant-mycorrhizal relationships. The impact of Au-NPs on the size and structure of microbial communities in soil; and soil enzymatic activities for β-glucosidase, urease, alkaline phosphatase, and dehydrogenase has been examined. In the present book, environmental safety and ethical issues related to the use of Au-NPs in plant and soil system have been also explored.

Overall, this book provides an up-to-date, most important and selected information associated with the plant response to Au-NPs. This book will be a good resource for upper-level students, faculty, scientists, and researchers working especially on plant biology, plant nanobiotechnology, plant biochemistry, plant microbiology, agriculture and other allied subjects and or science. I am grateful to all contributors for readily accepting my invitation, sharing their knowledge in specialized areas of research, and readily adjusting the suggestions for improving the shape of their contributions. Finally, my special thanks go to Shagufta, Zaara, Mehwish, and Huzaifa for providing their time and overall extended support to put everything together.

Wolaita, Ethiopia Azamal Husen

Contents

About the Editor

Azamal Husen is served as Full Professor and Head of the Department of Biology, University of Gondar, Ethiopia; and currently working as a Foreign Delegate at Wolaita Sodo University, Wolaita, Ethiopia. He is also working as an Adjunct Professor at Graphic Era (Deemed to be University), Dehradun, Uttarakhand, India. He also worked as Full Professor at Sankalchand Patel University, Visnagar, India; and Visiting Faculty of the Forest Research Institute and the Doon College of Agriculture and Forest at Dehra Dun, India. His research and teaching experience of 25 years encompasses Nanobiotechnology, Plant Physiology, Environmental Botany, Medicinal Plants; and Clonal Propagation. He has conducted research sponsored by the World Bank, the National Agricultural Technology Project, the Indian Council of Agriculture Research, the Indian Council of Forest Research Education, and the Japan Bank for International Cooperation. Husen has published extensively (over 300, which includes research papers, books, book chapters etc.) and served on the Editorial Board and as reviewer of reputed journals published by Elsevier, Frontiers Media, Taylor & Francis, Springer Nature, RSC, Oxford University Press, Sciendo, the Royal Society, CSIRO, PLOS, MDPI, John Wiley & Sons, and UPM Journals. He is on the advisory board of Cambridge Scholars Publishing, UK. He is a fellow of the Plantae group of the American Society of Plant Biologists, and a member of the International Society of Root Research, Asian Council of Science Editors, and International Natural Product Sciences. He has given numerous talks and presentations in many conferences and workshops. Husen has also given various orientation courses for academic staff, prepared different modules, developed BSc, MSc and PhD curriculum. Moreover, he has received a recognition award from University of Gondar, Ethiopia for a decade of service in teaching, research and community services. He is Editor-in-Chief of the *American Journal of Plant Physiology*, and a Series Editor of *Exploring Medicinal Plants* (Taylor & Francis Group, USA); *Plant Biology, Sustainability, and Climate Change* (Elsevier, USA); and *Smart Nanomaterials Technology* (Springer Nature, Singapore). He has achieved the distinguished honour of being recognized as one of the "World's Top 2% Scientists" for the years 2022, 2023, and 2024 by Stanford University, USA. This recognition has also been prominently featured in the Elsevier Data Repository. Moreover, he has also worked as a special issue guest editor on various topics such as Advances in Plant Metabolomics: Insights, Applications, and Integration (*Plant Physiology and Biochemistry*), Exploring Nanotechnology in Medicinal Plants (*Journal of Herbal Medicine*), Integrating Nanotechnology with Plant Compounds: Enhancing Abiotic Stress Resilience and Crop Improvement via Signaling Molecules (*Frontiers in Plant Science*), etc.

Plant Response to Gold Nanoparticles in Terms of Growth, Development, Production, and Protection: An Overview

Satya, Tahmeena Khan, Kulsum Hashmi, Saman Raza, Sakshi Gupta, and Seema Joshi

Abstract The specialized chemical and physical characteristics of nanoparticles make them quite different from bulk materials and therefore useful in nearly every aspect of life. A great deal of research has also gone into how they interact with living systems, as well as how they affect their physiology, morphology, and biochemistry. Nanoparticles (NPs) are widely used in agriculture to improve seed germination, encourage plant growth, and shield crops from biotic stress. Studies have shown that a variety of NPs from the environment can enter plants, accumulate there, and subsequently go up the food chain. Silver, gold, copper, titanium, iron, and zinc-based metallic nanoparticles are frequently utilized in agriculture to promote plant growth and yield. The potential of metallic NPs in protecting plants, promoting development, detecting diseases, and identifying pesticide and herbicide residues has been highlighted by recent studies. Different NPs bring about different changes in the plants. Gold nanoparticles (AuNPs) have gained immense popularity and are the most researched NPs due to their wide range of commercial applications, ease of synthesis, unique optical properties, chemical stability, and non-toxicity. The use of AuNPs in agriculture, specifically for plant growth and development, has been elucidated in this chapter. The interactions between AuNPs and the biota have also been described in detail in this chapter.

Keywords AuNPs · Gene regulation · Mechanistic insight · Biotic stress · Photosynthetic efficiency · Translocation

---

Satya · K. Hashmi · S. Raza · S. Gupta
Department of Chemistry, Isabella Thoburn College, Lucknow, India

T. Khan (✉) · S. Joshi
Department of Chemistry, Integral University, Lucknow, India
e-mail: tahminakhan30@yahoo.com

A. Husen (ed.), *Plant Response to Gold Nanoparticles*, Smart Nanomaterials Technology, https://doi.org/10.1007/978-981-97-7174-5_1

1 Introduction

Nanoparticles range in size from 1 to 100 nm [60, 158]. In nature, NPs can occur as a consequence of extensive forest fires, volcanic eruptions, or the weathering of rocks [18]. Nanotechnology was developed as a result of investigations of engineered nanoparticles, and it involves producing, utilizing, and studying NPs as well as modifying their shape, size, and structure [71, 143, 183]. Engineered NPs are widely used in almost every sphere of life. Various NPs are being used in the development of sensor technologies, optical devices, catalysis, bactericides, cosmetics, electronics, biological labelling, pharmaceutics, biomedicine, and many other aspects[16, 28, 62-70, 72, 73, 82, 107, 113, 140, 178]. Additionally, NPs have been utilized in agriculture for improved seed germination, plant growth, and to protect crops from biotic stress [67, 91].

As compared to other NPs, metallic NPs exhibit a large surface area, high surface-to-volume ratio and unique optical, electrical, chemical, physical and photoelectrochemical properties; besides these their synthesis and modification are also feasible [12]. The production and marketing of metallic NPs have significantly increased during the past several years. Metallic NPs are extensively used in medical imaging, biosensor diagnostics, therapeutics, pharmaceuticals, optoelectronics, chemical catalysis, drugs, and various other fields. Hence, as a result of their extensive production and applications, metallic NPs are inevitably released into the environment and may act as an exogenous stimulator for biological objects [49, 135, 179].

Research has demonstrated that a wide range of NPs can penetrate and accumulate in plants from the environment and then be transported through food chains [166, 192] (Fig. 1). The effects (positive or negative) of NPs on plants depend on their size, shape, concentration, coatings, self-aggregate tendency, and stability [157]. Plants have a higher probability of exposure to NPs present in their immediate environment because of their enormous size, high area of leaves, and stationary nature [37]. The main path through which plants can be exposed to NPs has been considered to be through roots, which may store a large portion of accumulated NPs that may cause physical or chemical toxic effects in plants [44]. Since plants act as the fundamental trophic level in the ecosystem and function as the basis of the food chain for many animals and also for humans, therefore, the interaction of plants with NPs is significantly important [106].

Several researchers have reported the toxic effects of metallic NPs on plants and their surroundings [154, 181, 211]. Among the metallic NPs, AuNPs are the most frequently synthesized, studied, and utilized for various applications. The extensive use of AuNPs has received considerable attention including interest in their synthesis by physical, chemical and biological methods [95]. Scientists have reported that AuNPs have a growth-promoting effect on plants. Despite some contradicting results, AuNPs have been observed to have adverse effects on plants at higher concentrations (more than 100 mg/L) [168], and at lower concentrations they can improve the growth and production of plants [109]. After treating plants with AuNPs, it has been

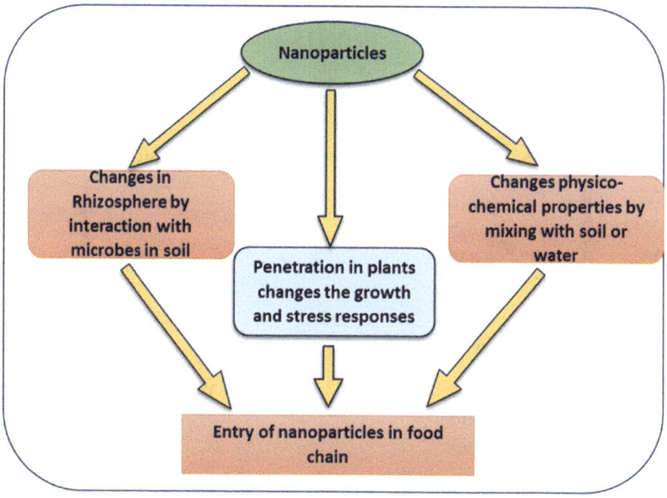

Fig. 1 Interaction of nanoparticles with plants

found that they can affect the reactive oxygen species (ROS) formation and immune responses [75, 114]. Hence, in plants, improved growth, developed primary and lateral roots, larger rosette diameters, and decreased oxidative stress responses have been observed as a result of the exposure of moderate levels of AuNPs. According to studies, exposure to AuNPs enhanced the antioxidant enzyme activities and free radical scavenging potential of plants as well as changed the expression of miRNAs that control several physiological, morphological, and metabolic processes [166].

1.1 Nanoparticles for Plant Growth and Development

As sessile organisms, plants have an incredible capacity to establish mechanisms that enable them to survive and grow in both favourable and unfavourable conditions. Scientists have been working for years to explore novel strategies that could benefit plants and improve their natural processes. NPs can increase plant metabolism [50, 159]. The use of NPs in agriculture is currently being explored because it helps to reduce the use of chemical fertilizers and significantly increase the growth of plants [84]. The role of NPs is well known for enhancing the strength of plants to resist different diseases and abiotic/biotic stresses [31, 195]. Abiotic and biotic stresses are represented in Fig. 2.

Metallic NPs have been shown to increase the proline level, nutrients and water consumption, accumulation of amino acids and the activity of various antioxidant enzymes like nitrate reductase, which improves the plant resistance [43]. NPs also influence the behaviour of genes related to stress tolerance [54]. The up-regulated genes become more active in dealing with metal exposure and oxidative stress,

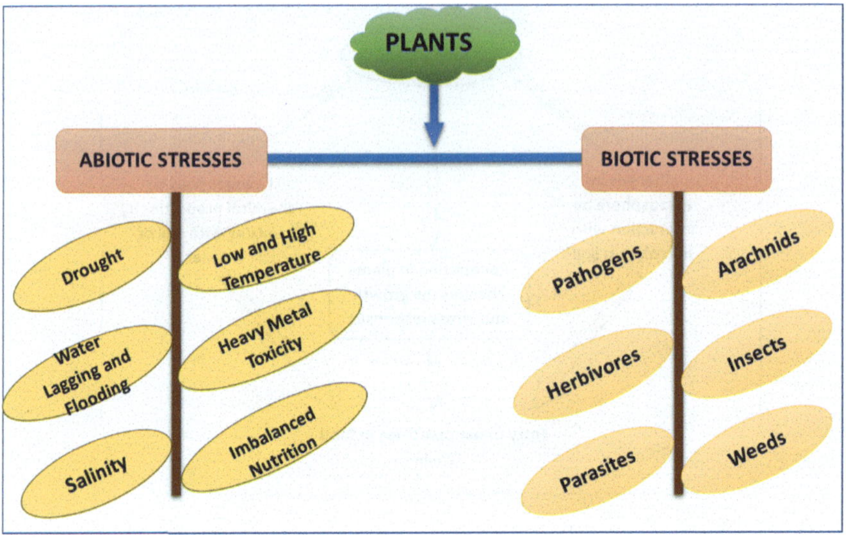

Fig. 2 Different types of abiotic and biotic stresses in plants

whereas, the down-regulated genes, like those involved in ethylene signalling, systemic acquired tolerance, and the regulation of growth and organ size through auxin, are associated with responses to pathogens and hormonal signalling [212]. For achieving the highest quality crop yield, it's widely recognized that the use of the recommended nutrient quantities should be maximum [38]. Nanofertilizers or nutrients encapsulated at the nano-scale may possess properties that prove highly beneficial to plants. They can improve the nutrient absorption, release nutrients as needed, and provide sustained release of chemical fertilizers for plant growth, thereby enhancing targeted activity [151]. NPs have been observed to influence various biochemical aspects linked to plant growth and development. They have been found to enhance nitrogen utilization efficiency and elevate the photosynthetic rate in significant crops like spinach, peanuts, and soybeans [20, 81, 100, 128, 141]. The interaction between NPs and plants leads to a range of physiological, morphological, biochemical, and genotoxic alterations in plant organisms [125]. NPs modify certain physiological processes within plants [191]. At the optimal concentration of NPs, changes in various physiological processes are induced, leading to enhanced seed germination and increased photosynthesis. Moreover, NPs influence the expression of specific genes and miRNAs, which positively affects stress tolerance and overall plant biomass (Fig. 3).

The effects induced by the NPs depend upon their characteristics, application methods, and the specific plant species involved [125]. The chemical composition and the concentration of NPs play a crucial role in determining their effects [10]. Scientists have employed diverse methods for applying NPs including foliar spray, soil application, and seed treatment, to assess their impact on seed germination and plant growth. Various studies have presented evidence that certain types of NPs can

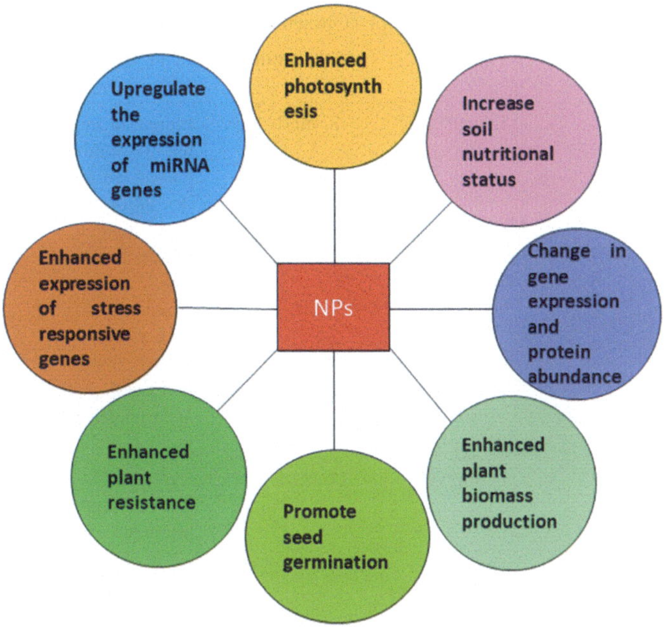

Fig. 3 Some processes through which NPs enhance plant growth and development

enhance seed germination [57] and promote plant growth and development when used at small concentrations [125, 130]. These NPs have been explored for their potential in mitigating crop diseases, leading to improved seed germination, vigour, growth, and increased yield. For instance, Magnesium nanoparticles (MgNPs) have been shown to influence seed germination in maize plants [80]. Aluminium nanoparticles (AlNPs) stimulate the growth of roots in radish and rape, although they showed no significant effect on cucumber [104]. Cerium dioxide nanoparticles (CeO_2 NPs) were found to enhance seed germination in Scots pine (*Pinus sylvestris*) [139]. The influence of CeO_2 NPs on germination of seeds of pine seedlings (*Pinus sylvestris L.*) has been studied, and they positively impacted the height and biomass of tomatoes [199]. Additionally, CeO_2 NPs have been reported to promote shoot biomass, plant growth, and grain yield in wheat. However, while Ce accumulation was observed in roots, there was no evidence of its transport into shoots [149]. Some recently developed NPs for plant growth and development are depicted in Table 1.

2 Deposition and Entry of Metallic NPs in Plants

Several studies have described the potential impacts of metallic NPs on plants [87, 126, 204]. The entry of NPs into the plant body occurs through various pathways [180]. Plants engage with atmospheric NPs primarily through their leaves, and leaf

Table 1 Some recently developed NPs for plant growth and development

NPs	Size	Plant	Concentration	Plant response to NPs	Reference
AgNPs	–	Fenugreek	40 mg/L	Increased shoot length, number of leaves, weight of dry shoot, photosynthetic pigments and indole acetic acid contents, antioxidant activity thus increased quality and quantity of yielded seeds	[156]
	–	Fenugreek	60 mg/L	Decreased photosynthetic pigments	[156]
ZnO NPs	25 nm	Drosophila Melanogaster	10–100 ppm	Increased inhibition of fungal pathogens *Alternaria solani*, Fusarium spp., and *Corynespora cassiicola*	[189]
	20–30 nm	Wheat (*Triticum aestivum* L.)	50 and 100 mg/L	Increased plant growth and chlorophyll contents; reduced oxidative stress and Cd-toxicity by 87% in normal and 81% in water deficiency	[3]
	>100 nm	Tomato (*Solanum lycopersicm* L. 'Linda')	40 mg/L	Increased plant growth and seedlings, decreased salinity stress	[61]

(continued)

Table 1 (continued)

NPs	Size	Plant	Concentration	Plant response to NPs	Reference
CuO NPs	94 nm	Eggplant (Solanum melongena)	1000 μg/ml	Antifungal activity against Fusarium oxysporum fungus, enhances plant growth, chlorophyll contents, plant biomass, and metabolic activity	[1]
	< 50 nm	Prunus domestica fruits	100 μg/mL	Reduced manifestations of grey mould symptoms	[111]
	50 nm	Tomato plants	250 mg/L	Reduced salinity stress, increased antioxidant activity, enzymatic activities,	[138]
SiO_2 NPs	–	*Oryza sativa*	20 and 30 mg/L	Improved plant growth, antioxidant activity and decreased Cd toxicity	[152]
	4–10 nm	*Oryza sativa*	5 mM	Enhanced grain weight and biomass	[105]
TiO_2 NPs	20 nm	*Oryza sativa*	25–750 mg/L	Enhance lateral root length and biomass of the plant	[209]
	6–7 nm	*Zea mays*	100 mg/L	Enhanced enzymatic activity, increased alanine, serine, glycine, threonine, glutamate and aspartate synthesis. Increased metabolism	[103]

(continued)

Table 1 (continued)

NPs	Size	Plant	Concentration	Plant response to NPs	Reference
	–	*Oryza sativa*	5–30 mg/L	Reduced Cd absorption, increased transpiration rate, elevated levels of chlorophyll contents and carotenoids, decreased electrolyte leakage, and reduced oxidative stress	[152]
Fe-0 (zero-valent) NPs	54 nm	*Arabidopsis thaliana*	500 mg/kg	Increased biomass, rate of photosynthesis and phosphorus level in rhizosphere	[206]
Fe_2O_3 NPs	40 nm	*Melissa officinalis*	5–40 μM	Increased chlorophyll content, decreased H_2O_2, MDA, and proline level	[120]
	20–40 nm	*Triticum aestivum cv* Cumhuriyet-75	500 mg/L	Enhanced plant height, root length, biomass, and chlorophyll contents	[7]
Al_2O_3 NPs	8–13 nm	*Lactuca sativa*	0.4 mg/mL	Increased biomass and level of phosphorus	[58]
	20 nm	*Lemna minor*	1–10,000 mg/L	Increased biomass, and root length	[86]

(continued)

Table 1 (continued)

NPs	Size	Plant	Concentration	Plant response to NPs	Reference
	30–60 nm	*Glycine max*	50 ppm	Increased plant growth and metabolism under flooding stress	[123]
CeO_2	10 nm	*Glycine max*	100 mg/kg	Enhanced plant growth	[25]
SWCNT	8 nm	Onion and Cucumber	325 and 1750 mg/L	Enhanced root length and plant growth	[24]
MWCNT	–	*Lycopersicon esculentum*	49 μg m/L	Increased nutrient uptake	[184]

stomata and hydathodes play a pivotal part in the foliar uptake of NPs. They traverse cell walls of palisade parenchyma, and leaf phloem, subsequently translocating to other parts of the plant. Other routes are aerial parts including cuticles, bark surfaces, and stigma. Epicuticular structures further facilitate NP deposition in epicuticular cavities, while trichomes on shoot surfaces enhance the likelihood of NP deposition. They may enter as NPs or metallic NPs undergo oxidation in the soil solution, converting them as ions, followed by subsequent reduction [150].

2.1 Deposition of Metallic NPs on Plant Surface

The deposition of metallic NPs on plants is a complex process, influenced by a multitude of factors. These include characteristics of the NPs in addition to factors like application routes, interactions with environmental components, and the unique physiology of each plant species.

2.1.1 Foliar Application

Metallic NPs can be applied directly to the leaves of plants as a spray or solution. This method allows for rapid absorption through stomata or directly through the leaf cuticle [10, 59]. The cuticle serves as the primary barrier on leaves, allowing only NPs smaller than 5 nm to enter, whereas NPs larger than 10 nm pass through stomata [154]. The transfer of NPs ranging between 10 and 50 nm is facilitated through the cytoplasm of adjacent cells. On the other hand, larger NPs between 50 and 200 nm are translocated between cells.

The efficacy of NP adsorption after foliar application is contingent on crucial factors. These include the method of application, NP size, concentration, and prevailing climate conditions. Trichomes, leaf exudates, and waxes influence the binding and retention of NPs on the leaf's surface [99]. These combined factors collectively determine the extent to which NPs are effectively absorbed by plants through foliar application. The foliar treatment of AgNPs on fenugreek plants led to an increase in total carbohydrate and protein content compared to control plants [156]. There are two established routes for the uptake of metallic NP solutions through the cuticle layers: a lipophilic pathway for non-polar solutes and an aqueous pore pathway for polar solutes. The stomatal pathway has been confirmed as a route for the foliar spray of NPs, allowing them to move from leaf surfaces to other tissues within plants [41]. In a study involving watermelon plants, four metal oxide NPs (sized between 24 and 47 nm) were sprayed on the leaves. The research found that smaller-sized NPs entered the leaves through the stomata and were detected in both the shoots and roots [196]. This underscores the potential for foliar application to influence the distribution and accumulation of nanoparticles within plants.

2.1.2 Soil Application

Metallic NPs can be mixed into the soil surrounding the plant. Through root uptake, they can be transported to different parts of the plant. This root uptake occurs through apoplastic and symplastic routes. The epidermal cell walls in plant roots form a complex matrix with pores that allow passage into the plant cell [36]. Certain NPs can disrupt the plasma membrane and induce the formation of new pores on the epidermal cell wall, facilitating the passage of large NPs [10]. In the apoplastic route, NPs passing through pores diffuse between the cell walls and the plasma membrane under osmotic pressure. This enables them to reach the endodermis, further facilitating their movement within the plant. Once internalized, NPs travel through different cells becoming integrated into the cytoplasm [29, 52].

2.2 Translocation of Metallic NPs into Plants

The translocation of metallic NPs depends on several factors including the physiological and structural characteristics of plant cells, interactions between nanomaterials and soil/leaves, and nature [79]. The plant cell wall acts as a barrier, regulating the entry of NPs into the cell and influencing their solubilization and passage based on their properties [89]. Metallic NPs taken up by plant roots are subsequently translocated to various tissues [187]. AuNPs have been found to accumulate in the shoots of *Oryza sativa*, while in the shoots of *Cucurbita pepo* and *Raphanus raphanistrum* [216], their accumulation was not reported. Plant roots exhibit a higher uptake rate for positively charged AuNPs. NPs selectively traverse the cell wall, entering the endothelial layer from the epidermis via either the symplast or exoplast pathway, and subsequently travel upwards through the xylem vessels [187]. The cortex comprises the endothelium and exodermis, with the endothelium connected to the pericycle.

This intricate process allows NPs to traverse the root and potentially reach other parts of the plant.

3 Biological Mechanisms of Metallic NPs for Plant Growth

Metallic NPs find applications in physical, chemical, biological, engineering, biomedical, and agricultural practices [150]. Commonly used metallic NPs are that of Ag, Au, Cu, Ti, Fe, and Zn. The biological mechanisms underlying the effects of metallic NPs on plant growth are complex and dependent upon the specific type of NP and plant species. There are some key mechanisms described below:

3.1 Stress Response Activation

Indeed, plants encounter various environmental stresses, and to bolster their defence against these stresses, plants employ a range of strategies, involving genetic, biochemical, and physiological pathways. They adapt by making molecular adjustments, often through significant alterations in gene expression [10]. In the initial response to abiotic stresses, plants undergo a transient elevation in cytoplasmic Ca (II), ROS, abscisic acid, and an upsurge in mitogen-activated protein kinase (MAPK) pathways. As stress persists, plants further engage mechanisms that involve the modulation of proteins responsible for shielding against cellular damage and the regulation of stress-specific gene expression [97] (Fig. 4). Recent advancements suggest that nanofertilizers can enhance crop yield in challenging environments. Additionally, nanomaterials influence the expression of stress-related genes. For example, the use of AgNPs in Arabidopsis microarray analysis reveals changes in genes important for stress tolerance [54]. In plants, responses induced by nanomaterials directly contribute to defence against different environmental stresses.

3.1.1 Activation of Abiotic Stress Response

Plant abiotic stress, stemming from factors like drought, high salinity, and heavy metals, poses a primary challenge to plant growth. This stress hampers crop yield by impeding normal plant development. Plants have evolved diverse physiological defence mechanisms to counteract these stresses and NPs offer a potential solution to alleviate abiotic stress [88].

Studies by Sun et al. demonstrated the positive impact of ZnO NPs on corn, enhancing photosynthetic rates. This research suggested that NPs can effectively mitigate plant drought stress [177]. The process of photosynthesis, vital for plant functioning, is particularly sensitive to chilling stress [119]. This stress adversely affects the plant's photosystem, leading to a decrease in chlorophyll content, CO_2

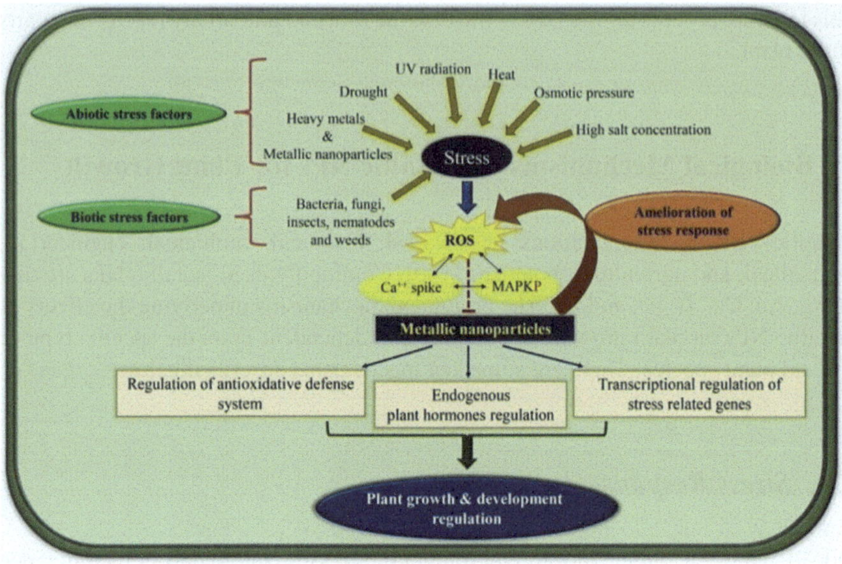

Fig. 4 Structured depiction of the general regulatory process of nanoparticles in mitigating various stress reactions in plants [186]

absorption, transpiration rate, and the degradation of the photosynthetic enzyme, Rubisco [77]. NPs exert their influence on the photosystem by enhancing the production of the Rubisco enzyme and improving the light absorption capacity of chloroplasts [77]. Additionally, they inhibit the production of ROS, further safeguarding the photosystem [9]. Exposure to metallic NPs also leads to an upregulation of the Rubisco and chlorophyll-binding protein genes [56], an increase in leaf pigments, and pronounced antioxidant enzyme activities [119], ultimately enhancing the plant's resilience to chilling stress.

Under cold stress conditions, plants respond by upregulating genes and enhancing the activities of enzymes. These enzymes are instrumental in scavenging ROS, thereby reducing oxidative stress. This leads to the repression of damaging processes such as lipid peroxidation, chlorophyll degradation, and hydrogen peroxide production, ultimately bolstering the plant's ability to tolerate stress [205]. Exposure to NPs in conjunction with cold stress has been demonstrated to enhance plant growth and various biochemical traits. Excessive salinity in plant growth environments can disrupt the nutritional balance of plants and impede their growth [116]. The introduction of NPs in agriculture offers a promising avenue to enhance the activity of enzymes crucial for salt tolerance in plants [2]. Research has demonstrated that ZnO NPs have the potential to alleviate salt stress in crops like cotton [74] and wheat [198].

The presence of heavy metals poses a greater threat to plant growth, causing disruptions in morphology, inhibiting growth, and inducing oxidative stress. To combat this stress, NPs can enhance the effectiveness of antioxidant systems. This

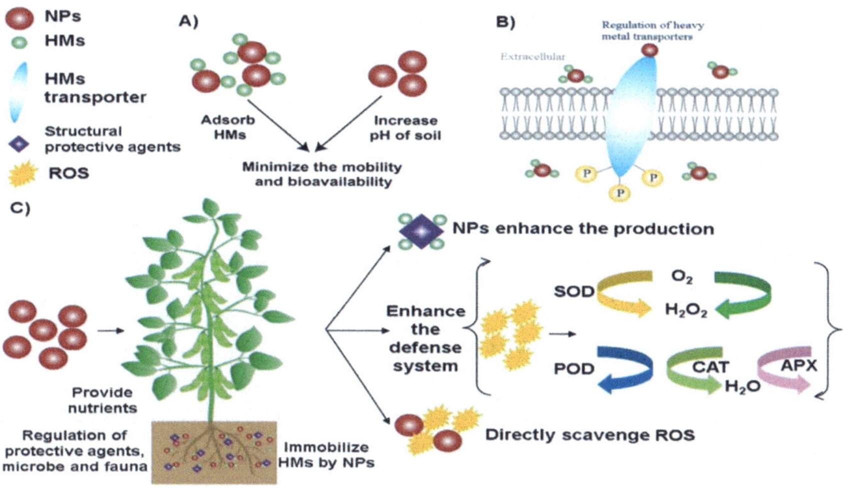

Fig. 5 Mechanism of NPs to mitigate heavy metal (HM) stress in plants [214]

is achieved through mechanism such as regulating the concentration of heavy metal ions in the soil, decreasing the expression of genes responsible for heavy metal ion transport in plants, and prompting the production of defence substances like organic acids, root exudates, and phytochelatin to counteract the impact of heavy metal ions [103]. The apoplastic barriers within plant roots serve a crucial physiological role in safeguarding the plant and regulating the movement of water, ions, and oxygen [27]. Metallic NPs can influence the formation of these barriers, potentially leading to a reduction in the accumulation of heavy metals within the root [153]. Additionally, specific NPs can regulate plant metal transport genes, thereby augmenting the plant's extracellular barrier to intercept heavy metals (Fig. 5).

3.1.2 Activation of Biotic Stress Response

NPs exert their inhibitory effects on fungi and bacteria through various mechanisms. This includes the formation of pores in the cell membrane, disruption of membrane potential, damage to the cell wall, direct attachment to the cell surface, induction of DNA damage, arrest of the cell cycle, inhibition of enzyme activity, and generation of ROS. Ultimately, these processes culminate in cell death. ROS generated by NPs inflict damage on cellular structures. ROS comprise both free radicals and non-radicals and when NPs accumulate in the membrane of bacteria or fungi, it alters membrane permeability and disrupts the proton motive force. Elevated concentrations of oxidative stress lead to DNA damage [170, 171].

3.2 Enhancement of Photosynthesis Efficiency

Within chloroplasts, energy from light is converted into a chemical form, which is then stored as sugar molecules. The application of metallic NPs on leaves significantly enhances chlorophyll content in plants resulting in increased light energy absorption and enhanced photosynthesis [173].

Metallic NPs, particularly those of TiO_2, have been extensively researched due to their photocatalytic properties. They can initiate oxidation–reduction reactions, facilitating charge transfer between light-harvesting complexes II and NPs [96]. In *A. thaliana*, plants treated with TiO_2 NPs exhibit a notable 3.83-fold increase in light-harvesting complex II (LHCII) content leading to the regulation of light energy distribution from photosystem I (PSI) to PSII by augmenting LHCII content, expediting the conversion of light energy to electronic energy, promoting photolysis of water, and facilitating evolution of oxygen [210].

3.3 Gene Regulation

The application of AgNPs in *Arabidopsis* led to the upregulation or downregulation of several genes [19]. The up-regulation was linked to the response to metals towards oxidative stress. Conversely, the down-regulated genes were associated with responses to pathogens and hormonal stimuli [19]. These responses induced by nanomaterials play a direct role in safeguarding plants against various stresses. MicroRNAs (miRNAs) can control various biological processes in plants and are pivotal in plant responses to stresses [47]. Studies in *A. thaliana* and *N. tabacum* have scrutinized the expression of miRNA genes. miRNAs have been identified as significant mediators of stress-responsive gene expression in the face of challenges like salt and water stress, through their regulatory influence on gene expression.

4 Significance of AuNPs in Plants

AuNPs have numerous commercial uses and are simple to synthesize in different shapes and sizes [52, 175], having specific optical characteristics, chemically stability [147], simple to quantify [207] and non-toxic [155] for therapeutic uses. Thus, there has been a lot of interest in the interactions between AuNPs and the biota [30]. The transport of AuNPs in entire plants or plant cells is significantly affected by their varied sizes, coatings, surface charges, and shapes [216]. The biomagnification and transfer of AuNPs in the food system increases the danger of human exposure. For example, Detritivores absorbed AuNPs from the soil and dispersed them across different tissues [85]. Another investigation demonstrated that AuNPs could

Table 2 Some AuNPs produced by plants

Plant species	Size (nm)	Shape	Reference
Vitex negundo leaf extract	70–90	Spherical	[162]
Platycodon grandiflorum leaf extract	~15	Spherical	[11]
Peganum harmala seed extract	43–52	Spherical	[121]
Coleus aromaticus leaf extract	~80	Spherical, rod, and triangular	[23]
Marsdenia tenacissima	~50	Spherical	[176]
Capsicum annuum var. grossum pulp extract	6–37	Triangle, hexagonal, and quasi-spherical	[208]
Papaver somniferum seed pulp extract	~77	Spherical	[194]
Plumeria alba flower extract	15–28	Spherical	[115]

be passed from the soil to invertebrates and subsequently to secondary consumers [190].

To explore how AuNPs transport in plant tissues, rape seeds were germinated for 10 days, after which they were incubated with a suspension of AuNPs. After one day of exposure, the analysis conducted using inductively coupled plasma-mass spectrometry (ICPMS) revealed that the levels of Au in the leaves and stems of the rape seedlings were greater compared to the control groups and after three days of exposure, the concentration of Au in both leaves and stems significantly increased but, extending the incubation period to 5 days did not lead to any alterations in the Au levels within the plant tissues. It was also observed that the transport of AuNPs inside plant tissue after three days of exposure to different concentrations of AuNPs solution. The results indicated that in contrast to control groups, treatments with AuNPs at different concentrations led to an increase in the concentration of Au in the stem tissues. Although the treatment with 10 nm AuNPs demonstrated a significant rise in the concentration of Au in the tissue of leaves while at higher concentrations specifically 20 and 40 nm AuNPs treatments, caused a slight increase in Au concentration in the leaves tissues [213]. Some AuNPs which are synthesized by plants are illustrated in Table 2.

5 Synthesis of AuNPs

AuNPs can be produced on a large scale by utilizing a variety of physical, chemical, and biological techniques to reduce the oxidation state of Au from (III) or (I) to (0), i.e. atomic gold (Fig. 6). NPs can be prepared by either the ‘Top-down’ or ‘Bottom-up’ approach. The top-down approach uses various physical means to break down suitable bulk material, like polymers, to NP size. In the bottom-up approach, NPs

Fig. 6 Synthesis of AuNPs

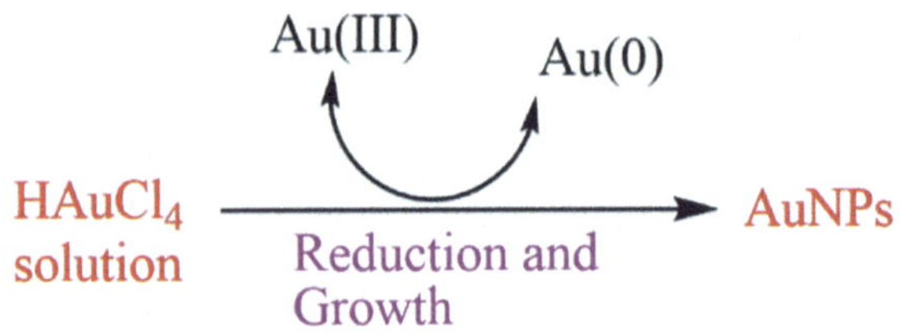

are made from smaller building blocks like atoms and molecules. This approach is used in chemical and biogenic methods of synthesis of NPs.

5.1 *Physical Methods*

AuNPs have been reported to be prepared by methods like laser irradiation [108], ultrasound [17], pyrolysis [188], thermolysis [165], and radiation-induced methods [185]. The limitation of this approach is that such NPs have surface imperfections which affect the characteristics of the NPs. Also, due to the use of special equipment and high energy consumption, this method is expensive.

5.2 *Chemical Methods*

In this method, common reducing agents like borohydride, citrate [92], ascorbic acid [98], or tannic acid [5] are used to reduce ionic Au, usually taken in the form of $HAuCl_4$ or its salt. This can be done in an aqueous or non-aqueous solvent as a reaction medium. After the reduction, a stabilizing agent like citrate or thiol is added to prevent agglomeration of colloidal Au particles formed. Using different types of such agents results in varying sizes and morphologies of the NPs formed [142]. For example, ascorbic acid has been used as both a reducing and stabilizing agent for the synthesis of AuNPs rapidly and efficiently, with the size of AuNPs ranging from 8 to 80 nm [112].

Both the above-discussed types of methods employed for the synthesis of NPs are associated with several drawbacks, particularly environmental concerns, which make them unsuitable for use.

5.3 *Biogenic Synthesis of AuNPs*

Biogenic synthesis of NPs is an environment-friendly alternative in which microorganisms, plants, and their extracts are used to synthesise NPs. Plants are a storehouse of phytochemicals produced by photosynthesis, food storage and metabolism, such as flavonoids, terpenoids, polysaccharides, phenols, and acids, like ascorbic acid; these phytochemicals are useful in NP synthesis as they function as reducing as well as stabilizing agents [131]. This property of plants together with their ability to absorb and accumulate metals from the soil makes them an excellent tool for the synthesis of metal NPs [164, 174]. The three steps of the biogenesis of NPs are activation, growth, and termination. In the activation step, the phytochemicals present in the plant cells reduce the metal ions to neutral metal atoms. In the growth phase, the metal atoms collect to form bigger crystals of nano size. The phytochemicals encapsulate the metal NPs, preventing their clumping, this stabilizes the NPs. The termination step involves the NPs taking final shape and size [26]. As biogenic synthesis follows several principles of Green Chemistry, it is a useful, environment-friendly, as well as energy- and cost-effective technique for the synthesis of NPs. Biogenic synthesis of NPs can be done either intracellularly, where the metal salt is accumulated in the plant cells and reduced by the plant's phytochemicals, or extracellularly, using plant extracts. It may be noted here that the size and structure of the metal NPs formed by biogenesis depend on several factors like the plant/plant extract used, incubation time, temperature, as well as the concentration and pH of the metal ion solution [167].

5.3.1 Intracellular Biogenesis of AuNPs

AuNPs were first prepared biogenically in the *Medicago sativa* (*alfalfa*) plant by growing the plant in an $AuCl_4$ solution, from which the plant absorbed and accumulated gold ions [48]. Later, several researchers prepared AuNPs in a similar manner from other plants, such as *Triticum aestivum* [14], *Avena sativa* [14], *Sesbania drummondii*, [163], *Chilopsis linearis* [48], etc. Microorganisms like bacteria, fungi, and algae have also been reported for the biogenesis of AuNPs. For example, AuNPs were prepared by using a strain of the bacteria Geobacillus using $HAuCl_4$ solution [32]. In another study, AuNPs were synthesized efficiently by the reduction of aqueous $AuCl_4$ using *Sargassum wightii* [169]. Several microorganisms have been used to prepare AuNPs, some of which are given in Table 3.

Although intracellular biogenesis is a convenient route for preparing NPs, it is associated with certain limitations like the tedious process of separation and purification of NPs synthesized. This has shifted the focus on extracellular biogenesis.

5.3.2 Extracellular Biogenesis Using Plant Extract

Several studies have shown that using plant extracts for biogenesis provides better control of the size and crystal structure of the NPs as compared to other methods [110]. It is also more convenient and cost-effective as no culture media need to be maintained and each part of the plant can be used, to prepare NPs of variable properties. [6, 78]. For example, the biogenic synthesis of AuNPs has been reported utilizing an aqueous extract of rose petals combined with an aqueous solution of $HAuCl_4$ at

Table 3 Microorganisms used to prepare AuNPs

Name of microorganism	Type	Metal ion solution	Time taken	Size of NP	Reference
Pseudomonas aeruginosa	Bacterium	$HAuCl_4$	–	48, 70, 84 nm	[42]
Rhodococcus sp.	Bacterium	$HAuCl_4$	–	60 ± 20 nm	[96]
Mariannaea sp. HJ	Fungus	$HAuCl_4$	12 h	37.4 nm	[136]
Sargassum muticum	Alga	$HAuCl_4$	30–60 min	4–6 nm	[127]
Geobacillus sp. strain ID17	Bacterium	$HAuCl_4$	16 h	5–50 nm	[32]
Stoechospermum marginatum	Alga	$HAuCl_4$	10 min	18–94 nm	[144]
Tetraselmis kochinensis	Alga	$HAuCl_4$	48 h	6–9 and 21–24 nm	[161]
Rhizopus oryzae	Fungus	$HAuCl_4$	24 h	10 nm	[35]
Escherichia coli DH5a	Bacterium	$HAuCl_4$	5 days	20–30 nm	[39]
Sargassum wightii	Alga	$HAuCl_4$	12 h	8–12 nm	[169]
Rhodococcuss sp.	Bacterium	$HAuCl_4$	24 h	5–15 nm	[4]
Verticillium	Fungus	$HAuCl_4$	3 days	NA	[122]

room temperature and incubated for a mere five minutes. FT-IR analysis showed that biomolecules with $-NH_2$, carbonyl group, and –OH groups were encasing AuNPs. TEM revealed that the generated NPs were multi-shaped. The NPs' face-centred cubic structure was revealed by XRD, and the average size was determined to be roughly 10 nm using the dynamic light scattering approach [129].

The biogenic production of AuNPs using the leaf extract of the sweetener *Stevia rebaudiana* has been reported in another study. In about 20 min, this extract and $HAuCl_4$ were combined to produce AuNPs. UV–visible spectrophotometry was used to track the reaction's progress, and HR-TEM and XRD revealed the creation of octahedral AuNPs, with sizes ranging from 8 to 20 nm, and confirmed the NPs' crystalline nature [117]. A few other examples include the synthesis of AuNPs using extracts of *Magnolia kabus* leaves [172] and black tea leaves [22], extracts of garlic, i.e. *Allium sativum* [148], and flower extract of *Plumeria alba* [124] among several others.

6 AuNPs and Plant Growth, Development, and Protection

It is necessary to conduct further research in the expanding field of NPs' effects on crop plants. In recent times, engineered NPs have received significant attention to improve crop yield. An annual harm of 157 billion dollars is caused by plant parasitic nematodes, one of the main agricultural pests. Research has shown that whereas AuNPs are present in tobacco, tomato, *alfalfa*, maize, and other plants, they are absent from the aerial sections of radish, pumpkin, barley, and wheat [45]. The kind of plant as well as the size and charge of the particles affect how well AuNPs penetrate. Plant roots absorb positively charged AuNPs, while negatively charged AuNPs can travel from roots to stems and leaves [102]. According to a study, when watermelon seedlings are sprayed with AuNPs (*Citrullus lanatus* var. *Lanatus*) [45], they enter leaves through stomata, and a phloem transport mechanism moves them from leaves to roots. [146]. Reduced AuNP dosages are expected to enhance growth and productivity in a variety of crops and model plants in lab settings, as well as being linked to seed germination and chlorophyll content. [109]. AuNPs' growth-promoting properties have also been noted. [95]. AuNPs have distinct physicochemical characteristics and are generally inert. A study provided evidence that AuNPs have efficient utility to manage nematodes that cause knots in tomato plants. After 3 h of incubation during which *Meloidogyne incognita* (Fig. 7) showed 100% mortality, a remarkable effect of AuNPs on *Meloidogyne incognita* J2 was found upon direct exposure in water. In addition to controlling nematodes known to cause knotting in roots, the AuNPs stimulated tomato plant development. When compared to the untreated control plants, the tomato plants (Fig. 8) treated with AuNP also showed a higher chlorophyll content [182].

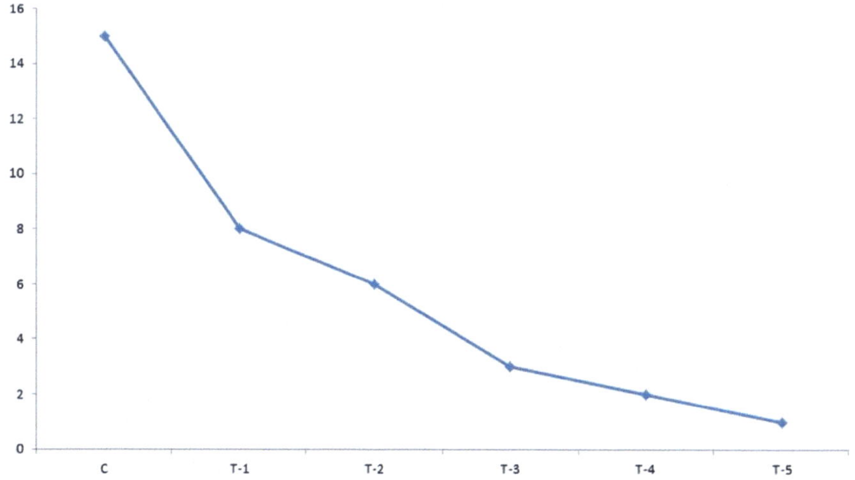

Fig. 7 Graph showing treatment of *M. incognia* with different concentrations of AuNPs [182]

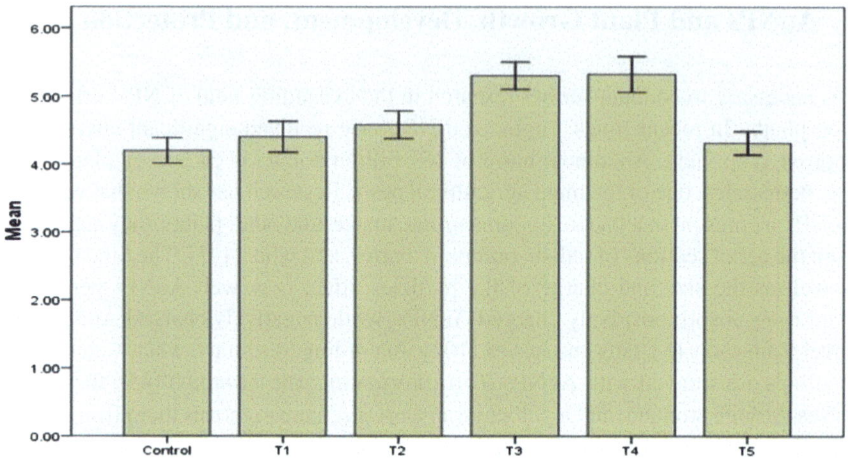

Fig. 8 Bar graph showing the profile of AuNPs' treated tomato plants [182]

Bioaccumulation and bio-magnification and bioinformation of engineered NPs need to be understood [150]. Regarding the accumulation and availability of NPs in food crops, very few research have been published. AuNPs are commonly employed in bombardment techniques to transfer genetic materials into plants since they have a high affinity for DNA and do not pose any harm to cells. [118]. AuNPs have a lower likelihood of entering the food chain through *Brassica* seed oil since they are linked to hydrophilic plant tissue compartments. In order to determine the impact of designed AuNPs on the growth and yield of *Brassica juncea* seeds, a study was conducted. Through foliar spraying, AuNPs at concentrations of 0, 10, 25, 50, and 100 ppm were administered. Plant height, stem diameter, number of branches and pods, and seed output are among the parameters that were found to be impacted by NP treatment. The number of leaves rose after treatment with AuNPs, and 10 ppm was found to be the ideal level of increase in seed output. Reducing and total sugar concentrations rose by up to 25 ppm. The redox status of the plant was also increased [15]. Leaf abscission was found to decrease due to slowed activity of ethylene. In *Arabidopsis thaliana*, AuNPs have also been shown to improve the antioxidant system and seed germination. In order to affect several physiological and metabolic processes in plants, they altered the quantities of miRNA [95]. In plants, NPs also function as intelligent delivery mechanisms. The expression of miRNAs, plant regulatory molecules, seed germination, growth, and antioxidant capacity of *Arabidopsis thaliana* were found to be significantly correlated with AuNP exposure [95]. Additionally, AuNPs enhanced the development of lettuce and cucumber seedlings [21]. Using X-ray diffraction, TEM, and UV–Vis, biosynthesized AuNPs from *Cassia auriculata* leaf extract were studied. The plant that produces biofuel, *Pennisetum galucum*, was used to test the effects of the produced AuNPs. The germination of the seeds and the growth of the seedlings produced positive outcomes. Increased plant biomass and improved germination have been related to significant economic relevance in horticulture and agriculture [134]. The outcomes matched those of other biologically produced AuNPs that improved the germination and growth of seedlings in crops that produced biofuel. Concerns with NP absorption, translocation, and biotransformation in plants are significant. For example, above 250 μM, AuNPs suppressed the development of *Arabidopsis thaliana* in liquid culture [105]. According to a different study, AuNPs did not inhibit plant growth or corn seed germination up to 160 mg/L [105]. Nonetheless, a plant's specific habitat, the charge and form of the NP, and the type of plant all influence how NPs work. Particularly for the characterization and quantification of NPs even at low NP concentration, single-particle inductively coupled plasma-mass spectrometry (SP-ICP-MS) has been identified as a developing approach. Gaining knowledge about how NPs are absorbed and accumulated in plant tissues is crucial. AuNPs, their size distribution, particle concentration, and dissolved Au content in tomato plant tissues were all simultaneously determined using an enzymatic digestion technique that was followed by SP-ICP-MS analysis. It was discovered that the Macerozyme R-10 enzyme could extract AuNPs from tomato plants without making them dissolve or clump together [33].

The effects of manufactured NPs on the environment are crucial to understand. Farm practices have investigated the use of AuNPs since they are thought to have a growth-promoting rather than a stress-inducing effect. Lower oxidative stress response induced by the immune-stimulatory PAMP flg22 in Arabidopsis seedlings has been linked to plants with longer roots, more lateral roots, and larger rosette diameters when exposed to moderate doses of AuNPs (Fig. 9) [46]. The study showed that AuNPs had a protective effect on plants' responses to oxidative stress without having any negative effects on them. Growth-promoting genes and proteins were up-regulated and oxidative stress responses were down-regulated, according to differential transcriptomics and proteomics.

AuNPs were found to affect the germination of *Arabidopsis thaliana* seeds. In comparison to the control, the total seed yield increased thrice when 24-nm AuNPs (10–80 mg/L) were added to the medium. Additionally, the length and diameter of the stem and roots increased [95]. Intriguing data were obtained by mass spectrometry and X-ray fluorescence in a study of entry of 5-, 10, and 15-nm AuNPs into the tissues of *Nicotiana tabacum* L. cv. *Xanthi* [85]. In addition to tobacco leaves, tobacco hornworm (*Manduca sexta*), which consumes tobacco leaves, was also shown to contain AuNPs in its tissues.

Certain bacteria, such as endophytes, epiphytes, and plant growth-promoting rhizobacteria (PGPR), are closely associated with plants. The PGPR is drawn towards the rhizosphere by the plant-mediated exudome. Numerous bacterial strains have been linked to the rhizosphere, including *Pseudomonas, Bacillus, Azospirillum, Alcaligenes, Azotobacter, Klebsiella, Arthrobacter, Enterobacter, Burkholderia,* and *Serratia* [93]. The rhizosphere offers a medium for chemical interactions between microbes and plants. According to a study, biogenic AuNPs have a concentration-dependent effect on the IAA synthesis of rhizospheric *Pseudomonas monteilii.*

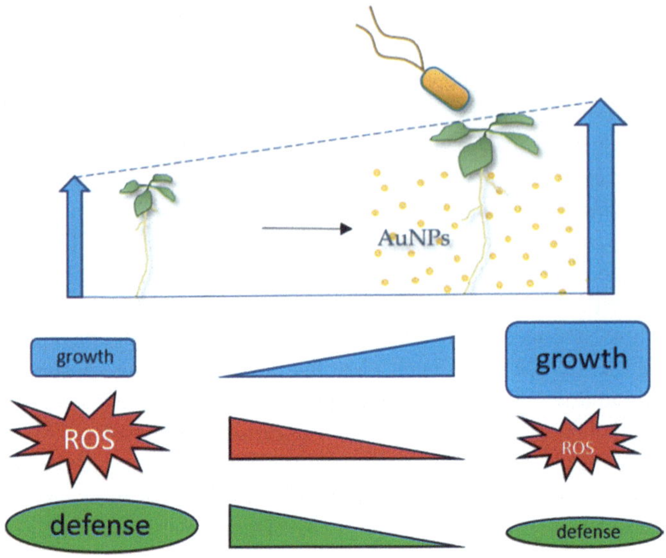

Fig. 9 Effects of AuNP on Arabidopsis seedlings [46]

Bacillus subtilis generated the AuNPs. The AuNPs obtained the spherical, hexagonal, and triangular geometries. AuNPs were utilized at varying concentrations to determine their effect on *Pseudomonas monteilii's* synthesis of IAA. In the presence of 50 μg/mL AuNPs, *P. monteilii* increased the generation of IAA, while *Vigna unguiculata* seedlings grown in the presence of 50 μg/mL AuNPs showed higher growth. The outcomes demonstrated that AuNPs can increase *P. monteilii's* capacity to produce IAAs. [133]. The composition of the essential oils that were extracted from the tissues changed significantly as a result of the micropropogation of lavender on media enhanced with AuNPs [203]. These substances have elicitor properties. A study was conducted to find out how oil gland formation and plant development in narrow-leaved lavender (*Lavandula angustifolia*) are impacted by the introduction of AuNPs into the culture media. On media treated with 1, 2, 5, 10, 20, and 50 mg dm^{-3} AuNPs, the plant shoots were propagated. A favourable impact on the plant's growth was discovered in vitro. The plants that were propagated on medium supplemented with 1 and 2 mg dm^{-3} NPs were found to have bigger trichome diameters on both sides of the leaf blade. [76]. Some effects of AuNPs are described in Table 4.

7 Future Prospects

Plants are impacted by NPs, which alter their morphology and physiology. NPs' effectiveness is influenced by their size, chemical makeup, surface charge, dosage, and other factors [13, 90]. Additionally, the plant species and plant portion that

Table 4 Effects of AuNPs on plants

Plant	Dosage of AuNP	Effect	References
Vigna unguiculata	100 μg/mL	Shoot enlargement	[132]
N. jatamansi	60 μM, 40 μM	Root enlargement	[83]
Vigna radiata	10, 100, 1000 mg/L	Increase in root and shoot length	[34]
Gloriosa superba *Allium cepa*	100, 500, 1000 μM	Increase in mitotic index	[8]
Arabidopsis thaliana	50 mg/L	Increase in biomass of suspension cell culture	[160]
Allium cepa	0.1–10 mg/L	Increase in chromosomal aberrations	[145]
Zea mays	5–15 ppm	Promotion of seed germination	[109]
Brassica juncea	0–400 mg/L	Root enlargement	[53]
Gloriosa superba	500–1000 μM	Enhanced seed germination, enhanced vegetative growth	[51]
Arabidopsis thaliana	10–80 mg/L	Promotion of seed germination	[95]
Nicotiana tabacum	30, 100 mg/L	Growth enhancement	[85]

they are applied to affect the efficacy [55]. The engineered NPs can bring about desired features and also deliver engineered DNA into plant cells. DNA can also be changed at the cellular level by NP-based DNA delivery, which can regulate the expression of changed genes and shorten the time needed to introduce desired genes into an alien creature. Additionally, the characteristics of seeds can be changed to provide particular qualities like colour, yield, and season growth. Thus, in addition to causing outward alterations in crops, nanotechnology can also alter internal expression patterns. Thus, the green revolution can be facilitated by the application of nanotechnology. Additionally, the detrimental effects of NPs on plant development and growth need to be addressed, especially at higher dosages. In order to assess AuNPs' potential harm in the environment and to understand the molecular pathways influencing NPs' impacts through controlled model systems, an ecotoxicological assessment of AuNPs is also required.

8 Conclusion

Due to their versatility in surface functionalization, AuNPs have been used extensively in a wide range of fields. The agriculture sector now has more opportunities due to recent advancements in nanotechnology. The use of AuNPs in agriculture, specifically in the growth and development of plants, has been expounded upon in this chapter. Moreover, the molecular details of AuNP entrance into plant cells are presented. It is well recognized that various plant species have diverse responses to NP treatment and that the same species might exhibit variable tolerance to different

NPs. Therefore, tolerance varies depending on the range of plant species examined. AuNP tolerance in soil should be taken into account, as well as how AuNP accumulation occurs in plants and how they enter the food chain. To increase the growing circumstances and yield of a wide variety of plant species, the best useable concentration of NPs needs to be investigated.

Acknowledgements The authors acknowledge the support extended by Prof. Abdul Rahman Khan, Head, Department of Chemistry, Integral University, Lucknow and Dr. (Mrs.) Vinita Prakash, Principal, Isabella Thoburn College, Lucknow.

References

1. Abdelaziz AM, Salem SS, Khalil AM, El-Wakil DA, Fouda HM, Hashem AH (2022) Potential of biosynthesized zinc oxide nanoparticles to control Fusarium wilt disease in eggplant (*Solanum melongena*) and promote plant growth. Biometals 35:601–616
2. Abideen Z, Hanif M, Munir N, Nielsen BL (2022) Impact of nanomaterials on the regulation of gene expression and metabolomics of plants under salt stress. Plants 11:691
3. Adrees M, Khan ZS, Hafeez M, Rizwan M, Hussain K, Asrar M, Alyemeni MN, Wijaya L, Ali S (2021) Foliar exposure of zinc oxide nanoparticles improved the growth of wheat (*Triticum aestivum* L.) and decreased cadmium concentration in grains under simultaneous Cd and water deficient stress. Ecotoxicol Environ Saf 208:111627
4. Ahmad A, Senapati S, Khan MI, Kumar R, Ramani R, Srinivas V, Sastry M (2003) Intracellular synthesis of gold nanoparticles by novel alkalotolerant actinomycetes, Rhodococcus sp. Nanotechnol 14:824–828
5. Ahmad T (2014) Reviewing the tannic acid mediated synthesis of metal nanoparticles. J Nanotechnol 1–11
6. Akhtar MS, Panwar J, Yun Y (2013) Biogenic synthesis of metallic nanoparticles by plant extracts. ACS Sustain. Chem Eng 1:591–602
7. Al-Amri N, Tombuloglu H, Slimani Y, Akhtar S, Barghouthi M, Almessiere M, Alshammari T, Baykal A6, Sabit H, Ercan I, Ozcelik S (2020) Size effect of iron (III) oxide nanomaterials on the growth, and their uptake and translocation in common wheat (*Triticum aestivum* L.). Ecotoxicol Environ Saf 194:110377
8. Alharbi NS, Bhakyaraj K, Gopinath K, Govindarajan M, Kumuraguru S, Mohan S, Kaleeswarran P, Kadaikunnan S, Khaled JM, Benelli G (2017) Gum-mediated fabrication of eco-friendly gold nanoparticles promoting cell division and pollen germination in plant cells. J Cluster Sci 28:507–517
9. Al-Huqail AA, Hatata MM, Al-Huqail AA, Ibrahim MM (2018) Preparation, characterization of silver phyto nanoparticles and their impact on growth potential of *Lupinus termis* L. seedlings. Saudi J Biol Sci 25:313–319
10. Ali S, Mehmood A, Khan N (2021) Uptake, translocation, and consequences of nanomaterials on plant growth and stress adaptation. J Nanomater 2021:1–17
11. Anbu P, Gopinath SCB, Jayanthi S (2020) Synthesis of gold nanoparticles using *Platycodon grandiflorum* extract and its antipathogenic activity under optimal conditions. Nanomater Nanotechnol 10:1–9
12. Anjum NA, Gill SS, Duarte AC, Pereira E, Ahmad I (2013) Silver nanoparticles in soil–plant systems. J Nanopart Res 15:1–26
13. Ansari N, Rahman QI, Khan T, Khan A, Khan R, Wagay JA, Husen A (2023). Mycogenic-assisted synthesis of nanoparticles and their efficient applications In: Husen A, Siddiqi KS (eds.). Advances in Smart Nanomaterials and their Applications. Elsevier

14. Armendariz V, Herrera I, Peralta-Videa JR, Jose-Yacaman M, Troiani H, Santiago P, Gardea-Torresdey JL (2004) Size controlled gold nanoparticle formation by *Avena sativa* biomass: Use of plants in nanobiotechnology. J Nanopart Res 6:377–382
15. Arora S, Sharma P, Kumar S, Nayan R, Khanna PK, Zaidi MGH (2012) Gold-nanoparticle induced enhancement in growth and seed yield of *Brassica juncea*. Plant growth Regul 66:303–310
16. Bachheti RK, Bachheti A, Husen A (2024) Metal and Metal-Oxide Based Nanomaterials (*Synthesis, Agricultural, Biomedical and Environmental Interventions*). Springer Nature Singapore Pte Ltd., 152 Beach Road, #21–01/04 Gateway East, Singapore 189721, Singapore
17. Baigent CL, Müller G (1980) A colloidal gold prepared with ultrasonics. Experientia 36:472–473
18. Bakshi S, He ZL, Harris WG (2015) Natural nanoparticles: implications for environment and human health. Crit Rev Environ Sci Technol 45:861–904
19. Banerjee J, Kole C (2016) Plant nanotechnology: an overview on concepts, strategies, and tools. Plant nanotechnology: principles and practices. Springer International Publishing, Cham, pp 1–14
20. Banijamali SM, Feizian M, Alinejadian Bidabadi A, Mehdipour E (2019) Evaluation uptake and translocation of iron oxide nanoparticles and its effect on photosynthetic pigmentation of chrysanthemum (*Chrysanthemum morifolium*)'Salvador.' J Ornament Plants 9:245–258
21. Barrena R, Casals E, Colón J, Font X, Sánchez A, Puntes V (2009) Evaluation of the ecotoxicity of model nanoparticles. Chemosphere 75:850–857
22. Begum NA, Mondal S, Basu S, Laskar RA, Mandal D (2009) Biogenic synthesis of Au and Ag nanoparticles using aqueous solutions of Black Tea leaf extracts. Colloids Surf B Colloid Surface B 71:113–118
23. Boomi P, Ganesan RM, Poorani G, Prabu HG, Ravikumare S, Jeyakanthan J (2019) Biological synergy of greener gold nanoparticles by using *Coleus aromaticus* leaf extract. Mater Sci Eng C 99:202–210
24. Cañas JE, Long M, Nations S, Vadan R, Dai L, Luo M, Ambikapathi R, Lee EH, Olazyk D (2008) Effects of functionalized and nonfunctionalized single-walled carbon nanotubes on root elongation of select crop species. Environ Toxicol Chem 27:1922–1931
25. Cao Z, Stowers C, Rossi L, Zhang W, Lombardini L, Ma X (2017) Physiological effects of cerium oxide nanoparticles on the photosynthesis and water use efficiency of Soybean (*Glycine max* L.). Environ Sci Nano 4:1086–1094
26. Chandirika JU, Annadurai G (2018) Biosynthesis and characterization of silver nanoparticles using leaf extract Abutilon indicum. Glob J Biotechnol Biochem 13:7–11
27. Chao-Dong Y, Xia Z, Guo-Feng L, Jun-Wei Z, Man-Zhu B, Zhi-Xiang Z (2013) Progress on the structure and physiological functions of apoplastic barriers in root. Bull Bot Res 33:114–119
28. Chawla S, Singh S, Husen A (2023) Smart Nanomaterials targeting pathological hypoxia. Springer Nature Singapore Pte Ltd., 152 Beach Road, #21–01/04 Gateway East, Singapore 189721, Singapore ISBN: 978-981-99-1717-4 https://doi.org/10.1007/978-981-99-1718-1
29. Chen J, Dou R, Yang Z, You T, Gao X, Wang L (2018) Phytotoxicity and bioaccumulation of zinc oxide nanoparticles in rice (*Oryza sativa* L.). Plant Physiol Biochem 130:604–612
30. Chithrani BD, Ghazani AA, Chan WC (2006) Determining the size and shape dependence of gold nanoparticle uptake into mammalian cells. Nano Lett 6:662–668
31. Choudhury FK, Rivero RM, Blumwald E, Mittler R (2017) Reactive oxygen species, abiotic stress and stress combination. Plant J 90:856–867
32. Correa-Llantén DN, Muñoz-Ibacache SA, Castro ME, Muñoz PA, Blamey JM (2013) Gold nanoparticles synthesized by Geobacillus sp. strain ID17 a thermophilic bacterium isolated from Deception Island, Antarctica. Microb Cell Factories 12:1–6
33. Dan Y, Zhang W, Xue R, Ma X, Stephan C, Shi H (2015) Characterization of gold nanoparticle uptake by tomato plants using enzymatic extraction followed by single-particle inductively coupled plasma–mass spectrometry analysis. Environ Sci Technol 49:3007–3014
34. Das S, Debnath N, Pradhan S, Goswami A (2017) Enhancement of photon absorption in the light-harvesting complex of isolated chloroplast in the presence of plasmonic gold nanosol—a

nanobionic approach towards photosynthesis and plant primary growth augmentation. Gold Bull 50:247–257
35. Das SK, Das AR, Guha AK (2009) Gold nanoparticles: microbial synthesis and application in water hygiene management. Langmuir 25:8192–8199
36. Deng YQ, White JC, Xing BS (2014) Interactions between engineered nanomaterials and agricultural crops: implications for food safety. J Zhejiang Univ Sci A 15:552–572
37. Dietz KJ, Herth S (2011) Plant nanotoxicology. Trends Plant Sci 16:582–589
38. Dimkpa C, Bindraban P, McLean JE et al (2017) Methods for rapid testing of plant and soil nutrients. In: Lichtfouse E (ed) Sustainable agriculture reviews. Springer Cham 1–43
39. Du L, Jiang H, Liu X, Wang E (2007) Biosynthesis of gold nanoparticles assisted by *Escherichia coli* DH5a and its application on direct electrochemistry of hemoglobin. Electrochem Commun 9:1165–1170
40. Du W, Tan W, Peralta-Videa JR, Gardea-Torresdey JL, Ji R, Yin Y, Guo H (2017) Interaction of metal oxide nanoparticles with higher terrestrial plants: physiological and biochemical aspects. Plant Physiol Biochem 110:210–225
41. Eichert T, Kurtz A, Steiner U, Goldbach HE (2008) Size exclusion limits and lateral heterogeneity of the stomatal foliar uptake pathway for aqueous solutes and water suspended nanoparticles. Physiol Plant 134:151–160
42. El-Newehy M, Thamer BM, Abdulhameed MM, Ahamed A, Hamshary HE (2023) Biofabricated gold nanoparticles with antibacterial and antibiofilm activities against foodborne bacterial pathogens using *Pseudomonas aeruginosa* metabolic extract. J Inorg Organomet Polym
43. Elsakhawy T, Omara AED, Alshaal T, El-Ramady H (2018) Nanomaterials and plant abiotic stress in agroecosystems. Environ Biodivers Soil Secur 2:73–94
44. Fabrega J, Luoma SN, Tyler CR, Galloway TS, Lead JR (2011) Silver nanoparticles: behaviour and effects in the aquatic environment. Environ Int 37:517–531
45. Feichtmeier NS, Walther P, Leopold K (2015) Uptake, effects, and regeneration of barley plants exposed to gold nanoparticles. Environ Sci Pollut Res Int 22:8549–8558
46. Ferrari E, Barbero F, Busquets-Fité M, Franz-Wachtel M, Köhler HR, Puntes V, Kemmerling B (2021) Growth-promoting gold nanoparticles decrease stress responses in *Arabidopsis* seedlings. Nanomaterials 11:3161
47. Frazier TP, Burklew CE, Zhang B (2014) Titanium dioxide nanoparticles affect the growth and microRNA expression of tobacco (*Nicotiana tabacum*). Funct Integr Genomics 14:75–83
48. Gardea-Torresdey JL, Rodriguez E, Parsons JG, Peralta-Videa JR, Meitzner G, Cruz-Jimenez G (2005) Use of ICP and XAS to determine the enhancement of gold phytoextraction by *Chilopsis linearis* using thiocyanate as a complexing agent. Anal Bioanal Chem 382:347–352
49. Geng M, Li L, Ai M, Jin J, Hu D, Song K (2022) Recent advances in metal-based nanoparticle-mediated biological effects in *Arabidopsis thaliana*: a mini review. Materials 15:4539
50. Giraldo JP, Landry MP, Faltermeier SM, McNicholas TP, Iverson NM, Boghossian AA, Revel NF, Hilmer AJ, Sen F, Brew JA, Strano MS (2014) Plant nanobionics approach to augment photosynthesis and biochemical sensing. Nat Mater 13:400–408
51. Gopinath K, Gowri S, Karthika V, Arumugam A (2014) Green synthesis of gold nanoparticles from fruit extract of *Terminalia arjuna*, for the enhanced seed germination activity of *Gloriosa superba*. J Nanostruc Chem 4:115
52. Grzelczak M, Perez-Juste J, Mulvaney P, Liz-Marzan LM (2008) Shape control in gold nanoparticle synthesis. Chem Soc Rev 37:1783–1791
53. Gunjan B, Zaidi MGH, Sandeep A (2014) Impact of gold nanoparticles onphysiological and biochemical characteristics of *Brassica juncea*. J Plant Biochem Physiol 2:2
54. Gupta SD, Agarwal A, Pradhan S (2018) Phytostimulatory effect of silver nanoparticles (AgNPs) on rice seedling growth: an insight from antioxidative enzyme activities and gene expression patterns. Ecotoxicol Environ Saf 161:624–633
55. Gutiérrez FJ, Mussons ML, Gatón P, Rojo R (2011) Nanotechnology and food industry. In: Benjamin, Valdez (Ed.) Scientific, health and social aspects of the food industry, InTech., (InTech., Croatia, 2011), pp 95–128

56. Hasanpour H, Maali-Amir R, Zeinali H (2015) Effect of TiO_2 nanoparticles on metabolic limitations to photosynthesis under cold in chickpea. Russ J Plant Physiol 62:779–787
57. Hatami M (2017) Stimulatory and inhibitory effects of nanoparticulates on seed germination and seedling vigor indices. Nanosci Plant–Soil Syst 357–385
58. Hayes KL, Mui J, Song B, Sani ES, Eisenman SW, Sheffield JB, Kim B (2020) Effects, uptake, and translocation of aluminum oxide nanoparticles in lettuce: a comparison study to phytotoxic aluminum ions. Sci Total Environ 719:137393
59. Hong J, Peralta-Videa JR, Rico C, Sahi S, Viveros MN, Bartonjo J, Zhao L, Gardea-Torresdey JL (2014) Evidence of translocation and physiological impacts of foliar applied CeO_2 nanoparticles on cucumber (*Cucumis sativus*) plants. Environ Sci Technol 48:4376–4385
60. Hossain Z, Mustafa G, Komatsu S (2015) Plant responses to nanoparticle stress. Int J Mol Sci 16:26644–26653
61. Hosseinpour A, Haliloglu K, Tolga Cinisli K, Ozkan G, Ozturk HI, Pour-Aboughadareh A, Poczai P (2020) Application of zinc oxide nanoparticles and plant growth promoting bacteria reduces genetic impairment under salt stress in tomato (*Solanum lycopersicum* L.'Linda'). Agriculture 10:521
62. Husen A (2019) Natural product-based fabrication of zinc oxide nanoparticles and their application, *In*: N*anomaterials* and Plant Potential (Eds. Husen A, Iqbal M) Springer International Publishing AG, Gewerbestrasse 11, 6330 Cham pp. 193–291 https://doi.org/10.1007/978-3-030-05569-1_7
63. Husen A (2020) Introduction and techniques in nanomaterials formulation. In: Nanomaterials for Agriculture and Forestry Applications (Eds. Husen, A, Jawaid M) Elsevier Inc. 50 Hampshire St., 5th Floor, Cambridge, MA 02139, USA, pp 1–14. https://doi.org/10.1016/B978-0-12-817852-2.00001-9
64. Husen A (2022) Engineered nanomaterials for sustainable agricultural production, soil improvement and stress management. Elsevier Inc., 50 Hampshire St., 5th Floor, Cambridge, MA 02139, USA https://doi.org/10.1016/C2021-0-00054-7
65. Husen A (2023a) Secondary metabolites based green synthesis of nanomaterials and their applications. Springer Nature Singapore Pte Ltd., 152 Beach Road, #21–01/04 Gateway East, Singapore 189721, Singapore. https://doi.org/10.1007/978-981-99-0927-8
66. Husen A (2023b) Nanomaterials from Agricultural and Horticultural Products. Springer Nature Singapore Pte Ltd., 152 Beach Road, #21-01/04 Gateway East, Singapore 189721, Singapore. https://link.springer.com/book/9789819934348
67. Husen A (2023c) Nanomaterials and nanocomposites exposures to plants (response, interaction, phytotoxicity and defense mechanisms). Springer Nature Singapore Pte Ltd., 152 Beach Road, #21-01/04 Gateway East, Singapore 189721, Singapore. https://doi.org/10.1007/978-981-99-2419-6
68. Husen A, Bachheti RK, Bachheti A (2023) Current trends in green nano-emulsions (food, agriculture and biomedical sectors). Springer Nature Singapore Pte Ltd., 152 Beach Road, #21-01/04 Gateway East, Singapore 189721, Singapore
69. Husen A, Iqbal M (2019) Nanomaterials and plant potential. Springer International Publishing AG, Gewerbestrasse 11, 6330 Cham, Switzerland https://doi.org/10.1007/978-3-030-05569-1
70. Husen A, Jawaid M (2020) Nanomaterials for agriculture and forestry applications. Elsevier Inc., 50 Hampshire St., 5th Floor, Cambridge, MA 02139, USA https://doi.org/10.1016/C2018-0-02349-X
71. Husen A, Siddiqi K (2014) Phytosynthesis of nanoparticles: concept, controversy and application. Nanoscale Res Lett 9:229
72. Husen A, Siddiqi KS (2014) Phytosynthesis of nanoparticles: concept, controversy and application. Nanoscale Res Lett 9(229):1–24. https://doi.org/10.1186/1556-276X-9-229
73. Husen A, Siddiqi KS (2023) Advances in smart nanomaterials and their applications. Elsevier Inc., 50 Hampshire St., 5th Floor, Cambridge, MA 02139, USA. https://doi.org/10.1016/C2021-0-02202-1
74. Hussein MM, Abou-Baker NH (2018) The contribution of nano-zinc to alleviate salinity stress on cotton plants. R Soc Open Sci 5:171809

75. Iqbal MS, Singh AK, Singh SP, Ansari MI (2020) Nanoparticles and plant interaction with respect to stress response. Nanomater Environ Biotechnol 1–15
76. Jadczak P, Kulpa D, Bihun M, Przewodowski W (2019) Positive effect of AgNPs and AuNPs in in vitro cultures of Lavandula angustifolia Mill. Plant Cell Tissue and Organ Culture (PCTOC) 139:191–197
77. Jahan S, Alias YB, Bakar AFB, Yusoff IB (2018) Toxicity evaluation of ZnO and TiO_2 nanomaterials in hydroponic red bean (*Vigna angularis*) plant: physiology, biochemistry and kinetic transport. J Environ Sci 72:140–152
78. Jamkhande PG, Ghule NW, Bamer AH, Kalaskar M (2019) Metal nanoparticles synthesis: an overview on methods of preparation, advantages and disadvantages, and applications. J Drug Deliv Sci Technol 53:101174
79. Janmohammadi M, Amanzadeh T, Sabaghnia N, Ion V (2016) Effect of nano-silicon foliar application on safflower growth under organic and inorganic fertilizer regimes. Botanica 22:53–64
80. Jayarambabu N, Kumari SB, Rao VK, Prabhu YT (2016) Enhancement of growth in maize by biogenic-synthesized MgO nanoparticles. Int J Pure Appl Zool 4:262–270
81. Jiang HS, Yin LY, Ren NN, Zhao ST, Li Z, Zhi Y, Shao H, Li W, Gontero B (2017) Silver nanoparticles induced reactive oxygen species via photosynthetic energy transport imbalance in an aquatic plant. Nanotoxicology 11:157–167
82. Jin-Chul K, Madhusudhan A, Husen A (2021) Smart nanomaterials in biomedical applications. Springer Nature Switzerland AG, Gewerbestrasse 11, 6330 Cham, Switzerland. https://doi.org/10.1007/978-3-030-84262-8
83. Joshi S, Dar AI, Acharya A, Joshi R (2022) Charged gold nanoparticles promote in vitro proliferation. In: Nardostachys jatamansi by differentially regulating chlorophyll content, hormone concentration, and antioxidant activity. Antioxidants 11(10):1962
84. Jośko I, Oleszczuk P (2013) Influence of soil type and environmental conditions on ZnO, TiO2 and Ni nanoparticles phytotoxicity. Chemosphere 92:91–99
85. Judy JD, Unrine JM, Bertsch PM (2011) Evidence for biomagnification of gold nanoparticles within a terrestrial food chain. Environ Sci Technol 45:776–781
86. Juhel G, Batisse E, Hugues Q, Daly D, van Pelt FN, O'Halloran J, Jansen MA (2011) Alumina nanoparticles enhance growth of Lemna minor. Aquat Toxicol 105:328–336
87. Karl-Josef D, Simone H (2011) Plant nanotoxicology. Trends Plant Sci 16:582–589
88. Khalid MF, Iqbal Khan R, Jawaid MZ, Shafqat W, Hussain S, Ahmed T, Rizwan M, Ercisli S, Pop OL, Alina Marc R (2022) Nanoparticles: the plant saviour under abiotic stresses. Nanomaterials 12:3915
89. Khan N, Bano A (2016) Modulation of phytoremediation and plant growth by the treatment with PGPR, Ag nanoparticle and untreated municipal wastewater. Int J Phytoremediation 18:1258–1269
90. Khan T, Rahman QI, Raza S, Zehra S, Ahmad N, Husen A (2023) Nanodimensional materials:an approach toward the biogenic synthesis. In: Husen A, Siddiqi KS (eds.). Advances in Smart Nanomaterials and their Applications. Elsevier
91. Khodakovskaya M, Dervishi E, Mahmood M, Xu Y, Li Z, Watanabe F, Biris AS (2009) Carbon nanotubes are able to penetrate plant seed coat and dramatically affect seed germination and plant growth. ACS Nano 3:3221–3227
92. Kimling J, Maier M, Okenve B, Kotaidis V, Ballot H, Plech A (2006) Turkevich method for gold nanoparticle synthesis revisited. J Phys Chem B 110:15700–15707
93. Kour D, Rana KL, Yadav N, Yadav AN, Kumar A, Meena VS, Singh B, Chauhan VS, Dhaliwal HS, Saxena AK (2019) Rhizospheric microbiomes: biodiversity, mechanisms of plant growth promotion, and biotechnological applications for sustainable agriculture. Plant growth promoting rhizobacteria for agricultural sustainability: from theory to practices 19–65
94. Kuang TY (2003) Mechanism and regulation of primary energy conversion process in photosynthesis. Science and Technology Press of Jiangsu Nanjing 22
95. Kumar V, Guleria P, Kumar V, Yadav SK (2013) Gold nanoparticle exposure induces growth and yield enhancement in *Arabidopsis thaliana*. Sci Total Environ 461:462–468

96. Kuyukina MS, Makarova MV, Ivshina IB, Kazymov KP, Osovetsky BM (2022) Biosynthesis and characterization of gold nanoparticles produced using *Rhodococcus Actinobacteria* at elevated chloroauric acid concentrations. Int J Mol Sci 23:12939
97. Landa P, Cyrusova T, Jerabkova J, Drabek O, Vanek T, Podlipna R (2016) Effect of metal oxides on plant germination: phytotoxicity of nanoparticles, bulk materials, and metal ions. Water Air Soil Pollut 227:448
98. Larm NE, Essner JB, Pokpas K, Canon JA, Jahed N, Iwuoha EI, Baker GA (2018) Room-temperature Turkevich method: formation of gold nanoparticles at the speed of mixing using cyclic oxocarbon reducing agents. J Phys Chem C 122:5105–5118
99. Larue C, Castillo-Michel H, Sobanska S, Cécillon L, Bureau S, Barthès V, Ouerdane L, Carriers M, Sarret G (2014) Foliar exposure of the crop *Lactuca sativa* to silver nanoparticles: evidence for internalization and changes in Ag speciation. J Hazard Mater 264:98–106
100. Lei Z, Mingyu S, Chao L, Liang C, Hao H, Xiao W, Xiaoqing L, Fan Y, Fengqing G, Fashui H (2007) Effects of nanoanatase TiO_2 on photosynthesis of spinach chloroplasts under different light illumination. Biol Trace Elem Res 119:68–76
101. Leyu AM, Debebe SE, Bachheti A, Rawat YS, Bachheti RK (2023) Green synthesis of gold and silver nanoparticles using invasive alien plant *Parthenium hysterophorus* and their antimicrobial and antioxidant activities sustainability 15:9456
102. Li H, Ye X, Guo X, Geng Z, Wang G (2016) Effects of surface ligands on the uptake and transport of gold nanoparticles in rice and tomato. J Hazard Mater 314:188–196
103. Lian J, Zhao L, Wu J, Xiong H, Bao Y, Zeb A, Tang J, Liu W (2020) Foliar spray of TiO_2 nanoparticles prevails over root application in reducing Cd accumulation and mitigating Cd-induced phytotoxicity in maize (*Zea mays* L.). Chemosphere 239:124794
104. Lin D, Xing B (2007) Phytotoxicity of nanoparticles: inhibition of seed germination and root growth. Environ Pollut 150:243–250
105. Liu C, Li F, Luo C, Liu X, Wang S, Liu T, Li X (2009) Foliar application of two silica sols reduced cadmium accumulation in rice grains. J Hazard Mater 161:1466–1472
106. Ma C, White JC, Dhankher OP, Xing B (2015) Metal-based nanotoxicity and detoxification pathways in higher plants. Environ Sci Technol 49:7109–7122
107. Madhusudhan A, Purohit SD, Prasad R, Husen A (2024) Functional smart nanomaterials and their theranostics approaches. Springer Nature Singapore Pte Ltd., 152 Beach Road, #21-01/04 Gateway East, Singapore 189721, Singapore
108. Mafuné F, Kohno J-Y, Takeda Y, Kondow T (2002) Full physical preparation of size-selected gold nanoparticles in solution: laser ablation and laserinduced size control. J Phys Chem B 106:7575–7577
109. Mahakham W, Theerakulpisut P, Maensiri S, Phumying S, Sarmah AK (2016) Environmentally benign synthesis of phytochemicals-capped gold nanoparticles as nanopriming agent for promoting maize seed germination. Sci Total Environ 573:1089–1102
110. Makarov VV, Love AJ, Sinitsyna OV, Makarova SS, Yaminsky IV, Taliansky ME, Kalinina NO (2014) "Green" nanotechnologies: synthesis of metal nanoparticles using plants. Acta Naturae 6:35–44
111. Malandrakis AA, Kavroulakis N, Chrysikopoulos CV (2019) Use of copper, silver and zinc nanoparticles against foliar and soil-borne plant pathogens. Sci Total Environ 670:292–299
112. Malassis L, Dreyfus R, Murphy RJ, Hough LA, Donnio B, Murray CB (2016) One-step green synthesis of gold and silver nanoparticles with ascorbic acid and their versatile surface post-functionalization. RSC Adv 6:33092–33100
113. Mandal AK, Ghorai S, Husen A (2023) Functionalized smart nanomaterials for point-of-care testing. Springer Nature Singapore Pte Ltd., 152 Beach Road, #21–01/04 Gateway East, Singapore 189721, Singapore
114. Marslin G, Sheeba CJ, Franklin G (2017) Nanoparticles alter secondary metabolism in plants via ROS burst. Front Plant Sci 8:832
115. Mata R, Bhaskaran A, Sadras SR (2016) Green-synthesized gold nanoparticles from *Plumeria alba* flower extract to augment catalytic degradation of organic dyes and inhibit bacterial growth. Particuology 24:78–86

116. Miao Y, Luo X, Gao X, Wang W, Li B, Hou L (2020) Exogenous salicylic acid alleviates salt stress by improving leaf photosynthesis and root system architecture in cucumber seedlings. Sci Hortic 272:109577
117. Mishra AN, Bhadauria S, Gaur MS, Pasricha R, Kushwah BS (2010) Synthesis of gold nanoparticles by leaves of zero-calorie sweetener herb (*Stevia Rebaudiana*) and their nanoscopic characterization by spectroscopy and microscopy. Int J Green Nanotech Phys Chem 1:118–124
118. Moaveni P, Karimi K, Valojerdi MZ (2011) The nanoparticles in plants: review paper. J Nano Struct Chem 2:59–78
119. Mohammadi R, Maali-Amiri R, Mantri NL (2014) Effect of TiO2 nanoparticles on oxidative damage and antioxidant defense systems in chickpea seedlings during cold stress. Russ J Plant Physiol 61:768–775
120. Mohasseli V, Farbood F, Moradi A (2020) Antioxidant defense and metabolic responses of lemon balm (*Melissa officinalis* L.) to Fe-nano-particles under reduced irrigation regimes. Ind Crops Prod 149:112338
121. Moustafa NE, Alomari AA (2019) Green synthesis and bactericidal activities of isotropic and anisotropic spherical gold nanoparticles produced using *Peganum harmala* L. leaf and seed extracts. Biotechnol Appl Biochem 66:664–672
122. Mukherjee P, Ahmad A, Senapati S, Khan MI, Kumar R, Ramani R, Srinivas V, Sastri M, Ajaykumar PV, Alam M, Parischa R (2001) Bioreduction of AuCl4 ions by the fungus, Verticillium sp. and surface trapping of the gold nanoparticles formed. Angew Chem Int 40:3585–3588
123. Mustafa G, Sakata K, Komatsu S (2015) Proteomic analysis of flooded soybean root exposed to aluminum oxide nanoparticles. J Proteomics 128:280–297
124. Nagaraj B, Malakar B, Divya TK, Krishnamurthy NB, Liny P, Dinesh R (2012) Environmental benign synthesis of gold nanoparticles from the flower extracts of Plumeria Alba Linn. (Frangipani) and evaluation of their biological activities. Int J Drug Dev Res 4:144–150
125. Nair R (2016) Effects of nanoparticles on plant growth and development. In: Plant nanotechnology: principles and practices. Springer India, New Delhi, pp 95–118
126. Nair R, Varghese SH, Nair BG, Maekawa T, Yoshida Y, Kumar DS (2010) Nanoparticulate material delivery to plants. Plant Sci 179:154–163
127. Namvar F, Azizi S, Ahmad MB, Shameli K, Mohamad R, Mahdavi M, Tahir P (2015) Green synthesis and characterization of gold nanoparticles using the marine macroalgae Sargassum muticum. Res Chem Intermed 41:5723–5730
128. Ngo QB, Dao TH, Nguyen HC, Tran XT, Van Nguyen T, Khuu TD, Huynh TH (2014) Effects of nanocrystalline powders (Fe, Co and Cu) on the germination, growth, crop yield and product quality of soybean (*Vietnamese species* DT-51). Adv Nat Sci Nanosci Nanotechnol 5:015016
129. Noruzi M, Zare D, Khoshnevisan K, Davoodi D (2011) Rapid green synthesis of gold nanoparticles using *Rosa hybrida* petal extract at room temperature. Spectroc Acta Pt A-Molec Biomolec Spectr 79:1461–1465
130. Pallavi Mehta CM, Srivastava R, Arora S, Sharma AK (2016) Impact assessment of silver nanoparticles on plant growth and soil bacterial diversity. 3 Biotech 6:1–10
131. Palomo JM, Filice M (2016) Biosynthesis of metal nanoparticles: novel efficient heterogeneous nanocatalysts. Nanomaterials 6:84
132. Panichikkal J, Krishnankutty RE (2022) Chitosan and gold nanoparticles supplementation for augmentation of indole-3-acetic acid production by rhizospheric *Pseudomonas aeruginosa* and plant growth enhancement. Curr Microbiol 79:185
133. Panichikkal J, Thomas R, John JC, Radhakrishnan EK (2019) Biogenic Gold Nanoparticle Supplementation to Plant Beneficial Pseudomonas monteilii was Found to Enhance its Plant Probiotic Effect. Curr Microbiol 76:503–509
134. Parveen A, Mazhari BBZ, Rao S (2016) Impact of bio-nanogold on seed germination and seedling growth in *Pennisetum glaucum*. Enzyme Microb Technol 95:107–111
135. Patra JK, Das G, Fraceto LF, Campos EVR, Rodriguez-Torres MDP, Acosta-Torres LS, Shin HS (2018) Nano based drug delivery systems: recent developments and future prospects. J Nanobiotechnol 16:1–33

136. Pei X, Qu Y, Shen W, Li H, Zhang X, Li S, Zhang Z, Li X (2017) Green synthesis of gold nanoparticles using fungus Mariannaea sp. HJ and their catalysis in reduction of 4-nitrophenol. Environ Sci Pollut Res 24:21649–21659
137. Pérez-de-Luque A (2017) Interaction of nanomaterials with plants: what do we need for real applications in agriculture? Front Environ Sci 5:12
138. Pérez-Labrada F, López-Vargas ER, Ortega-Ortiz H, Cadenas-Pliego G, Benavides-Mendoza A, Juárez-Maldonado A (2019) Responses of tomato plants under saline stress to foliar application of copper nanoparticles. Plants 8:151
139. Pinchuk AP, Likhanov AF, Babenko LP, Kryvtsova MV, Demchenko OA, Sherbakov OB, Lazarenko LM, Spivak MY (2017) The influence of cerium dioxide nanoparticles on germination of seeds and plastic exchange of pine seedlings (*Pinus sylvestris* L.). Biotechnologia Acta 63–71
140. Prasad R, Pandey R, Barman I (2015) Engineering tailored nanoparticles with microbes: quo vadis? Wiley Interdiscip Rev Nanomed Nanobiotechnol 8:316–330
141. Prasad TNVKV, Sudhakar P, Sreenivasulu Y, Latha P, Munaswamy V, Reddy KR, Sreeprasad TS, Sajanlal PR, Pradeep T (2012) Effect of nanoscale zinc oxide particles on the germination, growth and yield of peanut. J Plant Nutrition 35:905–927
142. Rac-Rumijowska O, Fiedot M, Suchorska-Wozniak P, Teterycz H (2017) Synthesis of gold nanoparticles with different kinds of stabilizing agents. In: 40th international spring seminar on electronics technology (ISSE) 1–6
143. Rai PK, Kumar V, Lee S, Raza N, Kim K-H, Ok YS, Tsang DCW (2018) Nanoparticle-plant interaction: Implications in energy, environment, and agriculture. Environ Int 119:1–19
144. Rajathi FAA, Parthiban C, Kumar VG, Anantharaman P (2012) Biosynthesis of antibacterial gold nanoparticles using brown alga, *Stoechospermum marginatum* (kützing). Spectrochim Acta Part A Mol Biomol Spectrosc 99:166–173
145. Rajeshwari A, Suresh S, Chandrasekaran N, Mukherjee A (2016) Toxicity evaluation of gold nanoparticles using an *Allium cepa* bioassay. RSC Adv 6:24000–24009
146. Raliya R, Franke C, Chavalmane S, Nair R, Reed N, Biswas P (2016) Quantitative understanding of nanoparticle uptake in watermelon plants. Front Plant Sci 7:1288
147. Raschke G, Kowarik S, Franzl T, Solnnichsen C, Klar TA, Feldmann J, Nichtl A, Kulrzinger K (2003) Biomolecular recognition based on single gold nanoparticle light scattering. Nano Lett 3:935–938
148. Rastogi L, Arunachalam J (2013) Green synthesis route for the size-controlled synthesis of biocompatible gold nanoparticles using aqueous extract of garlic (*Allium sativum*). Adv Mat Lett 4:548–555
149. Rico CM, Lee SC, Rubenecia R, Mukherjee A, Hong J, Peralta-Videa JR, Gardea-Torresdey JL (2014) Cerium oxide nanoparticles impact yield and modify nutritional parameters in wheat (*Triticum aestivum* L.). J Agric Food Chem 62:9669–9675
150. Rico CM, Majumdar S, Duarte-Gardea M, Peralta-Videa JR, Gardea-Torresdey JL (2011) Interaction of nanoparticles with edible plants and their possible implications in the food chain. J Agric Food Chem 59:3485–3498
151. Rico CM, Peralta-Videa JR, Gardea-Torresdey JL (2015) Chemistry, biochemistry of nanoparticles, and their role in antioxidant defense system in plants. *In*: Nanotechnology and Plant Sciences: Nanoparticles and Their Impact on Plants (Eds. Rai M et al). Springer International Publishing, pp 1–17
152. Rizwan M, Ali S, ur Rehman MZ, Malik S, Adrees M, Qayyum MF, Alamri SA, Alyemeni MN, Ahmad P (2019) Effect of foliar applications of silicon and titanium dioxide nanoparticles on growth, oxidative stress, and cadmium accumulation by rice (*Oryza sativa*). Acta Physiol Plant 41:35
153. Rossi L, Zhang W, Schwab AP, Ma X (2017) Uptake, uptake, accumulation, and in planta distribution of coexisting cerium oxide nanoparticles and cadmium in *Glycine max* (L.) Merr. Environ Sci Technol 51:12815–12824
154. Ruttkay-Nedecky B, Krystofova O, Nejdl L, Adam V (2017) Nanoparticles based on essential metals and their phytotoxicity. J Nanobiotechnol 15:1–19

155. Sabo-Attwood T, Unrine JM, Stone JW, Murphy CJ, Ghoshroy S, Blom D, Bertsch PM, Newman LA (2012) Uptake, distribution and toxicity of gold nanoparticles in tobacco (*Nicotiana xanthi*) seedlings. Nanotoxicology 6:353–360
156. Sadak MS (2019) Impact of silver nanoparticles on plant growth, some biochemical aspects, and yield of fenugreek plant (*Trigonella foenum-graecum*). Bull Natl Res Cent 43:1–6
157. Sami F, Siddiqui H, Hayat S (2020) Impact of silver nanoparticles on plant physiology: a critical review. Sustain Agric Rev 41: Nanotechnol Plant Growth Develop 111–127
158. Saranya S, Aswani R, Remakanthan A, Radhakrishnan EK (2019) Nanotechnology in agriculture. In: Nanotechnology for agriculture: advances for sustainable agriculture (Eds. Hussain CM et al). Springer International Publishing, pp 1–17 [satya yellow]
159. Satya HK, Gupta S, Singh N, Khan T, Joshi S (2023) Nanofabrication of metals and their compounds for effective medicinal and environmental applications (a review). Russ J Gen Chem 93:635–665
160. Selivanov NY, Selivanova OG, Sokolov OI, Sokolova MK, Sokolov AO, Bogatyrev VA, Dykman LA (2017) Effect of gold and silver nanoparticles on the growth of the *Arabidopsis thaliana* cell suspension culture. Nanotechnol Russia 12:116–124
161. Senapati S, Syed A, Moeez S, Kumar A, Ahmad A (2012) Intracellular synthesis of gold nanoparticles using alga *Tetraselmis kochinensis*. Mater Lett 79:116–118
162. Shanwaz MM, Shyam P (2023) Anti-bacterial effect and characteristics of gold nanoparticles (AuNps) formed with *Vitex negundo* plant extract. Appl Biochem Biotechnol 195:1630–1643
163. Sharma NC, Sahi SV, Nath S, Parsons JG, Gardea-Torresdey JL, Pal T (2007) Synthesis of plant-mediated gold nanoparticles and catalytic role of biomatrix-embedded nanomaterials. Environ Sci Technol 41:5137–5142
164. Shivaji S, Madhu S, Singh S (2011) Extracellular synthesis of antibacterial silver nanoparticles using psychrophilic bacteria. Process Biochem 46:1800–1807
165. Shubin Y, Plyusnin P, Sharafutdinov M (2012) In situ synchrotron study of Au-Pd nanoporous alloy formation by single-source precursor thermolysis. Nanotechnology 23:405302–405307
166. Siddiqi KS, Husen A (2016) Engineered gold nanoparticles and plant adaptation potential. Nanoscale Res Lett 11:1–10
167. Siddiqi KS, Husen A (2017) Recent advances in plant-mediated engineered gold nanoparticles and their application in biological system. J Trace Elem Med Biol 40:10–23
168. Siegel J, Záruba K, Švorčík V, Kroumanová K, Burketová L, Martinec J (2018) Round-shape gold nanoparticles: effect of particle size and concentration on *Arabidopsis thaliana* root growth. Nanoscale Res Lett 13:1–7
169. Singaravelu G, Arockiamary JS, Kumar VG, Govindaraju KA (2007) Novel extracellular synthesis of monodisperse gold nanoparticles using marine alga *Sargassum wightii* Greville. Colloids Surf B 57:97–101
170. Singh J, Vishwakarma K, Ramawat N, Rai P, Singh VK, Mishra RK, Kumar V, Tripathi DK, Sharma S (2019) Nanomaterials and microbes' interactions: A contemporary overview. 3 Biotech 9:1–4
171. Slavin YN, Asnis J, Häfeli UO, Bach H (2017) Metal nanoparticles: Understanding the mechanisms behind antibacterial activity. J NanobioTechnol 15:65
172. Song JY, Jang HK, Kim BS (2009) Biological synthesis of gold nanoparticles using *Magnolia kobus* and *Diopyros kaki* leaf extracts. Process Biochem 44:1133–1138
173. Sonkar S, Sarangi PK, Pandey B, Prakash A, Singh AK (2023) Response of plant photosynthesis to nanomaterials. In: Nanomaterial interactions with plant cellular mechanisms and macromolecules and agricultural implications (Eds. Rai M et al). Springer International Publishing 49–67
174. Stephen JR, Macnaughtont SJ (1999) Developments in terrestrial bacterial remediation of metals. Curr Opin Biotechnol 10:230–233
175. Su YH, Tu SL, Tseng SW, Chang YC, Chang SH, Zhang WM (2010) Influence of surface plasmon resonance on the emission intermittency of photoluminescence from gold nano-sea-urchins. Nanoscale 2:2639–2646

176. Sun B, Hu N, Han L, Pi Y, Gao Y, Chen K (2019) Anticancer activity of green synthesised gold nanoparticles from *Marsdenia tenacissima* inhibits A549 cell proliferation through the apoptotic pathway. Artif Cells Nanomed Biotechnol 47:4012–4019
177. Sun L, Song F, Zhu X, Liu S, Liu F, Wang Y, Li X (2021) Nano-ZnO alleviates drought stress via modulating the plant water use and carbohydrate metabolism in maize. Arch Agron Soil Sci 67:245–259
178. Taghiyari HR, Morrell JF, Husen A (2023) Emerging Nanomaterials (Opportunities and Challenges in Forestry Sectors). Springer Nature Switzerland AG, Gewerbestrasse 11, 6330 Cham, Switzerland. https://doi.org/10.1007/978-3-031-17378-3
179. Tan C, Cao X, Wu XJ, He Q, Yang J, Zhang X, Chen J, Zhao W, Han S, Nam GH, Sindoro M (2017) Recent advances in ultrathin two-dimensional nanomaterials. Chem Rev 117:6225–6331
180. Tarafdar JC, Xiong Y, Wang WN, Quinl D, Biswas P (2012) Standardization of size, shape and concentration of nanoparticle for plant application. Appl Biol Res 14:138–144
181. Taylor AF, Rylott EL, Anderson CW, Bruce NC (2014) Investigating the toxicity, uptake, nanoparticle formation and genetic response of plants to gold. PLoS ONE 9:e93793
182. Thakur RK, Dhirta B, Shirkot P (2018) Studies on effect of gold nanoparticles on *Meloidogyne incognitaand* tomato plants growth and development. BioRxiv 428144
183. Thul ST, Sarangi BK (2015) Implications of nanotechnology on plant productivity and its rhizospheric environment. In: Rai M et al. (Eds.) Nanotechnology and plant sciences: nanoparticles and their impact on plants. Springer International Publishing, pp 37–53
184. Tiwari DK, Dasgupta-Schubert N, Villaseñor Cendejas LM, Villegas J, Carreto Montoya L, Borjas García SE (2014) Interfacing carbon nanotubes (CNT) with plants: enhancement of growth, water and ionic nutrient uptake in maize (*Zea mays*) and implications for nanoagriculture. Appl Nanosci 4:577–591
185. Treguer M, De Cointet C, Remita H, Khatouri J, Mostafavi M, Amblard J, Belloni J, De Keyzer R (1998) Dose rate effects on radiolytic synthesis of gold-silver bimetallic clusters in solution. J Phys Chem B 102:4310–4321
186. Tripathi D, Singh M, Pandey-Rai S (2022) Crosstalk of nanoparticles and phytohormones regulate plant growth and metabolism under abiotic and biotic stress. Plant Stress 100107
187. Tripathi DK, Shweta SS, Singh S, Pandey R, Singh VP, Sharma NC, Prasad SM, Dubey NK, Chauhan DK (2017) An overview on manufactured nanoparticles in plants: uptake, translocation, accumulation and phytotoxicity. Plant Physiol Biochem 110:2–12
188. Tsai SC, Song YL, Tsai CS, Yang CC, Chiu WY, Lin HM (2004) Ultrasonic spray pyrolysis for nanoparticles synthesis. J Mater Sci 39:3647–3657
189. Tyagi PK, Mishra R, Khan F, Gupta D, Gola D (2020) Antifungal effects of silver nanoparticles against various plant pathogenic fungi and its safety evaluation on *Drosophila melanogaster*. Biointerface Res Appl Chem 10:6587–6596
190. Unrine JM, Shoults-Wilson WA, Zhurbich O, Bertsch PM, Tsyusko OV (2012) Trophic transfer of Au nanoparticles from soil along a simulated terrestrial food chain. Environ Sci Technol 46:9753–9760
191. Van Aken B (2015) Gene expression changes in plants and microorganisms exposed to nanomaterials. Curr Opin Biotechnol 33:206–219
192. Venzhik YV, Moshkov IE, Dykman LA (2021) Gold nanoparticles in plant physiology: principal effects and prospects of application. Russ J Plant Physiol 68:401–412
193. Verma SK, Das AK, Patel MK, Shah A, Kumar V, Gantait S (2018) Engineered nanomaterials for plant growth and development: a perspective analysis. Sci Total Env 630:1413–1435
194. Wali M, Sajjad AS, Sumaira S, Muhammad N, Safia H, Muhammad J (2017) Green synthesis of gold nanoparticles and their characterizations using plant extract of *Papaver somniferum*. Nano Sci Nano Technol 11:118
195. Wang P, Lombi E, Zhao FJ, Kopittke PM (2016) Nanotechnology: a new opportunity in plant sciences. Trends Plant Sci 21:699–712
196. Wang WN, Tarafdar JC, Biswas P (2013) Nanoparticle synthesis and delivery by an aerosol route for watermelon plant foliar uptake. J Nanopart Res 15:14117

197. Wang X, Xie H, Wang P, Yin H (2023) Nanoparticles in plants: uptake, transport and physiological activity in leaf and root. Materials 16:3097
198. Wang Z, Li H, Li X, Xin C, Si J, Li S, Li Y, Zheng X, Li H, Wei X, Zhang Z (2020) Nano-ZnO priming induces salt tolerance by promoting photosynthetic carbon assimilation in wheat. Arch Agron Soil Sci 66:1259–1273
199. Wang Z, Xie X, Zhao J, Liu X, Feng W, White JC, Xing B (2012) Xylem-and phloem-based transport of CuO nanoparticles in maize (*Zea mays* L.). Environ sci technol 46:4434–4441
200. Wassie AT, Ahmed IN, Bachheti A, Husen A, Bachheti RK (2024a) Plant extract-assisted green synthesis and characterization of copper nanocomposite. In: Bachheti RK, Bachheti A, Husen A (eds) metal and metal-oxide based nanomaterials. Smart Nanomater Technol. Springer, Singapore. pp 43–66 https://doi.org/10.1007/978-981-99-7673-7_3
201. Wassie AT, Ahmed, IN, Bachheti A, Husen A, Bachheti RK (2024c) Role of Zinc Oxide Nanomaterials for Photocatalytic Degradation of Environmental Pollutants. In: Bachheti RK, Bachheti A, Husen A (eds) Metal and Metal-Oxide Based Nanomaterials. Smart Nanomaterials Technology. Springer, Singapore. pp 287–311 https://doi.org/10.1007/978-981-99-7673-7_14
202. Wassie AT, Pandey DP, Bachheti A, Husen A, Vashishth DS, Bachheti RK (2024b) Use of green synthesized platinum nanoparticles for biomedical applications. In: Bachheti RK, Bachheti A, Husen A (eds) Metal and metal-oxide based nanomaterials. Smart nanomaterials technology. Springer, Singapore. pp 271–285. https://doi.org/10.1007/978-981-99-7673-7_13
203. Wesołowska A, Jadczak P, Kulpa D, Przewodowski W (2019) Gas chromatography–mass spectrometry (GC–MS) analysis of essential oils from AgNPs and AuNPs elicited *Lavandula angustifolia in vitro* cultures. Molecules 24:606
204. Xingmao M, Jane G-L, Yang D, Andrei K (2010) Interactions between engineered nanoparticles (ENPs) and plants: phytotoxicity, uptake and accumulation. Sci Total Environ 408:3053–3061
205. Xu J, Yang J, Duan X, Jiang Y, Zhang P (2014) Increased expression of native cytosolic Cu/Zn superoxide dismutase and ascorbate peroxidase improves tolerance to oxidative and chilling stresses in cassava (*Manihot esculenta Crantz*). BMC Plant Biol 14:1–14
206. Yoon H, Kang YG, Chang YS, Kim JH (2019) Effects of zerovalent iron nanoparticles on photosynthesis and biochemical adaptation of soil-grown *Arabidopsis thaliana*. Nanomaterials 9:1543
207. Yu L, Andriola A (2010) Quantitative gold nanoparticle analysis methods: a review. Talanta 82:869–875
208. Yuan CG, Huo C, Yu S, Gui B (2017) Biosynthesis of gold nanoparticles using Capsicum annuum var. grossum pulp extract and its catalytic activity. Phys E 85:9–26
209. Zahra Z, Waseem N, Zahra R, Lee H, Badshah MA, Mehmood A, Choi HK, Arshad M (2017) Growth and metabolic responses of rice (*Oryza sativa* L.) cultivated in phosphorus-deficient soil amended with TiO_2 nanoparticles. J Agric Food Chem 65:5598–5606
210. Ze Y, Liu C, Wang L, Hong M, Hong F (2011) The regulation of TiO2 nanoparticles on the expression of light-harvesting complex II and photosynthesis of chloroplasts of *Arabidopsis thaliana*. Biol Trace Elem Res 143:1131–1141
211. Zhai G, Walters KS, Peate DW, Alvarez PJ, Schnoor J (2014) Transport of gold nanoparticles through plasmodesmata and precipitation of gold ions in woody poplar. Environ Sci Technol Lett 1:146–151
212. Zhang CL, Jiang HS, Gu SP, Zhou XH, Lu ZW, Kang XH, Yin L, Huang J (2019) Combination analysis of the physiology and transcriptome provides insights into the mechanism of silver nanoparticles phytotoxicity. Environ Pollut 252:1539–1549
213. Zhang J, Pan L, Lv M, Aldalbahi A, Xie T, Li A, Tai R, Huang Q, Fan C, Zhao Y, Zhu Y (2015) Transportation and fate of gold nanoparticles in oilseed rape. RSC Adv 5:73827–73833
214. Zhou P, Adeel M, Shakoor N, Guo M, Hao Y, Azeem I, Li M, Liu M, Rui Y (2020) Application of nanoparticles alleviates heavy metals stress and promotes plant growth: an overview. Nanomaterials 24;11(1):26

215. Zhou P, Adeel M, Shakoor N, Guo M, Hao Y, Azeem I, Li M, Liu M, Rui Y (2021) Application of nanoparticles alleviates heavy metals stress and promotes plant growth: an overview. Nanomaterials 11:26
216. Zhu ZJ, Wang H, Yan B, Zheng H, Jiang Y, Miranda OR, Rotello VM, Xing B, Vachet RW (2012) Effect of surface charge on the uptake and distribution of gold nanoparticles in four plant species. Environ Sci Technol 46:12391–12398

Beneficial and Adverse Effects of Au-NPs on Plant

Noureddine Chaachouay, Lahcen Zidane, and Azamal Husen

Abstract The application of gold nanoparticles in several domains, such as agriculture, has garnered considerable interest recently. This chapter thoroughly examines the dual influence of gold nanoparticles on plants, investigating their advantageous and detrimental effects. Gold nanoparticles have been observed to positively impact plant growth by promoting seed germination, enhancing nutrient uptake, and stimulating overall plant development. These nanoparticles can potentially be used as nanocarriers to deliver agrochemicals to specific targets, promoting sustainable and efficient agricultural practices. Nevertheless, in addition to their anticipated advantages, the chapter examines the possible disadvantages and adverse effects of gold nanoparticles on plant physiology. Investigations have uncovered cases where nanoparticles have caused oxidative stress, changes in gene expression, and disturbances in biological functions. To effectively utilize the beneficial properties of gold nanoparticles in plant systems and minimize any potential hazards, it is crucial to have a comprehensive understanding of the complex interactions between them. The chapter provides an overview of recent research findings, the methodology used to evaluate the impacts of gold nanoparticles on plants, and the current understanding of their ecological consequences. Furthermore, it explores methods for enhancing the effectiveness of gold nanoparticle usage in agriculture, aiming to achieve the most significant advantages while minimizing any negative environmental impact.

Keywords Effects · Gold nanoparticles · Nanotechnology · Plants · Synthesis

N. Chaachouay (✉)
Agri-Food and Health Laboratory (AFHL), ENS of Berrechid, Hassan First University, Address: Po Box 382, 26000 Settat, Morocco
e-mail: nour.chay@gmail.com

L. Zidane
Plant, Animal Productions and Agro-Industry Laboratory, Department of Biology, Faculty of Sciences, Ibn Tofail University, B.P. 133 14000, Kenitra, Morocco

A. Husen
Wolaita Sodo University, PO Box 138, Wolaita, Ethiopia

Department of Biotechnology, Graphic Era (Deemed to Be University), Dehradun, Uttarakhand 248002, India

A. Husen (ed.), *Plant Response to Gold Nanoparticles*, Smart Nanomaterials Technology, https://doi.org/10.1007/978-981-97-7174-5_2

List of Abbreviations

APVMA	Australian pesticides and veterinary medicines authority
Au-NPs	Gold nanoparticles
DAWE	Department of agriculture, water and the environment
DNA	Deoxyribonucleic acid
EFSA	European food safety authority
EPA	Environmental protection agency
OECD	Organization for economic co-operation and development
ROS	Reactive oxygen species
SPR	Surface plasmon resonance
US	United States

1 Introduction

In recent years, nanotechnology has experienced a notable upsurge in attention and implementations, and gold nanoparticles (Au-NPs) have surfaced as multifunctional substances across diverse domains [71]. Au-NPs, which vary in dimensions from one hundred nanometers, are exceptional in their optical, electronic, and catalytic characteristics, rendering them advantageous in various applications [25]. Their applications include the pharmaceutical, electronic, catalytic, and sensing technology sectors [34, 36, 43, 64]. Within agriculture and environmental research, the growing use of Au-NPs has highlighted the importance of fully comprehending their effects on plants [34, 63]. Since nanoparticles can infiltrate the soil via multiple pathways, such as irrigation and the decomposition of items containing nanomaterials, it is essential to evaluate their possible impact on plant physiology, growth, and overall well-being [35, 54]. The significance of understanding the interactions between Au-NPs and plants lies in their potential impact on agricultural productivity and environmental sustainability [10]. Given that agriculture is a fundamental aspect of the worldwide food supply, any modification in plant behavior caused by Au-NPs could have significant consequences for crop output and food security [48]. Furthermore, it is crucial to comprehend the environmental destiny of Au-NPs to avert unforeseen ecological repercussions and establish acceptable recommendations for using nanomaterials. This investigation into the impacts of Au-NPs on plants enhances the overall comprehension of interactions between nanomaterials and plants and provides valuable insights for developing sustainable farming methods. By clarifying how Au-NPs affect plants, scientists can establish principles for their secure and efficient utilization, reducing potential hazards and optimizing the advantages of nanotechnology in agriculture and environmental control [17]. This research is becoming more critical as the incorporation of nanomaterials into different industries continues to expand, highlighting the necessity for a comprehensive comprehension of their effects on our ecosystems and food systems. Within this perspective, the current chapter examines the dual impact of Au-NPs on plants, encompassing both beneficial and detrimental effects. This study examines these nanoparticles' advantageous and damaging implications within the framework of plant interactions.

2 Synthesis and Characteristics of Au-NPs

2.1 Au-NP Synthesis Methods

The synthesis of Au-NPs can be achieved through various methods, and the choice of method often depends on the nanoparticles' desired size, shape, and surface properties [61]. These methods can be broadly classified into top-down and bottom-up approaches [32]. Top-down approaches involve breaking down bulk gold into smaller particles. Some commonly used top-down methods include laser ablation, mechanical milling, and electrochemical erosion [15, 81]. Bottom-up approaches involve building up Au-NPs from smaller components. Standard bottom-up methods include chemical reduction, photochemical reduction, and biological synthesis (Table 1).

Each method has its advantages and disadvantages, and the choice of method depends on the desired size, shape, and properties of the Au-NPs [32]. For example, chemical reduction is a widely used method that is relatively simple and inexpensive but may require toxic chemicals. Biological synthesis, on the other hand, is an eco-friendlier method that uses biological agents such as bacteria, fungi, and plants to synthesize Au-NPs.

2.2 Chemical and Physical Properties of Au-NPs

Gold nanoparticles (Au-NPs) represent nanoscale particles composed of gold atoms, and their interaction with plants is significantly influenced by a combination of chemical and physical properties. Some of the essential factors to consider are (Fig. 1).

- **Size and shape of Au-NPs:** In terms of chemical properties, the surface area-to-volume ratio increases as particle size decreases, affecting the reactivity and availability of gold atoms on the surface. Physically, different shapes, such as spheres, rods, or triangles, can yield varying effects, influencing cellular uptake and interactions with cellular structures in plants [16].
- **Surface chemistry:** Chemically, the surface of Au-NPs is often functionalized with molecules like citrate or polyethylene glycol to enhance stability and reduce agglomeration. These chemical modifications can impact interactions with plant cells and tissues. On a physical level, the presence of a stabilizing agent plays

Table 1 Different methods of Au-NP synthesis

Method	Description	Key references
Biological synthesis	Biological or green synthesis involves using physical entities, such as plant extracts or microorganisms, as reducing and stabilizing agents for forming Au-NPs. This method is considered environmentally friendly and can result in nanoparticles with unique properties	[14, 59]
Photochemical reduction	Photochemical methods use light energy to drive the reduction of gold ions, leading to the formation of nanoparticles. This approach often involves using photoactive agents or irradiation to induce the reduction process. Photochemical methods offer precise control of the reaction by adjusting light exposure	[31, 46]
Chemical reduction	Chemical reduction methods involve using chemical agents, such as citrate or other reducing agents, to reduce gold ions into nanoparticles. The Turkevich and Brust–Schiffrin methods are examples of chemical reduction approaches. These methods offer control over particle size and shape	[21, 23]
Electrochemical erosion	Electrochemical erosion is a process where an electric current erodes or dissolves bulk gold, producing nanoparticles. Electrochemical methods offer control over the size and shape of the resulting nanoparticles through adjustments in the electrochemical conditions	[22, 50]
Mechanical milling	Mechanical milling involves using mechanical forces, such as grinding or milling, to break down bulk gold into smaller particles—this physical process reduces gold particle size. However, control over size distribution may be challenging	[27, 30]
Laser ablation	In laser ablation, a laser beam is focused on a gold target in a liquid medium. The intense energy generated by the laser ablates or vaporizes the gold, leading to the formation of nanoparticles. These nanoparticles are then stabilized in the surrounding liquid	[8, 37, 44]

a role in the overall stability and dispersibility of Au-NPs in plant environments [7].

- **Charge of Au-NPs:** Chemically, Au-NPs can carry a positive or negative charge based on synthesis methods and surface modifications. This charge can affect interactions with plant cell membranes, influencing uptake and transport. Physically, electrostatic interactions between charged Au-NPs and plant cell surfaces play a role in adhesion and internalization [7, 21].
- **Agglomeration state:** Chemically, accumulation is influenced by surface chemistry and the surrounding environment. Agglomerated nanoparticles may exhibit different interactions with plant surfaces compared to individual ones. On a physical level, the size and stability of agglomerates can affect the penetration of Au-NPs through plant cell walls and membranes [16].

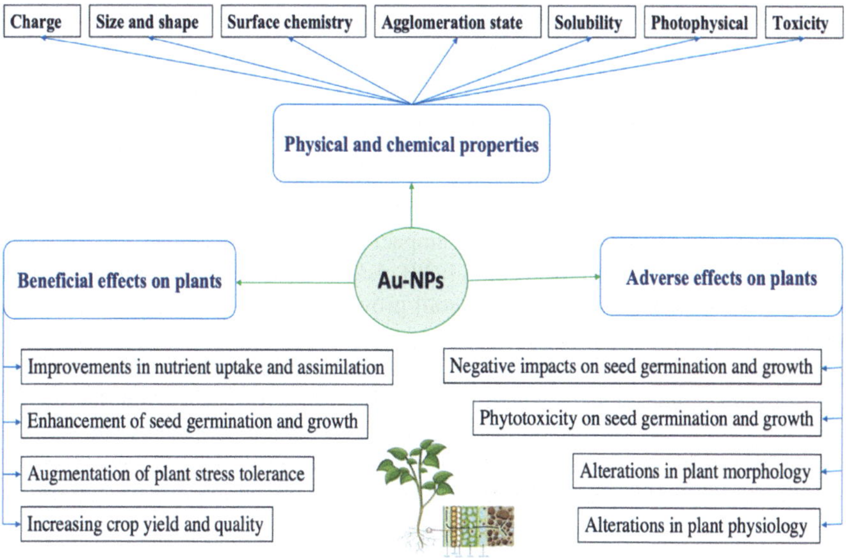

Fig. 1 Physical and chemical properties, beneficial and adverse effects of Au-NPs

- **Solubility of Au-NPs:** From a chemical standpoint, while gold is generally considered inert, the solubility of Au-NPs may play a role in their interaction with plant tissues. Au-NPs may release gold ions, influencing plant responses. Physically, the solubility of Au-NPs can impact their bioavailability and potential toxicity to plants [7].
- **Photophysical properties:** Chemically, they exhibit optical properties, including surface plasmon resonance (SPR). These properties can influence light absorption and scattering, potentially affecting plants' photosynthesis. On a physical level, the absorption and scattering of light by Au-NPs may have implications for plant growth and development, especially in light-sensitive processes [7].
- **Toxicity of Au-NPs:** Chemically, Au-NPs may exhibit toxicity depending on concentration and exposure duration [80]. The release of gold ions, if any, can contribute to toxicity. Physically, factors such as the physical form, size, and concentration of Au-NPs can impact their plant toxicity. Understanding these chemical and biological properties is essential for assessing the potential effects of Au-NPs in various applications related to plants, such as agriculture and phytoremediation [16].

3 Beneficial Effects of AU-NPS on Plants

3.1 Enhancement of Seed Germination and Growth

The application of Au-NPs in agriculture has demonstrated notable beneficial effects on the enhancement of seed germination and subsequent plant growth. Research findings suggest that Au-NPs can positively impact germination, leading to accelerated and more uniform seedling emergence [4, 28]. Additionally, these nanoparticles have been associated with promoting the overall growth of plants, influencing both root and shoot development. The enhanced germination and growth observed in the presence of Au-NPs could be attributed to their unique physicochemical properties, which may facilitate cellular processes and nutrient absorption [6]. As a result, the use of Au-NPs holds promise as a potential strategy to improve seedling establishment and foster robust plant growth, contributing to the advancement of sustainable agricultural practices [2]. However, it is crucial to continue research efforts to understand the mechanisms involved comprehensively and to address any potential environmental concerns associated with the application of nanoparticles in the context of plant development.

3.2 Improvements in Nutrient Uptake and Assimilation

Au-NPs have exhibited promising effects on improving nutrient uptake and assimilation in plants. Studies suggest that Au-NPs can positively influence the absorption of essential minerals from the soil, enhancing the plant's ability to take up nutrients crucial for growth and development [3, 67]. The unique surface properties of Au-NPs may facilitate interactions with plant root surfaces, promoting more efficient nutrient assimilation [12]. This heightened nutrient uptake can lead to healthier and more robust plants, with potential implications for increased agricultural productivity. As understanding the mechanisms underlying these effects continues to advance, the application of Au-NPs in agriculture can contribute to sustainable practices by optimizing nutrient utilization and promoting plant health [49]. However, ongoing research is essential to elucidate the specific pathways and conditions governing the interaction between Au-NPs and plant nutrient dynamics, ensuring safe and effective implementation in real-world agricultural settings.

3.3 Augmentation of Plant Stress Tolerance

The beneficial effects of Au-NPs on plant stress tolerance, particularly in the face of challenges such as drought, salinity, and heavy metal stress, have increased interest in scientific research. In drought stress, Au-NPs have demonstrated the ability to enhance the drought tolerance of plants [76]. This is often attributed to the nanoparticles' capacity to regulate water uptake and maintain cellular hydration levels and their potential role in activating stress-responsive genes and antioxidant systems within plant cells [57]. Similarly, in the case of salinity stress, Au-NPs have shown promise in mitigating the adverse effects by influencing ion homeostasis and minimizing the impact of excess salts on plant tissues [62]. Moreover, Au-NPs exhibit the potential to address heavy metal stress in plants, a common issue in contaminated soils [11]. These nanoparticles may contribute to reducing heavy metal toxicity by processes such as chelation or reduction, preventing the metals from exerting harmful effects on plant physiological processes [58]. The capacity of Au-NPs to enhance antioxidant defenses also plays a crucial role in mitigating the oxidative stress induced by heavy metals.

3.4 Potential Applications in Agriculture for Increased Crop Yield and Quality

The potential applications of Au-NPs in agriculture promise substantial crop yield and quality improvements. One key avenue through which Au-NPs may enhance agricultural productivity is by positively influencing various stages of plant development. Studies indicate that Au-NPs can expedite seed germination and promote seedling growth [56, 75]. This early boost in plant development can set the stage for increased crop yields. Furthermore, Au-NPs have been linked to improvements in nutrient uptake and assimilation [12]. By enhancing the plant's ability to absorb essential minerals from the soil, these nanoparticles contribute to better nutrition, leading to healthier and more robust crops [26]. The overall enhancement of plant health and growth can translate into increased biomass and, ultimately, higher yields.

Au-NPs also promise to mitigate the impact of environmental stressors, such as drought and salinity, as previously discussed [76]. By bolstering stress tolerance, these nanoparticles have the potential to protect crops from adverse conditions, ensuring more consistent yields even in challenging environments. The use of Au-NPs in agriculture aligns with the goal of sustainable farming practices [29]. By optimizing nutrient utilization, promoting stress resistance, and fostering overall plant health, Au-NPs could contribute to more efficient and environmentally friendly crop production [49]. However, as with any emerging technology, carefully considering factors such as nanoparticle concentration, application methods, and potential ecological impacts is crucial to ensure responsible and practical implementation in real-world agricultural settings.

3.5 *Mechanisms Underlying the Beneficial Effects*

In recent years, scientific research has delved into the advantageous impact of Au-NPs on plant growth and stress resilience. Despite the evolving nature of our understanding, key aspects have emerged, particularly in the burgeoning field of nanotechnology in agriculture [51]. The relatively nascent application of nanoparticles like Au-NPs prompts ongoing investigations, refining our comprehension of the fundamental mechanisms involved [82]. The crucial aspect is the enhanced nutrient uptake observed in plants treated with Au-NPs. The increased surface area of these nanoparticles facilitates the absorption of essential minerals and nutrients by plant roots from the soil, fostering overall growth.

Another noteworthy effect is the stimulation of enzyme activity by Au-NPs, influencing various enzymes crucial for plant growth and stress responses [9]. The heightened activity of these enzymes contributes to improved plant performance [75]. Additionally, plants treated with Au-NPs exhibit an augmented antioxidant defense system, indicating the potential of these nanoparticles to enhance a plant's ability to combat oxidative stress. Furthermore, Au-NPs can modulate gene expression and signal transduction in plants, positively influencing their development and stress tolerance [6]. The potential contribution of Au-NPs to better water use efficiency and their role as nano-fertilizers further underscores their significance in promoting sustainable plant growth. While these mechanisms present promising avenues for utilizing Au-NPs in agriculture, it is essential to consider the variability in effects based on factors such as nanoparticle concentration, size, and plant species [49]. Furthermore, the ecological implications and long-term effects necessitate thorough investigation [61]. The dynamic nature of nanotechnology in agriculture underscores the need for continuous research to unravel the intricacies of Au-NPs' beneficial effects on plant growth and stress tolerance.

3.6 *Adverse Effects oF Au-NPs on Plants*

3.6.1 Phytotoxicity and Negative Impacts on Seed Germination and Growth

The utilization of Au-NPs in diverse applications, including agriculture, has prompted apprehension regarding their potential detrimental effects on plants. Despite possessing unique properties that render them appealing for specific purposes, the interactions between gold nanoparticles and biological systems, particularly plants, still need to be understood. One primary concern revolves around the phytotoxicity of Au-NPs, where their presence can impede cellular structures or release ions, resulting in diminished plant growth, development, and overall health [69, 74].

Furthermore, studies have indicated that Au-NPs may hinder seed germination by disrupting the normal physiological and biochemical processes associated with this crucial stage of plant development [56, 67, 74]. The interference could affect water uptake, enzyme activity, and nutrient mobilization during germination. Additionally, the impact extends to root growth inhibition, diminishing nutrient uptake and compromising the plant's ability to anchor itself in the soil [66]. Oxidative stress induced by Au-NPs, characterized by the generation of ROS, poses another threat by causing damage to cellular components such as proteins, lipids, and DNA, potentially leading to cell death [13, 79]. The bioaccumulation and translocation of gold nanoparticles within plants, including edible parts, raise concerns about their transfer through the food chain, impacting human health. There is also apprehension that Au-NPs may induce genotoxic effects in plants, potentially resulting in DNA damage with long-term consequences on genetic diversity and resilience within plant populations.

Alterations in Plant Morphology and Physiology

Au-NPs possess the potential to elicit profound effects on plants, instigating notable alterations in both morphology and physiology based on existing knowledge up to the present [18]. Morphological changes associated with exposure to Au-NPs include root growth inhibition, where certain studies propose a reduction in root length and overall root development in plants [9]. Furthermore, shifts in shoot growth parameters, encompassing alterations in plant height, leaf size, and branching patterns, have been documented as responses to nanoparticle exposure [74]. These morphological variations underscore the intricate interactions between gold nanoparticles and plant structures.

Regarding physiological changes, the impact of Au-NPs extends to critical aspects of plant functioning. Photosynthetic activity, a fundamental process for plant life, is susceptible to nanoparticle influence, leading to modifications in chlorophyll content, photosynthetic rate, and efficiency [72]. Consequently, such changes can result in diminished overall plant productivity. Additionally, nanoparticle interference with nutrient uptake and plant transport mechanisms may induce nutrient deficiencies, adversely affecting plant health [33]. The induction of oxidative stress is another physiological consequence, as exposure to nanoparticles can upset the balance between ROS production and the plant's antioxidant defense system. This oxidative stress can manifest in cellular damage, potentially compromising plant cells' structural integrity and functionality [1, 39]. Beyond these physiological effects, exposure to gold nanoparticles is linked to genetic and molecular changes, influencing gene expression in various physiological processes and disrupting cellular signaling pathways crucial for average plant growth and development.

3.7 *Factors Influencing the Interactions*

3.7.1 Size, Shape, and Concentration of Au-NPs

The interactions between Au-NPs and plants are influenced by several key factors, with the size, shape, and concentration of Au-NPs playing a pivotal role. The size of Au-NPs is critical as it affects their uptake and transport within plant tissues [38]. Smaller nanoparticles may have increased mobility, potentially facilitating their entry into plant cells and influencing cellular responses. Additionally, the shape of Au-NPs can impact their interactions with plant structures. Nanoparticles with different shapes may exhibit distinct affinities for specific plant cell components, influencing their distribution and effects within the plant system [12]. Concentration is another crucial factor governing the interactions of Au-NPs with plants. The number of Au-NPs present in the environment can influence their uptake by plants and subsequent physiological effects [9]. High concentrations of Au-NPs may increase accumulation within plant tissues, potentially resulting in altered growth patterns or stress responses. Conversely, low concentrations may elicit minimal effects, emphasizing the importance of understanding the dose-dependent nature of Au-NP interactions with plants [75]. Investigations must consider these factors collectively to comprehensively assess the impact of Au-NPs on plant biology and ecology.

Soil Conditions and Plant Species-Specific Responses

The interactions between Au-NPs and plants are intricately influenced by various factors, with soil conditions and plant species-specific responses playing pivotal roles. Firstly, the pH of the soil emerges as a critical determinant. Soil acidity or alkalinity significantly affects the behavior of Au-NPs, influencing their stability and reactivity [20]. Additionally, soil composition, encompassing mineral content and organic matter, contributes to the dynamic interplay. Variations in soil components can impact the availability of nanoparticles, subsequently influencing their uptake by plant roots [5]. Secondly, soil moisture content is a crucial factor influencing the interactions between Au-NPs and plants. Fluctuations in soil moisture levels alter the mobility and transport of nanoparticles, impacting their distribution and accessibility to plant roots [68]. The moisture content of the soil is intimately linked to the plant's physiological state, and understanding this relationship is crucial for assessing the potential effects of Au-NPs on plant growth and development [66]. Lastly, the responses of different plant species to Au-NPs are diverse and species-specific. Factors such as root morphology, genetic variability, and metabolic pathways significantly determine how plants interact with nanoparticles [9]. Some plants may exhibit a higher affinity for uptake, while others may possess mechanisms to mitigate or sequester nanoparticles [66]. This inherent variability underscores the importance of considering plant species-specific responses when studying the interactions between Au-NPs and plants, as it contributes to a more nuanced understanding of the environmental implications associated with nanoparticle exposure in diverse ecosystems.

Duration of Exposure and Accumulation Patterns

The duration of exposure and accumulation patterns intricately influence the interactions of Au-NPs with plants. The term exposure refers to the length of time that plants are subjected to Au-NPs, and this temporal factor plays a crucial role in determining the extent of nanoparticle uptake and subsequent effects on plant biology [73]. More extended exposure periods may lead to increased accumulation of Au-NPs within plant tissues, potentially impacting various physiological processes. Additionally, accumulation patterns describe the distribution of Au-NPs within different plant organs and tissues over time [78]. Understanding these patterns is essential for assessing the potential translocation of Au-NPs from roots to shoots and their subsequent effects on plant growth and development. Researchers investigating the environmental implications of Au-NPs on plants must consider the dynamic interplay between exposure duration and accumulation patterns to evaluate the impact on plant health and ecosystem dynamics comprehensively.

3.8 Environmental Fate and Transport

The fate of Au-NPs in the soil–plant system is a multifaceted process influenced by various environmental factors [52]. When Au-NPs are introduced into the soil, their interactions with soil components play a crucial role. The surface properties of the nanoparticles, including size and charge, influence their adsorption onto soil particles [19]. Additionally, soil characteristics such as texture, organic matter content, and pH impact the mobility and transport of Au-NPs within the soil profile. Once in contact with plant roots, Au-NPs can be taken up through processes that are not fully elucidated, and their subsequent translocation within the plant varies across different species. Microbial interactions in the soil further complicate the fate of Au-NPs, as microorganisms may participate in transforming nanoparticles, altering their size, shape, and surface chemistry [24].

Environmental redox conditions and pH fluctuations can affect the stability of Au-NPs, while the presence of organic matter in the soil can influence their interactions [65]. The fate of Au-NPs also holds implications for human and environmental health. Concerns arise from the potential bioaccumulation of these nanoparticles in edible plant parts, posing risks for human exposure [24]. Beyond human health, the ecological impact of Au-NPs in the soil–plant system extends to soil health, microbial communities, and the overall ecosystem. The potential for bioaccumulation and translocation of Au-NPs within plants is a subject of increasing interest in environmental science. Studies have shown that Au-NPs, when present in soil or water, can be absorbed by plant roots and subsequently translocated to various plant tissues

[12, 42, 45]. This phenomenon raises concerns about the potential for bioaccumulation, where the nanoparticles may accumulate in higher concentrations within the plant than in the surrounding environment. The mechanisms governing the uptake and translocation of Au-NPs within plants are complex and can be influenced by factors such as nanoparticle size, surface coating, and plant species [47]. Understanding these processes is crucial for assessing the potential environmental impact of Au-NPs and developing strategies to mitigate any adverse effects on plant ecosystems. Accurate analytical techniques, such as transmission electron microscopy and inductively coupled plasma mass spectrometry, are crucial for monitoring the fate of Au-NPs and understanding their dynamics in the complex soil–plant environment.

3.9 Regulatory and Ethical Considerations

Au-NPs are nanosized particles of gold that have potential applications in various fields, including agriculture. However, using Au-NPs in agriculture may pose risks to human health and the environment, such as toxicity, bioaccumulation, and environmental fate [52]. Therefore, some regulations and guidelines aim to ensure the safe and sustainable use of Au-NPs in agriculture. For example, in Australia, the Australian Pesticides and Veterinary Medicines Authority (APVMA) has issued a guideline for regulating biological agricultural products, including a nanotechnology section [55]. The guideline provides information on the safety assessment, labeling, and registration requirements for natural products that contain or are intended to be used with nanotechnology.

Similarly, the US Environmental Protection Agency (EPA) has developed a nonpoint source program for agriculture that addresses the potential impacts of agricultural activities on water quality [40, 53]. The program guides how to prevent or reduce nutrient pollution from agricultural sources through best management practices. The Australian Department of Agriculture, Water and the Environment (DAWE) has also established animal welfare standards and guidelines for particular animal industries. The European Food Safety Authority (EFSA) evaluates the safety of Au-NPs as novel food ingredients in the European Union [77]. In Canada, Health Canada regulates nanomaterials in agriculture to ensure safety [60]. The Australian Pesticides and Veterinary Medicines Authority (APVMA) oversees Au-NPs used in agricultural products falling under the definition of pesticides [41]. International guidelines from organizations such as the Organization for Economic Co-operation and Development (OECD) and the Codex Alimentarius Commission may influence the regulation of Au-NPs in agriculture globally [70]. The standards and guidelines cover housing, feeding, handling, transport, slaughter, and disposal of animals. These regulations and procedures aim to protect human health and animal welfare while promoting the development and innovation of nanotechnology in agriculture.

The ethical considerations surrounding using Au-NPs in agriculture extend to potential impacts on the environment and human health. Environmental concerns include the risk of Au-NPs accumulating in soil and water ecosystems, posing potential threats to non-target organisms and overall ecosystem health. Questions about long-term effects on biodiversity and ecosystems necessitate thorough examination. On the human health front, concerns about potential exposure to Au-NPs through the consumption of treated agricultural products or environmental exposure prompted inquiries into their toxicity and bioaccumulative effects. Ethical considerations emphasize the importance of transparent communication regarding Au-NPs, ensuring that information about potential risks and benefits is accessible to the public and agricultural stakeholders. Striking a balance between technological innovation and the precautionary principle is essential, emphasizing thorough risk assessments, continuous monitoring, and adherence to evolving regulatory frameworks to mitigate any adverse effects of Au-NPs in agriculture. Ethical guidelines should prioritize public engagement, inclusivity in decision-making processes, and responsible innovation to address concerns and promote sustainable agricultural practices.

3.10 Future Perspectives

In future research directions, it is imperative to delve into the multifaceted applications of Au-NPs in agriculture. While existing studies have illuminated the potential of Au-NPs in enhancing crop productivity and stress tolerance, further investigation is warranted to comprehensively understand their long-term effects on soil health, plant physiology, and ecosystem dynamics [49, 76]. Unraveling the intricate mechanisms governing the interaction between Au-NPs and plant systems will contribute to developing targeted nanomaterial applications, minimizing unintended environmental consequences. Moreover, exploring innovative synthesis methods and surface modifications to tailor Au-NPs for specific agricultural challenges could open new avenues for sustainable crop management.

Strategic approaches must be devised to ensure the safe and sustainable use of Au-NPs in agriculture. Rigorous risk assessment protocols should be established to evaluate the potential ecological impacts and toxicity of Au-NPs, addressing concerns related to bioaccumulation and unintended consequences on non-target organisms. Additionally, guidelines for optimal application dosages and frequencies must be formulated to prevent overuse and minimize adverse effects. Collaboration between scientists, regulatory bodies, and agricultural stakeholders is crucial for developing a framework that balances the benefits of Au-NP technology with environmental safety. Implementing precision agriculture practices, incorporating real-time monitoring, and fostering education and awareness among farmers can further enhance the responsible utilization of Au-NPs, ensuring a sustainable integration of nanotechnology in modern agriculture.

3.10.1 Conclusion

The intricate relationship between Au-NPs and plant systems reveals a dichotomy of both advantageous and detrimental effects. On a positive note, Au-NPs exhibit potential benefits such as enhanced plant growth, improved nutrient absorption, and heightened stress resistance. These findings open avenues for innovative applications in agriculture to address global food security challenges. Nevertheless, the shadow of potential adverse consequences looms, with concerns ranging from soil health impacts to plant toxicity issues demanding careful consideration. As we navigate the complexities of Au-NP interactions with plants, a nuanced understanding emerges as a prerequisite for responsible utilization. Striking a delicate balance between exploiting the positive attributes of Au-NPs for agricultural advancement and mitigating their negative repercussions demands sustained research efforts. Ethical and environmental considerations should guide the integration of Au-NPs into agricultural practices, fostering sustainable solutions for the future. A comprehensive approach that combines scientific rigor, moral discernment, and environmental stewardship is imperative in this evolving landscape. Continued exploration of the underlying mechanisms and long-term implications will empower us to harness the potential benefits of Au-NPs while safeguarding the integrity of our ecosystems. Leveraging Au-NPs for plant enhancement requires a commitment to sustainable practices, ethical considerations, and ongoing scientific inquiry.

References

1. Aazami MA, Rasouli F, Ebrahimzadeh A (2021) Oxidative damage, antioxidant mechanism and gene expression in tomato responding to salinity stress under in vitro conditions and application of iron and zinc oxide nanoparticles on callus induction and plant regeneration. BMC Plant Biol 21(1):597. https://doi.org/10.1186/s12870-021-03379-7
2. Abbasi Khalaki M, Moameri M, Asgari Lajayer B, Astatkie T (2021) Influence of nano-priming on seed germination and plant growth of forage and medicinal plants. Plant Growth Regul 93(1):13–28. https://doi.org/10.1007/s10725-020-00670-9
3. Achari GA, Kowshik M (2018) Recent developments on nanotechnology in agriculture: plant mineral nutrition, health, and interactions with soil microflora. J Agric Food Chem 66(33):8647–8661. https://doi.org/10.1021/acs.jafc.8b00691
4. Acharya P, Jayaprakasha GK, Crosby KM, Jifon JL, Patil BS (2019) Green-synthesized nanoparticles enhanced seedling growth, yield, and quality of onion (*Allium cepa* L.). ACS Sustain Chem Eng 7(17):14580–14590. https://doi.org/10.1021/acssuschemeng.9b02180
5. Ahmed B, Rizvi A, Ali K, Lee J, Zaidi A, Khan MS, Musarrat J (2021) Nanoparticles in the soil–plant system: a review. Environ Chem Lett 19(2):1545–1609. https://doi.org/10.1007/s10311-020-01138-y
6. Ali S, Mehmood A, Khan N (2021) Uptake, translocation, and consequences of nanomaterials on plant growth and stress adaptation. J Nanomater 2021:1–17
7. Al-wardy RA, Rahi SK (2021) The physical properties and applications of gold nanoparticles (Au NPs). Samarra J Pure Appl Sci 3(1):74–86

8. Amendola V, Polizzi S, Meneghetti M (2006) Laser ablation synthesis of gold nanoparticles in organic solvents. J Phys Chem B 110(14):7232–7237. https://doi.org/10.1021/jp0605092
9. Aqeel U, Aftab T, Khan MMA, Naeem M, Khan MN (2022) A comprehensive review of impacts of diverse nanoparticles on growth, development and physiological adjustments in plants under changing environment. Chemosphere 291:132672
10. Aromal SA, Philip D (2012) Green synthesis of gold nanoparticles using Trigonella foenum-graecum and its size-dependent catalytic activity. Spectrochim Acta A Mol Biomol Spectrosc 97:1–5
11. Arsenov D, Beljin J, Jović D, Maletić S, Borišev M, Borišev I (2023) Nanomaterials as endorsed environmental remediation tools for the next generation: eco-safety and sustainability. J Geochem Explor 253:107283
12. Avellan A, Yun J, Zhang Y, Spielman-Sun E, Unrine JM, Thieme J, Li J, Lombi E, Bland G, Lowry GV (2019) Nanoparticle size and coating chemistry control foliar uptake pathways, translocation, and leaf-to-rhizosphere transport in wheat. ACS Nano 13(5):5291–5305. https://doi.org/10.1021/acsnano.8b09781
13. Balkrishna A, Kumar A, Arya V, Rohela A, Verma R, Nepovimova E, Krejcar O, Kumar D, Thakur N, Kuca K (2021) Phytoantioxidant functionalized nanoparticles: a green approach to combat nanoparticle-induced oxidative stress. Oxid Med Cell Longev 2021:1–20
14. Bharadwaj KK, Rabha B, Pati S, Sarkar T, Choudhury BK, Barman A, Bhattacharjya D, Srivastava A, Baishya D, Edinur HA (2021) Green synthesis of gold nanoparticles using plant extracts as beneficial prospect for cancer theranostics. Molecules 26(21):6389
15. Chaachouay N, Azeroual A, Benkhnigue O, Zidane L (2023) Alkaloids: a suitable precursor for nanomaterials synthesis, and their various applications. In: Husen A (eds) Secondary metabolites based green synthesis of nanomaterials and their applications. Smart nanomaterials technology. Springer, Singapore. https://doi.org/10.1007/978-981-99-0927-8_2
16. Chaachouay N, Zidane L (2024) Role of carbon nanotubes for herbicide detection and remediation. In: Bachheti A, Bachheti RK, Husen A (eds) Carbon-based nanomaterials. Smart nanomaterials technology. Springer, Singapore. https://doi.org/10.1007/978-981-97-0240-4_8
17. Chaachouay N, Douira A, Zidane L (2021) COVID-19, prevention and treatment with herbal medicine in the herbal markets of Salé Prefecture, North-Western Morocco. Eur J Integr Med 42:101285. https://doi.org/10.1016/j.eujim.2021.101285
18. Chaudhary RG, Bhusari GS, Tiple AD, Rai AR, Somkuvar SR, Potbhare AK, Lambat TL, Ingle PP, Abdala AA (2019) Metal/metal oxide nanoparticles: toxicity, applications, and future prospects. Curr Pharm Des 25(37):4013–4029
19. Darlington TK, Neigh AM, Spencer MT, Guyen OTN, Oldenburg SJ (2009) Nanoparticle characteristics affecting environmental fate and transport through soil. Environ Toxicol Chem 28(6):1191–1199. https://doi.org/10.1897/08-341.1
20. De Sousa BT, De Oliveira JL, Caixeta Oliveira H, De Castro VLSS (2022) Balancing the benefits to agriculture and adverse ecotoxicological impacts of inorganic nanoparticles. In Fernandes Fraceto L, Pereira De Carvalho HW, De Lima R, Ghoshal S, Santaella C (Éds.) Inorganic nanopesticides and nanofertilizers (pp 1–51). Springer International Publishing. https://doi.org/10.1007/978-3-030-94155-0_1
21. De Souza CD, Nogueira BR, Rostelato MEC (2019) Review of the methodologies used in the synthesis gold nanoparticles by chemical reduction. J Alloy Compd 798:714–740
22. Deng M, Zhang X, Zhang Z, Xin Z, Song Y (2014) A gold nanoparticle ink suitable for the fabrication of electrochemical electrode by inkjet printing. J Nanosci Nanotechnol 14(7):5114–5119
23. Dumur F, Guerlin A, Dumas E, Bertin D, Gigmes D, Mayer CR (2011) Controlled spontaneous generation of gold nanoparticles assisted by dual reducing and capping agents. Gold Bulletin 44(2):119–137. https://doi.org/10.1007/s13404-011-0018-5
24. Dwivedi AD, Dubey SP, Sillanpää M, Kwon Y-N, Lee C, Varma RS (2015) Fate of engineered nanoparticles: implications in the environment. Coord Chem Rev 287:64–78
25. Elahi N, Kamali M, Baghersad MH (2018) Recent biomedical applications of gold nanoparticles: a review. Talanta 184:537–556

26. Elemike EE, Uzoh IM, Onwudiwe DC, Babalola OO (2019) The role of nanotechnology in the fortification of plant nutrients and improvement of crop production. Appl Sci 9(3):499
27. El-Eskandarany MS, Al-Hazza A, Al-Hajji LA, Ali N, Al-Duweesh AA, Banyan M, Al-Ajmi F (2021) Mechanical milling: a superior nanotechnological tool for fabrication of nanocrystalline and nanocomposite materials. Nanomaterials 11(10):2484
28. El-Moneim DA, Dawood MFA, Moursi YS, Farghaly AA, Afifi M, Sallam A (2021) Positive and negative effects of nanoparticles on agricultural crops. Nanotechnol Environ Eng 6(2):21. https://doi.org/10.1007/s41204-021-00117-0
29. Fiol DF, Terrile MC, Frik J, Mesas FA, Álvarez VA, Casalongué CA (2021) Nanotechnology in plants: recent advances and challenges. J Chem Technol Biotechnol 96(8):2095–2108. https://doi.org/10.1002/jctb.6741
30. Gorrasi G, Sorrentino A (2015) Mechanical milling as a technology to produce structural and functional bio-nanocomposites. Green Chem 17(5):2610–2625
31. Grzelczak M, Liz-Marzán LM (2020) The relevance of light in the formation of colloidal metal nanoparticles. Colloid Synthesis Plasmon Nanometals 515–535
32. Hammami I, Alabdallah NM (2021) Gold nanoparticles: Synthesis properties and applications. J King Saud Univ-Sci 33(7):101560
33. Hong J, Wang C, Wagner DC, Gardea-Torresdey JL, He F, Rico CM (2021) Foliar application of nanoparticles: mechanisms of absorption, transfer, and multiple impacts. Environ Sci Nano 8(5):1196–1210
34. Husen A (2017) Gold nanoparticles from plant system: synthesis, characterization and their application, In: Nanoscience and plant–soil systems Vol.–48 (Eds. Ghorbanpourn M, Manika K, Varma A) Springer International Publishing AG, Gewerbestrasse 11, 6330 Cham pp 455–479. https://doi.org/10.1007/978-3-319-46835-8_17
35. Husen A (2023) Nanomaterials and nanocomposites exposures to plants (Response, Interaction, Phytotoxicity and Defense Mechanisms). Springer Nature Singapore Pte Ltd., 152 Beach Road, #21-01/04 Gateway East, Singapore 189721, Singapore. https://doi.org/10.1007/978-981-99-2419-6
36. Husen A, Rahman QI, Iqbal M, Yassin MO (2019) Plant-mediated fabrication of gold nanoparticles and their applications. In: N*anomaterials* and plant potential (Eds. Husen A, Iqbal M) Springer International Publishing AG, Gewerbestrasse 11, 6330 Cham, pp 71–110 https://doi.org/10.1007/978-3-030-05569-1_3
37. Imam H, Elsayed K, Ahmed MA, Ramdan R (2012) Effect of experimental parameters on the fabrication of gold nanoparticles via laser ablation. https://www.scirp.org/html/2-1190063_20398.htm
38. Ishida T, Murayama T, Taketoshi A, Haruta M (2020) Importance of size and contact structure of gold nanoparticles for the genesis of unique catalytic processes. Chem Rev 120(2):464–525. https://doi.org/10.1021/acs.chemrev.9b00551
39. Kumar V, Sharma M, Khare T, Wani SH (2018) Impact of nanoparticles on oxidative stress and responsive antioxidative defense in plants. In Nanomaterials in plants, algae, and microorganisms, pp 393–406. Elsevier. https://www.sciencedirect.com/science/article/pii/B9780128114872000177
40. Kumari R, Suman K, Karmakar S, Lakra SG, Saurav GK, Mahto BK (2023) Regulation and safety measures for nanotechnology-based agri-products. Front Genome Editing 5:1200987
41. Ludlow K (2010) Australian pesticides and veterinary medicines authority. Encyclop Nanosci Soc 1:38–39
42. Lv J, Christie P, Zhang S (2019) Uptake, translocation, and transformation of metal-based nanoparticles in plants: recent advances and methodological challenges. Environ Sci Nano 6(1):41–59
43. Madhusudhan A, Bandi R, Sharma G, Dadigala R, Husen A, Kim JC (2022) Current trends in engineered gold nanoparticles for cancer therapy. In: Kim JC, Alle M, Husen A (eds) Smart nanomaterials in biomedical applications. Nanotechnol Life Sci. Springer, Cham. pp 1–40. https://doi.org/10.1007/978-3-030-84262-8_1

44. Mafuné F, Kohno J, Takeda Y, Kondow T, Sawabe H (2001) Formation of gold nanoparticles by laser ablation in aqueous solution of surfactant. J Phys Chem B 105(22):5114–5120. https://doi.org/10.1021/jp0037091
45. Malejko J, Godlewska-Żyłkiewicz B, Vanek T, Landa P, Nath J, Dror I, Berkowitz B (2021) Uptake, translocation, weathering and speciation of gold nanoparticles in potato, radish, carrot and lettuce crops. J Hazard Mater418:126219
46. McGilvray KL, Fasciani C, Bueno-Alejo CJ, Schwartz-Narbonne R, Scaiano JC (2012) Photochemical strategies for the seed-mediated growth of gold and gold-silver nanoparticles. Langmuir 28(46):16148–16155. https://doi.org/10.1021/la302814v
47. Mishra V, Mishra RK, Dikshit A, Pandey AC (2014) Interactions of nanoparticles with plants : An emerging prospective in the agriculture industry. In: Emerging technologies and management of crop stress tolerance, pp 159–180. Elsevier. https://www.sciencedirect.com/science/article/pii/B9780128008768000084
48. Mittal AK, Chisti Y, Banerjee UC (2013) Synthesis of metallic nanoparticles using plant extracts. Biotechnol Adv 31(2):346–356
49. Mittal D, Kaur G, Singh P, Yadav K, Ali SA (2020) Nanoparticle-based sustainable agriculture and food science: recent advances and future outlook. Front Nanotechnol 2:579954
50. Muhammad A, Hajian R, Yusof NA, Shams N, Abdullah J, Woi PM, Garmestani H (2018) A screen printed carbon electrode modified with carbon nanotubes and gold nanoparticles as a sensitive electrochemical sensor for determination of thiamphenicol residue in milk. RSC Adv 8(5):2714–2722
51. Mukhopadhyay SS (2014) Nanotechnology in agriculture: Prospects and constraints. Nanotechnol Sci Appl 63. https://doi.org/10.2147/NSA.S39409
52. Murali M, Gowtham HG, Singh SB, Shilpa N, Aiyaz M, Alomary MN, Alshamrani M, Salawi A, Almoshari Y, Ansari MA (2022) Fate, bioaccumulation and toxicity of engineered nanomaterials in plants: current challenges and future prospects. Sci Total Environ 811:152249
53. Parry R (1998) Agricultural phosphorus and water quality: A U.S. environmental protection agency perspective. J Environ Quality 27(2):258–261. https://doi.org/10.2134/jeq1998.00472425002700020003x
54. Patra AK, Adhikari T, Bhardwaj AK (2016) Enhancing crop productivity in salt-affected environments by stimulating soil biological processes and remediation using nanotechnology. In Dagar JC, Sharma PC, Sharma DK, AK Singh (Éds.) Innovative Saline Agriculture, pp 83–103. Springer India. https://doi.org/10.1007/978-81-322-2770-0_4
55. Pesticides A (2015) Nanotechnologies for pesticides and veterinary medicines: regulatory considerations. The Manager, Public Affairs. Australian Pesticides and Veterinary—Medicines Authority; PO Box 6182; KINGSTON ACT 2604 Australia. https://apvma.gov.au/sites/default/files/publication/15626-nanotechnologies-pesticides-veterinary-medicines_regulatory-considerations_july2015.doc
56. Rahman MS, Chakraborty A, Mazumdar S, Nandi NC, Bhuiyan MNI, Alauddin SM, Khan IA, Hossain MJ (2020) Effects of poly (vinylpyrrolidone) protected platinum nanoparticles on seed germination and growth performance of *Pisum sativum*. Nano-Struct Nano-Objects 21:100408
57. Rajput VD, Minkina T, Kumari A, Harish Singh VK, Verma KK, Mandzhieva S, Sushkova S, Srivastava S, Keswani C (2021) Coping with the challenges of abiotic stress in plants: new dimensions in the field application of nanoparticles. Plants 10(6):1221
58. Rehman AU, Nazir S, Irshad R, Tahir K, ur Rehman K, Islam RU, Wahab Z (2021) Toxicity of heavy metals in plants and animals and their uptake by magnetic iron oxide nanoparticles. J Mol Liquids321:114455
59. Salem SS, Fouda A (2021) Green synthesis of metallic nanoparticles and their prospective biotechnological applications: an overview. Biol Trace Elem Res 199(1):344–370. https://doi.org/10.1007/s12011-020-02138-3
60. Saner MA, Heafey E, Bowman DM (2012) Mapping the regulatory environment for nanomaterials in Canada. Nanotech L Bus 9:343
61. Sarfraz N, Khan I (2021) Plasmonic gold nanoparticles (AuNPs): properties, synthesis and their advanced energy, environmental and biomedical applications. Chem –Asian J 16(7):720–742. https://doi.org/10.1002/asia.202001202

62. Saritha GNG, Anju T, Kumar A (2022) Nanotechnology-Big impact : how nanotechnology is changing the future of agriculture? J Agric Food Res 100457
63. Siddiqi KS, Husen A (2016) Engineered gold nanoparticles and plant adaptation potential. Nanoscale Res Lett 11(400):1–10
64. Siddiqi KS, Husen A (2017) Recent advances in plant-mediated engineered gold nanoparticles and their application in biological system. J Trace Elem Med Biol 40:10–23
65. Sharma VK, Sayes CM, Guo B, Pillai S, Parsons JG, Wang C, Yan B, Ma X (2019) Interactions between silver nanoparticles and other metal nanoparticles under environmentally relevant conditions: a review. Sci Total Environ 653:1042–1051
66. Singh A, Sharma R, Rawat S, Singh AK, Rajput VD, Fedorov Y, Minkina T, Chaplygin V (2022) Nanomaterial-plant interaction : Views on the pros and cons. In: Toxicity of nanoparticles in plants, pp 47–68. Elsevier. https://www.sciencedirect.com/science/article/pii/B9780323907743000155
67. Singla R, Kumari A, Yadav SK (2019) Impact of nanomaterials on plant physiology and functions. In: Husen A, Iqbal M (Éds.) Nanomaterials and plant potential, pp 349–377. Springer International Publishing. https://doi.org/10.1007/978-3-030-05569-1_14
68. Su Y, Ashworth V, Kim C, Adeleye AS, Rolshausen P, Roper C, White J, Jassby D (2019) Delivery, uptake, fate, and transport of engineered nanoparticles in plants: a critical review and data analysis. Environ Sci Nano 6(8):2311–2331
69. Sz\Holl\Hosi R, Molnár Á, Kondak S, Kolbert Z (2020) Dual effect of nanomaterials on germination and seedling growth: Stimulation vs phytotoxicity. Plants 9(12):1745
70. Tarhan Ö (2020) Safety and regulatory issues of nanomaterials in foods. Handbook Food Nanotechnology 655–703
71. Thipe VC, Karikachery AR, Çakılkaya P, Farooq U, Genedy HH, Kaeokhamloed N, Phan D-H, Rezwan R, Tezcan G, Roger E (2022) Green nanotechnology—An innovative pathway towards biocompatible and medically relevant gold nanoparticles. J Drug Delivery Sci Technol 70:103256
72. Tighe-Neira R, Carmora E, Recio G, Nunes-Nesi A, Reyes-Diaz M, Alberdi M, Rengel Z, Inostroza-Blancheteau C (2018) Metallic nanoparticles influence the structure and function of the photosynthetic apparatus in plants. Plant Physiol Biochem 130:408–417
73. Tomaszewska-Sowa M, Lisiecki K, Pańka D (2022) Response of rapeseed (*Brassica napus* L.) to silver and gold nanoparticles as a function of concentration and length of exposure. Agronomy 12(11):2885
74. Verma DK, Patel S, Kushwah KS (2021) Effects of nanoparticles on seed germination, growth, phytotoxicity and crop improvement. Agric Rev 42(1):1–11
75. Verma SK, Das AK, Patel MK, Shah A, Kumar V, Gantait S (2018) Engineered nanomaterials for plant growth and development: a perspective analysis. Sci Total Environ 630:1413–1435
76. Wahab A, Batool F, Muhammad M, Zaman W, Mikhlef RM, Naeem M (2023) Current knowledge, research progress, and future prospects of phyto-synthesized nanoparticles interactions with food crops under induced drought stress. Sustainability 15(20):14792
77. Wang M, Li S, Chen Z, Zhu J, Hao W, Jia G, Chen W, Zheng Y, Qu W, Liu Y (2021) Safety assessment of nanoparticles in food: current status and prospective. Nano Today 39:101169
78. Wang P, Lombi E, Sun S, Scheckel KG, Malysheva A, McKenna BA, Menzies NW, Zhao F-J, Kopittke PM (2017) Characterizing the uptake, accumulation and toxicity of silver sulfide nanoparticles in plants. Environ Sci Nano 4(2):448–460
79. Xuan L, Ju Z, Skonieczna M, Zhou P, Huang R (2023) Nanoparticles-induced potential toxicity on human health: Applications, toxicity mechanisms, and evaluation models. MedComm 4(4):e327. https://doi.org/10.1002/mco2.327
80. Yah CS (2013) The toxicity of Gold Nanoparticles in relation to their physiochemical properties. Biomed Res (0970-938X), 24(3). https://www.researchgate.net/profile/Clarence-Yah/publication/279700491_The_toxicity_of_Gold_Nanoparticles_in_relation_to_their_physiochemical_properties/links/55f40cb408ae63926cf2669a/The-toxicity-of-Gold-Nanoparticles-in-relation-to-their-physiochemical-properties.pdf

81. Yanagimoto T, Yu Y-T, Kaneko K (2012) Microstructure and CO gas sensing property of Au/SnO2 core–shell structure nanoparticles synthesized by precipitation method and microwave-assisted hydrothermal synthesis method. Sens Actuators B Chem 166:31–35
82. Zhang J, Claverie J, Chaker M, Ma D (2017) Colloidal metal nanoparticles prepared by laser ablation and their applications. ChemPhysChem 18(9):986–1006. https://doi.org/10.1002/cphc.201601220

Beneficial and Adverse Effects of Gold Nanoparticles on Rhizosphere Biology

Atul Loyal, S. K. Pahuja, Dev Vart, Kavita, Vipul Sharma, Rakesh K. Srivastava, Mayur Mukut Murlidhar Sharma, Mandeep Redhu, Divya Kapoor, Pankaj Sharma, and Azamal Husen

Abstract Nanoparticles have recently garnered significant interest across various disciplines such as physics, chemistry, material science, medicine, and biology. This heightened attention is a consequence of their distinctive electronic, magnetic, optical, mechanical, and chemical properties. Among the plethora of nanoparticles, gold nanoparticles stand out as particularly valuable, finding extensive applications in both industrial and medical contexts. The rhizosphere is home to a complex and dynamic microbial community, including bacteria, fungi, archaea, and other microorganisms. These microorganisms play crucial roles in nutrient cycling, plant growth promotion, and protection against pathogens. The rhizosphere is a hotspot for interactions between plants and microorganisms. Au-NPs in various studies have shown

A. Loyal · S. K. Pahuja · D. Vart · Kavita · V. Sharma
Department of Genetics and Plant Breeding, CCS Haryana Agricultural University, Hisar, Haryana 125004, India
e-mail: sharmap@uic.edu

R. K. Srivastava
International Crops Research Institute for the Semi-Arid Tropics (ICRISAT), Patancheru 502324, India

M. M. M. Sharma
Department of Agriculture and Industry, Kangwon National University, Chuncheon, Korea

M. Redhu
Department of Plant, Soil and Agricultural System, Southern Illinois University, Carbondale, IL 61901, USA

D. Kapoor · P. Sharma (✉)
Department of Microbiology, CCS Haryana Agricultural University, Hisar, Haryana 125004, India
e-mail: sharmap@uic.edu

P. Sharma
Department of Ophthalmology & Visual Sciences, University of Illinois Chicago, Chicago 60612, USA

A. Husen
Wolaita Sodo University, PO Box 138, Wolaita, Ethiopia

Department of Biotechnology, Graphic Era (Deemed to Be University), Dehradun, Uttarakhand 248002, India

A. Husen (ed.), *Plant Response to Gold Nanoparticles*, Smart Nanomaterials Technology, https://doi.org/10.1007/978-981-97-7174-5_3

an increase in microfauna growth when the particle size greater than 45 nm has bacterial growth and also improved germination rate and seedling vigour. Improved rhizosphere led to better plant growth and development. But Au-NPs are proven to be lethal when applied in larger concentration and in smaller sizes (10–15 nm). In order to recommend the use Ai-NPs for crop area there should be clarity regarding beneficial and adverse effects. This chapter will discuss the positive and detrimental aspects of the use of Au-NPs on rhizosphere growth and diversity.

Keywords Au-NPs · Nanoparticles · Rhizosphere · Microbiome · Symbiosis

1 Introduction

Nanomaterials are the particles whose size is 10^{-9} (nano); or in other terms, the particles single/joined/agglomerate whose size is less than 100 nm at least in one dimension falls under the category of nanoparticles [38]. Nanoparticles have recently garnered significant interest across various disciplines such as physics, chemistry, material science, medicine, and biology. This heightened attention is a consequence of their distinctive electronic, magnetic, optical, mechanical, and chemical properties. Among the plethora of nanoparticles, gold nanoparticles stand out as particularly valuable, finding extensive applications in every sector [22, 24, 42, 46, 73, 74]. Gold stands out as a stable metal, displaying notable resistance to oxidation. Technologies utilizing gold nanoparticles offer solutions to significant environmental challenges by addressing issues like eco-friendly production methods, pollution management, and water purification. Fundamental requirement for the utilization of gold nanoparticles in biomedical applications is their necessity to be non-toxic and compatible with both in vivo and in vitro environments [51]. Gold is one of the stable metals and stands resistant to oxidation. Among the various nanomaterials, nanoparticles have gained considerable attention across multiple disciplines, including physics, chemistry, materials science, medicine, and biology. Within this diverse group, gold nanoparticles (AuNPs) are particularly notable for their stability and resistance to oxidation, properties that make them highly desirable for applications in numerous fields. Gold nanoparticles offer promising solutions to environmental challenges. They play crucial roles in eco-friendly production methods, pollution management, and water purification. In biomedical applications, gold nanoparticles must be non-toxic and biocompatible for successful use in both in vivo and in vitro environments. The inherent stability of gold as a metal and its resistance to oxidation further contribute to its value in these areas.

The term rhizosphere was coined in 1904 it can be described as the narrow region of soil that is directly influenced by root secretions, root exudates, and associated microorganisms in the vicinity of plant roots The study of the rhizosphere, encompassing its biology, diversity, and interactions with plants, holds significant promise for potential applications aimed at fostering eco-friendly and sustainable practices in agriculture and environmental biotechnology and nanotechnology [61]. Plants release a variety of organic compounds, known as root exudates, into the soil. These exudates consist of sugars, amino acids, organic acids, and other substances. These compounds can serve as a carbon source for microorganisms and influence the soil microbial community. The rhizosphere is home to a complex and dynamic microbial community, including bacteria, fungi, archaea, and other microorganisms [19]. These microorganisms play crucial roles in nutrient cycling, plant growth promotion, and protection against pathogens. The microbial population within this zone comprises beneficial, harmful, and a limited number of potentially detrimental microorganisms. The rhizosphere is a hotspot for interactions between plants and microorganisms [56–63]. Beneficial microbes can form symbiotic relationships with plants, promoting growth and enhancing nutrient uptake. On the other hand, pathogenic microorganisms may also colonize the rhizosphere, leading to plant diseases. In this context, the positive microorganisms refer to beneficial microbes, the negative ones denote plant pathogens, and the potentially harmful ones are identified as opportunistic human pathogens [61]. The persistent application of nanoparticles within agricultural systems underscores the importance of evaluating both their beneficial and adverse impacts on rhizosphere biology. With each crop cycle, the repeated introduction of nanoparticles leads to their accumulation in the soil, which could pose significant risks, including potential toxicity to rhizosphere microorganisms [71–75]. Gold nanoparticles, in particular, are gaining traction in agricultural practices due to their innovative applications and multifunctional roles. However, understanding the interactions between gold nanoparticles and the rhizosphere is essential, as their benefits—such as improved nutrient availability and enhanced disease resistance—must be carefully weighed against possible risks to the microbial balance within this critical zone. This chapter delves into the dual nature of gold nanoparticles in the rhizosphere, exploring both their positive contributions and potential negative effects. By examining how AuNPs influence microbial communities, nutrient cycling, and plant-microbe interactions, we aim to provide insights into the broader implications of AuNP use for sustainable agricultural practices.

2 Properties of Au-NPs

Au-NPs exhibit unique physical, chemical, and optical properties, which make them useful in a wide range of applications including catalysis, sensing, imaging, drug delivery, and biomedicine.

2.1 Size, Shape, and Surface Characteristics

The size of Au-NPs plays a pivotal role in dictating their physical, chemical, and optical properties. Typically ranging from 1 to 100 nm, the size of AuNPs can be precisely controlled during synthesis. As the size decreases, the surface area to volume ratio increases, leading to enhanced reactivity and unique quantum effects that differ from bulk gold. This phenomenon influences various applications, as smaller nanoparticles are often favoured in biomedical contexts such as drug delivery and imaging, while larger nanoparticles may find utility in catalytic reactions. The tunability of Au-NP size allows researchers to tailor their properties to suit specific requirements across diverse fields, ranging from materials science to biotechnology.

The size of Au-NPs is a critical parameter that influences various properties:

2.1.1 Surface Area to Volume Ratio

As the size of Au-NPs decreases, their surface area to volume ratio increases dramatically. This high surface area is advantageous for applications such as catalysis, where surface interactions play a crucial role in chemical reactions. Numerous studies have explored the significance of surface area to volume ratio in nanomaterials, including AuNPs, which can offer insights into its implications like in catalysis, sensing, detection, drug delivery, and other biomedical applications. For example, Au-NPs were used in the catalytic reduction of 4-nitrophenol with sodium borohydride ($NaBH_4$) and it was found that the reactivity increases with increasing particle diameter keeping the surface area constant for the investigated model reaction [34].

2.1.2 Plasmonic Effects

Plasmonic effects in AuNPs refer to the phenomenon of localized surface plasmon resonance (LSPR), which arises from the collective oscillation of conduction electrons on the nanoparticle surface when stimulated by incident light. This unique optical property is highly dependent on the size, shape, and composition of the nanoparticles. When the frequency of incident light matches the natural frequency of the plasmons, resonance occurs, leading to the intense absorption and scattering of light by the Au-NPs. The plasmonic effects result in strong enhancement of the local electromagnetic field around the nanoparticles, making them highly sensitive to changes in their immediate environment. This sensitivity is exploited in various applications such as sensing, imaging, and spectroscopy. By tuning the size and shape of Au-NPs, researchers can precisely control the plasmonic resonance frequencies, allowing for tailored interactions with light across the visible and near-infrared spectra. Plasmonic effects have significant implications in fields such as biosensing, where AuNPs functionalized with specific biomolecules can detect target analytes with high sensitivity and selectivity, making them valuable tools in biological and medical research.

2.1.3 Quantum Effects

These effects observed in AuNPs arise due to the confinement of electrons within a nanoscale volume, leading to alterations in their electronic and optical properties compared to bulk gold. As the size of AuNPs decreases, these quantum effects become more pronounced and influential. One of the primary quantum effects is the quantum size effect, where the electronic band structure of the nanoparticles is modified. In bulk materials, the electronic band structure is continuous, but in nanoparticles, discrete energy levels emerge due to quantum confinement. This results in a shift in the energy levels and optical absorption spectra of AuNPs compared to bulk gold. Additionally, the quantization of electron motion leads to changes in conductivity, optical absorption, and luminescence properties. These effects are exploited in various applications such as catalysis, sensing, and biomedical imaging, where the unique properties of AuNPs at the nanoscale offer advantages over bulk materials. The ability to manipulate and control these quantum effects makes AuNPs valuable building blocks in nanotechnology and materials science.

2.2 *Methods of Synthesis*

AuNPs can be synthesized using various methods, each offering different levels of control over size, shape, and properties. Some common methods of Au-NP synthesis are given below.

2.2.1 Chemical Reduction

It stands as one of the foremost methods for synthesizing AuNPs, leveraging the reduction of gold ions (Au^{3+}) into metallic gold nanoparticles in the presence of reducing and stabilizing agents. Typically, sodium citrate or sodium borohydride serves as the reducing agent, while stabilizing agents such as citrate or cetyltrimethyl ammonium bromide (CTAB) help control the size and stability of the resulting nanoparticles. The reaction occurs under controlled conditions of temperature, pH, and concentration, allowing for the precise modulation of the size and shape of the AuNPs. As the reducing agent acts upon the gold ions, nucleation and subsequent growth of the nanoparticles occur, forming a colloidal suspension of AuNPs. Chemical reduction offers simplicity, scalability, and versatility, making it a widely adopted method for synthesizing AuNPs in research and industrial settings across various fields, including catalysis, biomedicine, and nanotechnology. For example, in the Turkevich method, AuNPs are synthesized by reducing chloroauric acid (HAuCl4) with a mild reducing agent such as sodium citrate in an aqueous solution [14].

2.2.2 Seed-Mediated Growth

This is also a widely employed method for the synthesis of AuNPs with controlled size and shape. This approach involves the initial formation of small AuNP seeds, typically through the reduction of gold ions using a strong reducing agent such as sodium borohydride. These seed nanoparticles act as nucleation sites for the subsequent growth of larger nanoparticles in the presence of additional gold precursor ions and a growth-controlling agent. By adjusting the concentration of reactants, reaction temperature, and reaction time, researchers can precisely control the size, shape, and aspect ratio of the resulting AuNPs. Seed-mediated growth offers versatility in producing a variety of nanoparticle shapes including nanorods, nanostars, and nanocubes, which find applications in catalysis, sensing, imaging, and biomedical research. For instance, Gold Nanorods were synthesized at low concentrations of CTAB by seed-mediated method [81]. This method has the ability to tailor AuNPs' morphologies with high precision makes it a valuable tool in nanotechnology for designing materials with desired properties and functionalities.

2.2.3 Microemulsion Method

This is the most versatile approach for the synthesis of AuNPs offering precise control over their size, shape, and distribution. This method relies on the formation of stable water/oil emulsions stabilized by surfactants or amphiphilic molecules. Within these emulsions, gold precursors are dissolved in the water phase, while the reducing agent is dissolved in the oil phase. Under controlled reaction conditions, such as temperature and pH, the reduction of gold ions at the water–oil interface leads to the nucleation and growth of AuNPs. The confined environment of the microemulsion system facilitates uniform growth and limits nanoparticle aggregation, resulting in the production of monodisperse AuNPs with narrow size distributions. In a study, self-assembled water-in-oil (W/O) microemulsions have been reported as a suitable route for the synthesis of size-controlled gold nanoparticles [52]. The microemulsion method is widely used for its ability to produce AuNPs with tailored properties suitable for various applications, including catalysis, sensing, imaging, and biomedical research. Moreover, its compatibility with scalable production processes makes it an attractive choice for industrial applications in nanotechnology and materials science.

2.2.4 Green Synthesis

This represents a sustainable approach to AuNPs production, utilizing natural compounds such as plant extracts, microbes, or biomolecules as reducing and stabilizing agents. Unlike conventional synthesis methods that often rely on toxic chemicals, green synthesis approaches offer environmental friendliness and biocompatibility. By harnessing the reducing potential of biological entities, these methods promote the formation of AuNPs under mild reaction conditions, typically at ambient temperature and pressure. The choice of green synthesis route depends on factors such as the availability of raw materials, desired nanoparticle properties, and scalability. For example, in a study, preparation of AuNPs has been done inwardly or outside by microbes, but outside the cell, handling is facile in terms of insulation, and reported that modified nanocellulose is a preferred technique for gold nanoparticle extraction [21]. Furthermore, green synthesis strategies yield AuNPs with biocompatible surface coatings, making them suitable for biomedical applications such as drug delivery, imaging, and therapy. Through ongoing research and innovation, green synthesis methods continue to play a vital role in advancing the field of nanotechnology while minimizing environmental impact.

2.2.5 Template-Assisted Synthesis

This method is mostly used for the controlled fabrication of AuNPs with specific sizes, shapes, and structures. In this approach, templates such as micelles, polymers, or biological molecules act as scaffolds or guides for the nucleation and growth of AuNPs. The template provides a structured environment where gold precursors can selectively deposit and subsequently form nanoparticles with desired geometries dictated by the template's morphology. By adjusting the properties of the template, including its size, shape, and surface chemistry, researchers can precisely control the characteristics of the synthesized AuNPs. This method offers a high degree of tunability and enables the production of AuNPs with intricate architectures and tailored properties suitable for diverse applications in catalysis, sensing, imaging, and biomedicine. Template-assisted synthesis continues to be an area of active research, with ongoing efforts focused on developing novel templates and optimizing synthesis conditions to expand the capabilities of Au-NP fabrication.

3 Beneficial Effects of Au-NPs on Rhizosphere Biology

Rhizosphere is the region surrounding the plant roots up to a certain depth. These are the hotspot for plant microbes' interaction and their colonization [15]. They create a microenvironment around the plant roots and help in plant nutrition and development in numerous ways; such nitrogen fixation by leguminous and free-living bacteria, phosphorus fixation, and making the nutrients in available form from unavailable form. In addition, these are also involved in various enzyme-dependent reaction and fasten the process of germination. Some microbes provide cross-protection against pathogenic bacteria or fungi such as bacillus have N fixation and biocontrol property *Sporisorium scitamineum* and *Ceratocystis paradoxa* in maize [76]. The activity of these microbes may get enhanced or reduced to very low level it all depends on the external micro-environment. The change in chemical or physical properties of soil due to any external stimuli may alter the behaviour or performance of microfauna.

Au-NPs are a kind of external stimuli with very minute particle size/diameter having the ability to get attached to the cell surface or penetrate the cell wall.

Asadishad et al. [6] in their research observed the impact of Au-NPs on soil enzyme activity, focusing on two surface coatings: citrate and polyvinylpyrrolidone (PVP). Different nominal diameters (5, 50, and 100 nm) of PVP-coated Au-NPs were tested at low concentration (0.1 mg Au-NPs kg^1 soil), revealing an enhanced stimulation of soil enzyme activity with decreasing PVP-Au-NPs size. 50 nm sizes citrate-coated Au-NPs consistently increased soil enzyme activity after 30 days compared to PVP-coated Au-NPs. At higher concentrations (100 mg Au-NPs kg^1 soil), soils treated with citrate-coated Au-NPs exhibited significantly distinct community compositions compared to PVP Au-NPs treated soils. This underscores the influence of nanoparticle surface coatings on soil microbial communities. The abundance of bacteria from important soil groups, such as Actinobacteria and Proteobacteria, notably increased after exposure to Au-NPs, with citrate-coated Au-NPs having a more substantial impact over 30 days. While [72] observed that as the size of the nanoparticles increased from 15 to 45 nm and beyond the lethality of nanoparticles decreased and eventually it promoted growth. There was an increase in growth of *P. fluorescens* and *B. subtilis* as the size of Au-NPs increased. Meanwhile, there was an increase in the growth of *P. elgii* with respect to increase in the concentration of Au-NPs. There are limited studies that exclusively explore the impact of Au-NPs on microfauna. Yet effect of Au-NPs on germination seedling and impact on seedling growth and development have been studied and are listed in Table 1.

Beneficial Effects: Highlight positive impacts such as improved nutrient uptake, enhanced plant growth, and effects on soil microorganisms.

4 Adverse Effects on Rhizosphere Biology and Plant Health

Just like every coin has two sides Au-NPs are not so different. There are some cons that are inevitable such as phytotoxicity, alterations in soil microbial communities, and broader ecological implications. The phytotoxicity of nanoparticles (NPs) is reportedly determined by a number of significant factors, including NPs' chemical and physical properties, production of reactive oxygen species (ROS), solubility of NPs, dose delivery, density, and plant and microbial fauna thriving in that soil ecosystem. It is imperative to take into account each of these factors when performing nanotoxicological experiments. Exposure to nanoparticles results in negative effects on various physiological and biochemical events in diverse microbial species which influences their growth stages and their integration with plants. Nanoparticles can exert toxicity on plants as well as rhizosphere through two distinct mechanisms: (i) a toxic effect based on their chemical composition, such as the release of (toxic) ions, and (ii) the induction of stimuli or stress attributed by surface area and shape of the nanoparticles [25]. Cell wall serve as a principal site of interaction for extraneous particles so entry of NPs is challenging as well. The precise mechanism through which nanoparticles invade plant and microbe cells is still unexplained. However,

Table 1 Review of literature highlighting beneficial effects of Au nanoparticles on rhizosphere

Crop	Stage of application	Method of synthesis	Source	Dose	Effect	References
Pearl millet	Seed	Green synthesis	*Cassia auriculata*	20 and 50 μg/ml	Positive effects were observed on percentage of seed germination and growth of seedlings	[50]
Rice	Seed and seedlings	Green synthesis	*Tiliacora triandra*	0, 10, 100, 500, 1000, 2000 mg L^{-1})	The germination percentages were high, a slight decrease in root and shoot lengths relative to the control was observed	[49]
Maize	Seed	Green synthesis	Galanga rhizome extracts	5, 10 and 15 ppm	Enhanced both emergence percentage and seedling vigour index	[43]
Wheat	Seed and seedlings	Chemical Synthesis	Reducing of chloro-auric acid solution with a solution of tri-sodium citrate	10, 20 and 30 mg/L	Plants treated with 10 mg/L of Au NPs produced a significantly higher fresh weight in comparison to control plants Plants were sprayed with 20 mg/L and 30 mg/L of Au NPs, the lowest fresh weight was recorded, which was also obviously less than that of the control plants	[45]

(continued)

Table 1 (continued)

Crop	Stage of application	Method of synthesis	Source	Dose	Effect	References
Wheat and Oat	Seed	Chemical Synthesis	Citrate reduction method	100 and 200 μl/ml	Enhanced seed germination percent, increased shoot and root length, relative water content of seedlings compared to control	[26]
Barley	Seed	Chemical synthesis	Reduction of Au^{III}	1, 3, 5, 8 and 10 mg/L	Seed germination was not affected by gold nanoparticles within this range	[16]
Mustard	Seed and seedlings	Chemical Synthesis	Reduction of $HAuCl_4$ by tri-sodium citrate salt	0, 10, 25, 50 and 100 ppm	Percentage germination was positively affected up to 25 ppm concentration and increase in leaf number, plant height, stem diameter and number of branches per plant in seedlings	[5]
Mung bean	Seed and seedlings	Chemical Synthesis	Using aspartic acid, citric acid, or bovine serum albumin (BSA)	10,100, 1000, μg/L, 100 and 1000 μg/ml	Average root length and shoot length of seedlings will be increased	[11]
Lentil	Seed	Green synthesis	*Zingiber officinale*	0, 10, 25, 50 and 100 ppm	All growth parameters were increased at 5 ppm and 10 ppm, while at high concentrations the effect is reversed	[1]

(continued)

Table 1 (continued)

Crop	Stage of application	Method of synthesis	Source	Dose	Effect	References
French Bean	Seedlings	Chemical form	Purchased	5 and 50 mg/L	Seedling biomass accumulation, transpiration, shoot and root elongation all showed insignificant impact of AuNPs at 5 mg/L,	[41]
Cucumber and lettuce	Seed	Chemical synthesis	From trisodium citrate	62, 100 and 116 μg mL^{-1}	Significant reduction effect on germination index	[8]
Onion	Seed	Green synthesis	Onion extract	1.31 μg/g	Enhanced seed germination and emergence	[3]
Watermelon	Seed	–	nano-Au acquired from National Center for Nanoscience and Technology (NCNST)	50×10^{-7}, 100×10^{-7}, and 200×10^{-7} mg/L	Improved root growth, root length and zero toxicity toward root elongation under all the concentrations, lower germination rate and increased germination energy and germination index	[79]
Gloriosa superba	Seed	Green synthesis	*Terminalia arjuna,*	500 and 1,000 μM	Increment in the frequency of germination index	[18]

(continued)

Table 1 (continued)

Crop	Stage of application	Method of synthesis	Source	Dose	Effect	References
S. marianum	Seed	Chemical synthesis	–	30 μg/ml	Positive influence on seed germination	[58]
Eruca sativa	Seed	Chemical synthesis	Using polyethyleneimine and $HAuCl_4$	30 μg/ml	Au showed inhibition in start and then showed positive response after 28 days, reduced the root and shoot length	[82]
Arabidopsis thaliana	Seed	Green synthesis	*Syzygium cumini*	10 and 80 μg/ml	Improved seed germination rate	[32]

NPs have the ability to exacerbate alterations in cellular structures, molecules, and even defence mechanisms [58]. Shukla et al. [72] reported that small-sized Au-NPs (12 nm) when treated induced cell death in *P. fluorescens* and *B. subtilis*. There are numerous means to determine phytotoxicity of NPs. A phytotoxicity test based on seed germination and root elongation is fast and precise and characterized by its ease of use, cost-effectiveness, sensitivity, and suitability for unstable compounds [31]. Lin and Xing [9] and Battke et al. [35] have documented the detrimental impacts of Au-NPs on the growth of plants. The observed adverse impacts on biomass production and growth of plants could potentially be elucidated by the adsorption of Au-NPs onto the cell wall surfaces of the primary root. This adsorption leads to a reduction in pore size, which subsequently obstructs water transport capacity. Ultimately, this mechanism culminates in attenuated plant growth and weight [7]. The toxicity of Au nanoparticles was evaluated by [8] utilizing standard toxicity assays. Precisely, anaerobic toxicity, bioluminescent (*Photobacterium phosphoreum*), and germination (cucumber and lettuce) tests were executed. Zero or minimal toxicity of Au nanoparticles was observed in every case. Feichtmeier et al. [16] evaluated the phytotoxicity of Au-NPs and discovered that prolonged exposure to elevated concentrations of Au-NPs over a short period of time results in irreversible damage to the roots of the plants and indirectly causes irreversible damage to the microfauna harbouring in the roots of plants. There are limited studies which exclusively studied the adverse impact of Au-NPs on rhizosphere but rhizosphere cannot be separated from roots. Because either they have symbiotic relationship with plants or in commensalism with plants. So, those having detrimental effect on plant growth, development, and germination will have an adverse impact rhizosphere as well. The research conducted by Zaka et al. [82] revealed that Au exhibited an inhibitory effect and reduced germination frequency (53%), reduced root and shoot length of *E. sativa* primarily due to its toxic effects. In contrast, Mahakham et al. [43] reported that nano priming enhanced the growth and physiology of seedlings while causing no toxicity symptoms. Siegel et al. [75] examined the impact of varying concentrations (1, 10, and 100 mg/L) and diameters (10, 14, and 18 nm) of Au-NPs on the root growth of *A. thaliana*. A substantial reduction in both the number and the length of lateral roots was observed subsequent to giving of Au-NPs solutions containing higher concentrations of particles, irrespective of their specific size. In the study performed by Manaf et al. [45] greater concentrations of Au nanoparticles (20 mg/L and 30 mg/L) had an adverse impact on the amount of chlorophyll 'a' and 'b' in wheat seedlings. The accumulation of a considerably reduced quantity of chlorophyll 'a' in comparison to the control provides evidence for the phytotoxic impact of Au-NPs. The activity of enzymes was also considerably less which was further reduced by the rise in the concentration of Au-NPs as compared to that of the control plant.

A critical effect analysed on Au nanoparticles was genotoxicity, which can be produced by direct interaction with DNA or by means of reactive oxygen species and the generation of stress due to oxidation [17]. Genotoxicity studies have been conducted on the most frequently utilized model plant, *Allium cepa*, focusing on oxidative stress, mitotic indices (MI), and chromosome aberrations (CA). When the roots of *A. cepa* were treated with citrate-coated Au nanoparticles measuring 15,

30, and 40 nm in diameter, at concentrations varied from 0.1 to 10 $mgmL^{-1}$ the frequency of CAs was increased in the root ends with the rise in both concentration and size [53]. The mitotic indices in treated root species exhibited a direct correlation with the concentration of NPs while demonstrating an inverse relationship with their size. The quantity of ROS produced was also increased with the dose [10]. In a study performed by Lovecka et al. [37], tobacco plants (*Nicotiana tabacum cv. Wisconsin*) were subjected to varying concentrations of Au-NPs (12–15 nm and 22–25 nm). These plants were either grown hydroponically for short-term exposure or submerged in soil for a long-term experiment assessment in order to determine the genotoxicity of the Au-NPs. The outcomes of both trials revealed that the damage caused by both sizes of Au-NPs on DNA was significantly dependent on concentration and exhibited major changes of DNA damage in the leaves for the highest concentration (30 $mgkg^{-1}$).

4.1 Implications for Soil Health and Microbial Communities

Agricultural applications of excessive quantities of various NPs are currently being implemented, with soils serving as a potential sink. Additionally, soil characteristics like soil organic material and soil pH can significantly impact the toxic effects of nanoparticles (NPs) on microorganisms [54, 57]. The majority of Au-NPs that have the potential to enter the food chain and are therefore hazardous to humans are absorbed by plants from the soil. Limited pertinent data exists regarding the quantity of Au-NPs discharged into the environment. Nevertheless, it is evident that the primary source of gold NPs, as perceived by plants, originates from wastewater inflowing sewage treatment facilities and subsequently settling in biosolids which are subsequently applied as fertilizer to agricultural soils. Soil microbial diversity serves as an extremely quick indicator of a decline in soil health and is imperative to the proper functioning of the soil ecosystem. Consequently, the evaluation of soil health via microbial diversity is presently the subject of considerable interest. With the exception of a few beneficial functions including trace element transport, bioavailability, and speciation, nanoparticles have been found to be toxic to microbes in general [29]. NPs, particularly those composed of metals and metal oxides, lack biodegradability and their accumulation in soil could have detrimental effects on the functional and genetic diversity of the soil's microbial community. NPs have the ability to invade microbial cells via penetrating cells or endocytosis. Au nanoparticles themselves or those coated with innocuous stabilizers (*e.g.*, PVP, citrate) are often considered as biologically inert [4]. Antimicrobial activity of gold nanoparticles is particularly reliant on purification and surface modification (e. g. capping and coating) methods [48]. Furthermore, research examining the impact of Au nanoparticles on the structure of microbial communities showed insignificant shifts in microbial counts and metabolic activities and only modest variations in community structure [44, 60]. While extensive research suggests that Au-NPs lack substantial antimicrobial activity on their own, it has been observed that they have the ability to enhance the antimicrobial effects of certain mild antimicrobials [80].

On the contrary, the reactive oxygen species production mechanism of gold (Au), silver (Ag), silicon (Si), nickel (Ni) NPs in aqueous suspension was investigated by Zhang et al. [83]. The findings suggested that Au NPs exclusively produced singlet oxygen, which permeated cells and exerted an antibacterial impact. Gold nanoparticles (AuNPs) have been observed to exhibit greater efficacy against Gram negative bacteria, like *Pseudomonas aeruginosa* and *E. coli* as compared to Gram positive bacteria such as *Staphylococcus aureus* and *Micrococcus luteus.* [47]. It has been shown that NPs (Ag, Au) with a particle size of approximately 2 nm are capable of penetrating cells and exhibiting direct interactions with numerous bacterial strains [2]. The enhanced antimicrobial activity of NPs can be attributed to their high surface-to-volume ratio, which may result in the production of more active oxygen species per unit weight. Furthermore, 5 nm AuNPs were employed to induce an antimicrobial response against species of *Staphylococcus aureus* and *E. coli* bacteria. It was reported that 90–95% of all bacterial colonies were reduced due to the effectiveness of AuNPs. Moreover, the most diminutive nanoparticles exhibit a more potent antimicrobial impact when contrasted with larger sizes. This phenomenon is attributed to their enhanced capability to interact seamlessly with bacterial cells. As per Asadishad et al. [6], polyvinylpyrrolidone and citrate-coated and Au nanoparticles exhibit a stimulating effect on microbial community composition and soil enzyme activity, including cellobiohydrolase, acid phosphatase, β 2 1,4-xylosidase, β 2 1,4-N-acetylglucosaminidase, and β 2 1,4-glucosidase, when present at small particle sizes (5 and 50 nm) and low concentrations. Conversely, showed significant inhibition when the soil is enriched with higher concentrations and larger sizes (100 nm) of Au NPs.

4.2 Fate of Au-NPs

The environmental fate of Au-NPs (gold nanoparticles) refers to the processes and mechanisms by which these nanoparticles behave and persist in the environment, particularly in soil ecosystems. Understanding the environmental fate of Au-NPs is crucial for assessing their potential impact on ecosystems, human health, and for developing sustainable applications. The persistence of Au-NPs in soil is influenced by various factors and processes directly or indirectly such as (a) edaphic factors and interaction with soil components (b) aggregation and Sedimentation (c) leaching and transport (d) biological transformation (e) persistence. Edaphic factors play major role in determining the fate of NPs across various conditions. The Au-NPs in soil are influenced by factors like Soil composition, pH, organic matter content, and the presence of other chemicals. These factors can affect the rate of aggregation, mobility, and stability of Au-NPs in the soil matrix. In addition, the ions present in the clay portion of soil and organic matter are responsible for the aggregation of the NPs. Sedimentation causes the settling of NPs and leads to their unequal distribution and concentration across the various horizons of soil. Au-NPs have the capability to engage with both soil minerals and organic substances, resulting in their adsorption or coating. These interactions have the potential to impact the chemical and physical characteristics of the nanoparticles. Within the soil, microorganisms contribute to the biological transformation of Au-NPs, where certain microbes may assist in the reduction or oxidation of these nanoparticles, thereby modifying their chemical state and reactivity. Another thing to be taken care of is leaching of nanoparticles to deeper root zones or soil profile. As the size of NPs is very minute, they can easily leach down with the flow of water and get deposited in deeper root zones, and that soil becomes almost impossible to ameliorate.

5 Conclusion

The various studies indicate that the introduction of nanomaterials like Au-NPs into soils can significantly influence soil enzyme activities and microbial communities, thereby affecting nutrient cycling processes in the soil. The results underscore the importance of considering nanoparticle characteristics and concentrations in understanding their potential environmental impacts. Despite showing promising results in various studies and parameters such as enhanced seed germination, growth, and development. It is essential to consider the potential long-term effects and risks associated with their presence in soil ecosystems. Monitoring the persistence and interactions of Au-NPs in different environmental conditions is critical for sustainable nanoparticle applications and environmental management. There are chances of bioaccumulation and vertical transfer of their toxic effects may cause serious blow to the food web/chain. In general, ensuring the nontoxic utilization of nanotechnology in agricultural products necessitates a cooperative endeavour involving researchers, producers, regulatory bodies, and consumers. Continued vigilance and evaluation of the potential risks linked to nanotechnology-based agricultural products are required to ensure their safety and efficacy.

References

1. Abd El-Aziz AR, Al-Othman MR (2019) Gold nanoparticles biosynthesis using zingiber officinale and their impact on the growth and chemical composition of lentil (lens culinaris medic). Pak J Bot 51(2):443–450
2. Abdussalam-Mohammed W (2020) Comparison of chemical and biological properties of metal nanoparticles (Au, Ag), with metal oxide nanoparticles (ZnO-NPs) and their applications. Adv J Chem Sect A. 3(2):111–236
3. Acharya P, Jayaprakasha GK, Crosby KM, Jifon JL, Patil BS (2019) Green-synthesized nanoparticles enhanced seedling growth, yield, and quality of onion (*Allium cepa* L.). ACS Sustain Chem Eng 7(17):14580–14590
4. Amin RM, Mohamed MB, Ramadan MA, Verwanger T, Krammer B (2009) Rapid and sensitive microplate assay for screening the effect of silver and gold nanoparticles on bacteria. Nanomed 4(6):637–643
5. Arora S, Sharma P, Kumar S, Nayan R, Khanna PK, Zaidi MGH (2012) Gold-nanoparticle induced enhancement in growth and seed yield of Brassica juncea. Plant Growth Regul 66:303–310
6. Asadishad B, Chahal S, Cianciarelli V, Zhou K, Tufenkji N (2017) Effect of gold nanoparticles on extracellular nutrient-cycling enzyme activity and bacterial community in soil slurries: role of nanoparticle size and surface coating. Environ Sci Nano 4(4):907–918
7. Asli S, Neumann PM (2009) Colloidal suspensions of clay or titanium dioxide nanoparticles can inhibit leaf growth and transpiration via physical effects on root water transport. Plant, cell & Environ 32(5):577–584
8. Barrena R, Casals E, Colón J, Font X, Sánchez A, Puntes V (2009) Evaluation of the ecotoxicity of model nanoparticles. Chemosphere 75(7):850–857
9. Battke F, Leopold K, Maier M, Schmidhalter U, Schuster M (2008) Palladium exposure of barley: uptake and effects. Plant Biol 10(2):272–276
10. Burketová L, Martinec J, Siegel J, Macůrková A, Maryška L, Valentová O (2022) Noble metal nanoparticles in agriculture: impacts on plants, associated microorganisms, and biotechnological practices. Biotechnol Adv 58:107929
11. Das S, Debnath N, Pradhan S, Goswami A (2017) Enhancement of photon absorption in the light-harvesting complex of isolated chloroplast in the presence of plasmonic gold nanosol—a nanobionic approach towards photosynthesis and plant primary growth augmentation. Gold Bull. 50(3):247–257
12. Dlamini SP, Akanmu AO, Babalola OO (2022) Rhizospheric microorganisms: The gateway to a sustainable plant health. Front Sustain Food Syst 6
13. Dlamini SP, Akanmu AO, Babalola OO (2022) Rhizospheric microorganisms: The gateway to a sustainable plant health. Frontiers Sust Food Syst 4(6):925802
14. Dong J, Carpinone PL, Pyrgiotakis G, Demokritou P, Moudgil BM (2020) Synthesis of precision gold nanoparticles using Turkevich method. KONA Powder J 37:224–232
15. Dutta J, Bora U (2019) Rhizosphere microbiome and plant probiotics. New Futur. Dev. Microb. Biotechnol. Bioeng. Microbes Soil, Crop Environ Sustain. Elsevier B.V.
16. Feichtmeier NS, Walther P, Leopold K (2015) Uptake, effects, and regeneration of barley plants exposed to gold nanoparticles. Environ Sci Pollut Res 22:8549–8558
17. Ghosh M, Ghosh I, Godderis L, Hoet P, Mukherjee A (2019) Genotoxicity of engineered nanoparticles in higher plants. Mutat Res Genet Toxicol Environ Mutagen 842:132–145
18. Gopinath K, Gowri S, Karthika V, Arumugam A (2014) Green synthesis of gold nanoparticles from fruit extract of *Terminalia arjuna*, for the enhanced seed germination activity of *Gloriosa superba*. J. Nanostructure Chem. 4:1–11
19. Hakim S, Naqqash T, Nawaz MS, Laraib I, Siddique MJ, Zia R, Mirza MS, Imran A (2021) Rhizosphere engineering with plant growth-promoting microorganisms for agriculture and ecological sustainability. Front Sustain Food Syst 5:1–23
20. Hammami I, Alabdallah NM (2021) Gold nanoparticles: synthesis properties and applications. J King Saud Univ Sci 33(7):101560

21. Hammami I, Alabdallah NM (2021) Gold nanoparticles: Synthesis properties and applications. J king Saud Univer Sci 33(7):101560
22. Husen A (2017) Gold nanoparticles from plant system: synthesis, characterization and their application, In: Nanoscience and plant–soil systems Vol.–48 (Eds. Ghorbanpourn M, Manika K, Varma A). Springer International Publishing AG, Gewerbestrasse 11, 6330 Cham pp. 455–479
23. Husen A (2023) Nanomaterials and Nanocomposites Exposures to Plants (Response, Interaction, Phytotoxicity and Defense Mechanisms). Springer Nature Singapore Pte Ltd., 152 Beach Road, #21-01/04 Gateway East, Singapore 189721, Singapore https://doi.org/10.1007/978-981-99-2419-6
24. Husen A, Rahman QI, Iqbal M, Yassin MO (2019) Plant-mediated fabrication of gold nanoparticles and their applications. In: Husen A, Iqbal M (Eds.) N*anomaterials* and plant potential. Springer International Publishing AG, Gewerbestrasse 11, 6330 Cham, pp 71–110. https://doi.org/10.1007/978-3-030-05569-1_3
25. Jia G, Wang H, Yan L, Wang X, Pei R, Yan T, Zhao Y, Guo X (2005) Cytotoxicity of carbon nanomaterials: single-wall nanotube, multi-wall nanotube, and fullerene. Environ. Sci. Technol. 39(5):1378–1383
26. Joshi A, Nayyar H, Dharamvir K, Verma G (2018) Detection of gold nanoparticles signal inside wheat (*Triticum aestivum* L) and oats (*Avena sativa*) seedlings. In AIP conference proceedings (Vol. 1953, No. 1). AIP Publishing
27. Kapoor D, Sharma P, Sharma MM, Yadav S, Husen A (2024) Exploring soil microbiota and their role in plant growth, stress tolerance, disease control and nutrient immobilizer. Biocatal Agricult Biotechnol 11:103358
28. Kapoor D, Yadav S, Sharma MM, Sharma P (2023) Interaction between metal nanoparticles and PGPR on the plant growth and development. InNanomaterials and Nanocomposites Exposures to Plants: Response, Interaction, Phytotoxicity and Defense Mechanisms. Springer, Singapore pp. 327–351
29. Khan T, Ullah N, Khan MA, Nadhman A (2019) Plant-based gold nanoparticles; a comprehensive review of the decade-long research on synthesis, mechanistic aspects and diverse applications. Adv Colloid Interface Sci 272:102017
30. Khot LR, Sankaran S, Maja JM, Ehsani R, Schuster EW (2012) Applications of nanomaterials in agricultural production and crop protection: a review. Crop Prot 35:64–70
31. Kordon HA (1992) Seed viability and germination: A multi-purpose experimental system. J Biol Educ 26(4):247–251
32. Kumar V, Guleria P, Kumar V, Yadav SK (2013) Gold nanoparticle exposure induces growth and yield enhancement in Arabidopsis thaliana. Sci Total Environ 461:462–468
33. Kumari R, Suman K, Karmakar S, Lakra SG, Saurav GK, Mahto BK (2023) Regulation and safety measures for nanotechnology-based agri-products. Front Genome ed 5:1200987
34. Liang C, Cheong JY, Sitaru G, Rosenfeldt S, Schenk AS, Gekle S, Kim ID, Greiner A (2022) Size-dependent Catalytic behavior of gold nanoparticles. Adv. Mater. Interfaces 9(4):2100867
35. Lin D, Xing B (2007) Phytotoxicity of nanoparticles: inhibition of seed germination and root growth. Environ Polln 150(2):243–250
36. Lovecká P, Macůrková A, Záruba K, Hubáček T, Siegel J, Valentová O (2021) Genomic damage induced in Nicotiana tabacum L. plants by colloidal solution with silver and gold nanoparticles. Plants 10(6):1260
37. Lovecká P, Macůrková A, Záruba K, Hubáček T, Siegel J, Valentová O (2021) Genomic damage induced in Nicotiana tabacum L. plants by colloidal solution with silver and gold nanoparticles. Plants 10(6):1260
38. Loyal A, Pahuja SK, Sharma P, Malik A, Srivastava RK, Mehta S (2023) Potential environmental and human health implications of nanomaterials used in sustainable agriculture and soil improvement. Eng Nanomater Sustain Agric Prod Soil Improv Stress Manag 387–412
39. Loyal A, Pahuja SK, Rani N, Pooja, Srivastava RK, Sharma P (2023) Titanium oxide nanoparticles: plant response, interaction, phytotoxicity, and defence mechanisms. In: Nanomaterials and Nanocomposites Exposures to Plants: Response, Interaction, Phytotoxicity and Defense Mechanisms. Springer, Singapore pp. 263–284

40. Ma X, Quah B (2016) Effects of surface charge on the fate and phytotoxicity of gold nanoparticles to phaseolus vulgaris. J Food Chem Nanotechnol 2(1):57–65
41. Ma X, Quah B (2016) Effects of surface charge on the fate and phytotoxicity of gold nanoparticles to phaseolus vulgaris. J Food Chem Nanotechnol 2(1):57–65
42. Madhusudhan A, Bandi R, Sharma G, Dadigala R, Husen A, Kim JC (2022) Current trends in engineered gold nanoparticles for cancer therapy. In: Kim JC, Alle M, Husen A (eds) Smart nanomaterials in biomedical applications. nanotechnology in the life sciences. Springer, Cham, pp 1–40 https://doi.org/10.1007/978-3-030-84262-8_1
43. Mahakham W, Theerakulpisut P, Maensiri S, Phumying S, Sarmah AK (2016) Environmentally benign synthesis of phytochemicals-capped gold nanoparticles as nanopriming agent for promoting maize seed germination. Sci Total Environ 573:1089–1102
44. Maliszewska I (2016) Effects of the biogenic gold nanoparticles on microbial community structure and activities. Ann Microbiol 66:785–794
45. Manaf A, Wang X, Tariq F, Jhanzab HM, Bibi Y, Sher A, Razzaq A, Fiaz S, Tanveer SK, Qayyum A (2021) Antioxidant enzyme activities correlated with growth parameters of wheat sprayed with silver and gold nanoparticle suspensions. Agronomy 11(8):1494
46. Mishra VK, Kumar A (2009) Impact of metal nanoparticles on the plant growth promoting rhizobacteria. Dig J Nanomater Biostruct 4:587–592
47. Mohammed AA (2015) Biosynthesis and size of silver nanoparticles using Aspergillus niger ATCC 16404 as antibacterial activity. IJCMAS. 4(2):522–528
48. Nam SH, Shin YJ, Lee WM, Kim SW, Kwak JI, Yoon SJ, An YJ (2015) Conducting a battery of bioassays for gold nanoparticles to derive guideline value for the protection of aquatic ecosystems. Nanotoxicology 9(3):326–335
49. Ndeh NT, Maensiri S, Maensiri D (2017) The effect of green synthesized gold nanoparticles on rice germination and roots. Adv Nat Sci: Nanosci Nanotechnol 8(3):035008
50. Parveen A, Mazhari BBZ, Rao S (2016) Impact of bio-nanogold on seed germination and seedling growth in *Pennisetum glaucum*. Enzyme Microb Technol 95:107–111
51. Rad AG, Abbasi H, Afzali MH (2011) Gold nanoparticles: synthesising, characterizing and reviewing novel application in recent years. Phys Procedia 22:203–208
52. Rajapantulu A, Bandyopadhyaya R (2021) Formation of gold nanoparticles in water-in-oil microemulsions: experiment, mechanism, and simulation. Langmuir 37(22):6623–6631
53. Rajeshwari A, Suresh S, Chandrasekaran N, Mukherjee A (2016) Toxicity evaluation of gold nanoparticles using an *Allium cepa* bioassay. RSC Adv 6(29):24000–24009
54. Rajput VD, Minkina T, Kumari A, Harish Singh VK, Verma KK, Mandzhieva S, Sushkova S, Srivastava S, Keswani C (2021) Coping with the challenges of abiotic stress in plants: new dimensions in the field application of nanoparticles. Plants 10(6):1221
55. Rani K, Sharma P, Kumar S, Wati L, Kumar R, Gurjar DS, Kumar D, Kumar R (2019) Legumes for sustainable soil and crop management. Sustainable management of soil and environment, pp. 193–215
56. Rani K, Rani A, Sharma P, Dahiya A, Punia H, Kumar S, Sheoran S, Banerjee A (2022) Legumes for agroecosystem services and sustainability. In: Advances in Legumes for Sustainable Intensification. Academic Press, pp. 363–380
57. Reidy B, Haase A, Luch A, Dawson KA, Lynch I (2013) Mechanisms of silver nanoparticle release, transformation and toxicity: a critical review of current knowledge and recommendations for future studies and applications. Materials 6(6):2295–2350
58. Salman Khan M, Zaka M, Haider Abbasi B, Rahman L, Shah A (2016) Seed germination and biochemical profile of *Silybum marianum* exposed to monometallic and bimetallic alloy nanoparticles. IET Nanobiotechnol 10(6):359–366
59. Sangwan S, Sharma P, Wati L, Mehta S (2023) Effect of chitosan nanoparticles on growth and physiology of crop plants. In: Engineered nanomaterials for sustainable agricultural production, soil improvement and stress management. Academic Press pp. 99–123
60. Schlich K, Díaz C, Gomez Pizarro B, Knopf B, Schlinkert R, Wege FF, Jural A, Hund-Rinke K (2021) Microbial population dynamics in model sewage treatment plants and the fate and effect of gold nanoparticles. Toxics 9(3):54

61. Sharma P, Mukut M, Sharma M, Malik A (2021) Plant growth-promoting microbes for sustainable biotic and abiotic stress management. plant growth-promoting microbes. Sustain Biot Abiotic Stress Manag
62. Sharma MM, Kapoor D, Rohilla R, Loyal A, Sharma P, Husen A (2024) Role of essential minerals and rhizosphere in plant under changing environmental conditions. In: Essential Minerals in Plant-Soil Systems, Elsevier 235–252
63. Sharma P, Patra A, Singh B, Mehta S (2023) Microbial rejuvenation of soils for sustainable agriculture. In: Sustainable Agriculture Reviews 60: Microbial Processes in Agriculture. Springer Nature Switzerland, Cham 293–323
64. Sharma P, Sharma MM, Malik A, Vashisth M, Singh D, Kumar R, Singh B, Patra A, Mehta S, Pandey V (2021) Rhizosphere, rhizosphere biology, and rhizospheric engineering. In: Plant growth-promoting microbes for sustainable biotic and abiotic stress management. Springer International Publishing, Cham 577–624
65. Sharma MM, Sharma P, Kapoor D, Beniwal P, Mehta S (2021) Phytomicrobiome community: an agrarian perspective towards resilient agriculture. Plant Performance Under Environmental Stress: Hormones, Biostimulants and Sustainable Plant Growth Management 493–534
66. Sharma P, Sharma MM, Anamika, Kapoor D, Rani K, Singh D, Barkodia M (2020) Role of microbes for attaining enhanced food crop production. Microbial biotechnology: basic research and applications, pp. 55–78
67. Sharma P, Pandey V, Sharma MM, Patra A, Singh B, Mehta S, Husen A (2021) A review on biosensors and nanosensors application in agroecosystems. Nanoscale Res Lett 16:1–24
68. Sharma P, Sangwan S, Mehta S (2023) Emerging role of phosphate nanoparticles in agriculture practices. InEngineered Nanomaterials for Sustainable Agricultural Production, Soil Improvement and Stress Management, Academic Press pp. 71–97
69. Sharma MM, Kapoor D, Rohilla R, Sharma P (2023) Nanomaterials and their toxicity to beneficial soil microbiota and fungi associated plants rhizosphere. InNanomaterials and Nanocomposites Exposures to Plants: Response, Interaction, Phytotoxicity and Defense Mechanisms. Springer, Singapore, pp 353–380
70. Sharma MM, Kapoor D, Loyal A, Kumar R, Sharma P, Husen A (2024) Environmental Toxicity of Engineered Carbon Nanoparticles. In: Carbon-Based Nanomaterials: Synthesis, Agricultural, Biomedical, and Environmental Interventions. Springer, Singapore, pp. 337-353
71. Sharma MM, Kapoor D, Sharma P, Husen A (2024) Role of Carbon Nanomaterials in the Prevention of Plant Disease. In: Carbon-Based Nanomaterials: Synthesis, Agricultural, Biomedical, and Environmental Interventions. Springer, Singapore, pp. 105–121
72. Shukla SK, Kumar R, Mishra RK, Pandey A, Pathak A, Zaidi M, Srivastava SK, Dikshit A (2015) Prediction and validation of gold nanoparticles (GNPs) on plant growth promoting rhizobacteria (PGPR): A step toward development of nano-biofertilizers. Nanotechnol Rev 4:439–448
73. Siddiqi KS, Husen A (2016) Engineered gold nanoparticles and plant adaptation potential. Nanoscale Res Lett 11(400):1–10
74. Siddiqi KS, Husen A (2017) Recent advances in plant-mediated engineered gold nanoparticles and their application in biological system. J Trace Elem Med Biol 40:10–23
75. Siegel J, Záruba K, Švorčík V, Kroumanová K, Burketová L, Martinec J (2018) Round-shape gold nanoparticles: effect of particle size and concentration on Arabidopsis thaliana root growth. Nanoscale Res Lett 13:1–7
76. Singh RK, Singh P, Li HB, Song QQ, Guo DJ, Solanki MK, Verma KK, Malviya MK, Song XP, Lakshmanan P, Yang LT, Li YR (2020) Diversity of nitrogen-fixing rhizobacteria associated with sugarcane: A comprehensive study of plant-microbe interactions for growth enhancement in Saccharum spp. BMC Plant Biol 20:1–21
77. Singh SP, Mishra A, Shyanti RK, Singh RP, Acharya A (2021) Silver nanoparticles synthesized using Carica papaya leaf extract (AgNPs-PLE) causes cell cycle arrest and apoptosis in human prostate (DU145) cancer cells. Biol Trace Elem Res 199:1316–1331
78. Singh S, Sangwan S, Sharma P, Devi P, Moond M (2021) Nanotechnology for sustainable agriculture: an emerging perspective. J Nanosci Nanotechnol 21(6):3453–3465

79. Wan Y, Li J, Ren H, Huang J, Yuan H (2014) Physiological investigation of gold nanorods toward watermelon. J Nanosci Nanotechnol 14(8):6089–6094
80. Wanarska E, Maliszewska I (2020) Gold nanoparticles in an enhancement of antimicrobial activity. Physicochem Probl Miner 56
81. Wei MZ, Deng TS, Zhang Q, Cheng Z, Li S (2021) Seed-mediated synthesis of gold nanorods at low concentrations of CTAB. ACS Omega 6(13):9188–9195
82. Zaka M, Abbasi BH, Rahman LU, Shah A, Zia M (2016) Synthesis and characterisation of metal nanoparticles and their effects on seed germination and seedling growth in commercially important Eruca sativa. IET Nanobiotech 10(3):134–140
83. Zhang W, Li Y, Niu J, Chen Y (2013) Photogeneration of reactive oxygen species on uncoated silver, gold, nickel, and silicon nanoparticles and their antibacterial effects. Langmuir 29(15):4647–4651

Genotoxicity of Gold Nanoparticles in Plants and Underlying Mechanisms

Maharudra Pratap Singh, Pawan Kumar, Akhilesh Kumar Singh, Om Prakash, Ashna Parveen, Ahmad Gazali, Prakash Kumar Sarangi, Laxuman Sharma, and Anand Prakash

Abstract The use of gold nanoparticles (Au-NPs) has gained popularity in various industries, but concerns about their environmental impact have emerged. This chapter explores the genotoxic potential of Au-NPs on plant genetic material and the mechanisms behind these effects. Understanding the genotoxic potential is crucial for assessing environmental safety and developing mitigation strategies to protect plant life. Studies show that Au-NPs can cause DNA damage, mutations, and genetic instability, possibly due to their ability to produce reactive oxygen species and cause oxidative stress in plant cells. This chapter explores the link between Au-NPs and plant genotoxicity, highlighting potential risks in the use of nanomaterials in agriculture. It suggests that understanding these mechanisms can help develop strategies to reduce genotoxic effects, promote responsible and sustainable farming practices, and create safer alternatives. The findings can also help formulate regulatory policies and guidelines for nanomaterial use in agriculture, promoting a more informed and responsible approach.

M. P. Singh · P. Kumar · A. K. Singh · O. Prakash · A. Gazali · A. Prakash (✉)
Department of Biotechnology, School of Life Sciences, Mahatma Gandhi Central University, Motihari, Bihar 845401, India
e-mail: akhileshsingh@mgcub.ac.in

A. K. Singh
e-mail: akhileshsingh@mgcub.ac.in

A. Parveen
Amity Institute of Biotechnology, Amity University Uttar Pradesh, Lucknow Campus, Lucknow 226028, India

P. K. Sarangi
College of Agriculture, Central Agricultural University, Imphal 795004, India

L. Sharma
Department of Horticulture, Sikkim University, Gangtok 737102, India

A. Parveen
Department of Biotechnology & Microbiology, Meerut Institute of Engineering and Technology, N.H. 58, Delhi-Roorkee Highway, Baghpat Road Bypass Crossing, Meerut 250005, India

A. Husen (ed.), *Plant Response to Gold Nanoparticles*, Smart Nanomaterials Technology, https://doi.org/10.1007/978-981-97-7174-5_4

Keywords Green synthesis · Au-NPs · Genotoxicity · DNA damage · Oxidative stress · ROS · Antioxidant · Engineered nanomaterial · Oxidative damage

1 Introduction

Nanotechnology, the study of matter manipulation at the nanoscale, has advanced significantly in recent years, resulting in the development and utilization of nanomaterials in different sectors. Gold nanoparticles (Au-NPs), or tiny gold particles, are one of them that has attracted a lot of attention because of their special chemical and physical characteristics and wide range of uses [20, 22, 36, 39, 58, 59]. The genotoxic effects of Au-NPs on plants have emerged as a focus of interest for researchers studying the interactions among nanoparticles and living organisms. These genotoxic effects can have serious consequences for plant growth, development, and ecosystem health. Researchers can reduce any negative effects and maximize the benefits of nanotechnology in agriculture and environmental remediation by developing strategies based on an understanding of the mechanisms by which Au-NPs interact with plants. Genotoxicity refers to the damage to an organism's genetic material, including DNA, due to exposure to specific substances [58]. Genotoxicity can lead to mutations, genetic instability, and negatively impact plant growth, development, and reproductive success. Au-NPs induce oxidative stress through their surface characteristics, producing reactive oxygen species (ROS) that can overpower antioxidant defenses and damage DNA, proteins, and lipids. Further research is needed to fully understand genotoxicity mechanisms and potential risks in agriculture and environmental applications. Nanotechnology has led to the widespread use of nanoparticles in various industries, including agriculture. However, concerns have been raised about the potential environmental and health effects of nanoparticles, particularly their genotoxicity in plants. Understanding the mechanisms by which nanoparticles interact with plant cells and their resulting genotoxic effects is crucial for assessing their risks and developing appropriate safety measures [53]. Siddiqi and Husen [58] have suggested that internalization of Au-NPs in *Arabidopsis thaliana* can lead to genotoxic effects. These findings highlight the need for regulatory measures to ensure the safe use of nanoparticles in agriculture and other industries. Au-NPs can disrupt plant cell membranes, releasing intracellular contents and leading to genotoxicity. They also play a role in modulating gene expression and epigenetic modifications, influencing genes related to stress response, DNA repair, and cell cycle regulation in plants. These findings underscore the multifaceted nature of genotoxicity caused by Au-NPs and the need for further research to fully understand their impact on plant biology. The study reveals that Au-NPs, or nanoparticles, can cause genotoxicity through direct physical interactions with DNA molecules, oxidative stress, and modulation of epigenetic mechanisms [39]. Exposure to Au-NPs in maize and soybean plants can lead to changes in DNA methylation patterns and histone acetylation levels, causing genotoxic effects. This has significant implications for the ecological sustainability of Au-NPs and their use in agriculture. Long-term exposure to Au-NPs can negatively impact plant growth and reproduction, leading to reduced plant populations and altered ecosystem dynamics [67]. Understanding these long-term effects is crucial for assessing the safety and responsible application of nanotechnology in agriculture, developing guidelines for safe use and disposal, and developing strategies to mitigate the negative impacts of Au-NPs [39].

2 Genotoxicity

Genotoxicity is the damage caused by an agent to genetic information within a cell, a crucial aspect of environmental and agricultural research. It involves adverse effects on genetic material, such as DNA damage and mutations, which can lead to heritable changes, potentially causing cancer, birth defects, or other genetic disorders. Genotoxic substances can disrupt cellular processes and genomic integrity, making it essential to understand the mechanisms and consequences of genotoxicity in plants [11]. Genotoxicity can be divided into three types: mutagenicity, which involves the ability of a substance to induce mutations in genetic material, clastogenicity, which involves the induction of chromosomal fragmentation or breakage, and aneugenicity, which is characterized by the induction of numerical chromosomal aberrations, such as aneuploidy, where cells possess an abnormal number of chromosomes. Understanding these types of genotoxicity is crucial for assessing the potential impact of pollutants and stressors on plant populations [67]. Plants are exposed to various genotoxic agents in their natural environment, including chemical pollutants, radiation, and biological factors. Pesticides, heavy metals, air pollutants, ultraviolet radiation, and certain pathogens can induce genotoxic effects in plants. These agents interact with plant DNA, leading to mutations, chromosomal aberrations, and other genetic alterations. DNA damage can affect plant growth and development, impacting individual plants and ecosystems [25]. The genotoxicity is a result of complex interactions between genotoxic agents and cellular machinery. Direct interactions result in DNA strand breaks, while indirect mechanisms generate reactive oxygen species (ROS), causing oxidative damage. These processes can lead to mutations, chromosomal aberrations, and other genetic alterations. Plants' repair mechanisms attempt to fix these damages, but under continuous exposure or high concentrations, these systems may become overwhelmed, leading to persistent genetic damage [58]. To assess genotoxicity in plants, a combination of molecular, cytogenetic, and physiological approaches is needed. Molecular markers like DNA damage and repair-related genes help understand genetic damage, cytogenetic techniques like chromosomal aberration provide insights into DNA fragmentation and physiological indicators like antioxidant enzyme changes indicate cellular response. Genotoxicity in plants is determined using various methods, including cytogenetic assays like the comet assay and micronucleus test, PCR and DNA sequencing, and gene expression patterns related to DNA repair and stress response. These methods help identify specific mutations and changes in DNA sequence, providing valuable insights into plant genotoxic effects [24]. Genotoxicity in plants can cause significant heritable changes that can impact offspring and potentially lead to long-term consequences for plant populations. Genotoxicity can negatively impact plants' ability to adapt to environmental stress, impacting their survival and crop yield. This can also pose a threat to food security in agriculture. Genetic mutations can lead to altered physiological processes, reduced fitness, and reproductive success. Over time, this damage can lead to population decline and reduced genetic diversity, making plant species more vulnerable to environmental challenges and climate change [53]. The study of genotoxicity in plants is evolving, necessitating further research to address gaps and challenges. This includes developing standardized protocols for genotoxicity assessment, exploring chronic exposure effects, and investigating the interaction of multiple genotoxic agents. Integrating omics approaches like genomics and transcriptomics can provide a comprehensive understanding of plant molecular responses to genotoxic stress. Genotoxicity in plants is crucial for environmental health and agricultural sustainability. The interaction between genotoxic agents and plant DNA presents challenges for researchers and managers. Advancements in genotoxicity mechanisms and robust assessment methods can protect plant populations, ecosystems, and human health [39].

3 Gold Nanoparticles

Gold, a precious metal that had special significance in ancient China and Egypt, was once believed to have healing properties. Research on gold nanoparticles began in the mid-nineteenth century with Faraday's findings in 1857. Zsigmondy's work on colloidal gold synthesis allowed for greater precision in manipulating nanoparticle properties [55]. The global market for gold nanoparticles is expected to reach 6.33 billion USD by 2025 due to their widespread use in biomedicine, biosensing, diagnostics, biofertilizers, and cosmetics [17]. Gold nanoparticles are being explored for targeted drug delivery, cancer therapy, and imaging in biomedicine, as they can be functionalized with specific molecules to improve treatment efficacy and minimize side effects. Gold nanoparticles are used in biosensing and diagnostics due to their unique optical properties, which enable the detection and quantification of various analytes like proteins, DNA, and pathogens. They are also used in biofertilizers to improve plant nutrient uptake and in cosmetics for their anti-aging and skin-enhancing properties [12]. Gold nanoparticles are chemically inert, allowing for stable and long-lasting performance in various environments. Their low intrinsic cytotoxicity makes them biocompatible and suitable for biomedical applications. Their plasmonic properties have opened up new possibilities for inducing biological responses and enhancing imaging techniques. These advantages make gold nanoparticles a valuable tool in the scientific community and industry [45].

3.1 AuNPs Biosynthesis

Traditionally, chemical and physical methods have been used for the synthesis of gold nanoparticles. Gold nanoparticle production often involves hazardous reagents and toxic chemicals, which can pose risks to human health and the environment. Physical techniques like laser irradiation require high energy inputs and can produce unwanted byproducts. Controlling growth and preventing particle aggregation is challenging due to factors like laser intensity and duration. Alternative methods are needed to address these drawbacks, aiming for safer and more environmentally friendly approaches to reduce hazardous reagent use, minimize environmental harm, and ensure high-quality nanoparticle production without contamination [46, 52]. The "green" method of AuNPs synthesis offers several advantages over traditional chemical and physical methods. It uses non-toxic agents and renewable materials, eliminating harmful reagents and reducing environmental pollution [20]. The low energy expenditure required for the synthesis process contributes to its sustainability, and the ecological safety ensures that the resulting AuNPs are environmentally friendly and compatible with various applications [7]. The "green" method of AuNPs synthesis is a sustainable and cost-effective approach that uses non-toxic agents, renewable materials, and low energy expenditure. This method is gaining popularity due to its safety, environmental benefits, and low energy expenditure. It eliminates the need for additional stabilizers and reducing agents, reducing the use of harmful chemicals. Renewable materials like plant extracts or biopolymers from agricultural waste can be used in the biological synthesis of AuNPs, making them abundant, easily accessible, and having a lower environmental impact. The "green" method also requires no temperature or pressure increase, making it more accessible and cost-effective. This method is becoming increasingly popular due to its environmental benefits [27]. The synthesis of AuNPs involves adding biological extracts from bacteria, fungi, or plants to a solution containing HAuCI4 salt. The process begins with the reduction of Au3 + ions to Au0, followed by agglomeration and stabilization to form AuNPs. The success of the synthesis depends on various bio-compounds involved. The pH of the solution plays a crucial role in the reduction process, with higher values favoring reduction and lower values causing aggregation. The size, shape, and stability of synthesized AuNPs are influenced by reaction temperature, reactant concentration, and reaction time. Agglomeration affects the optical, magnetic, and catalytic properties of AuNPs, while stabilization prevents further agglomeration and maintains their size, shape, and dispersibility. These steps are essential for controlling the characteristics of synthesized AuNPs and ensuring their suitability for specific applications [41] (Fig. 1).

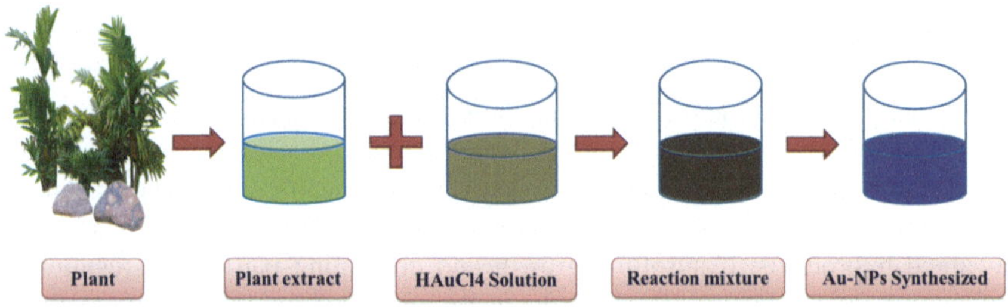

Fig. 1 Scheme of green synthesis of Au-NPs

3.2 *AuNPs Morphology*

Gold nanoparticles have found extensive applications in various fields due to their unique electrical and optical properties. In the field of medicine, they have been used for targeted drug delivery, cancer therapy, and imaging. In electronics, gold nanoparticles are utilized in the development of sensors, displays, and electronic devices. They also play a crucial role in catalysis, enabling efficient chemical reactions. Additionally, gold nanoparticles have been employed in environmental remediation to remove pollutants from water and soil [43]. The versatility of gold nanoparticles makes them highly valuable in numerous industries. While gold nanoparticles have shown great potential for various agricultural applications, it is crucial to thoroughly assess their biosafety. The extensive use of these nanoparticles hinges on understanding their potential health impacts and ensuring their genotoxicity and biocompatibility. This assessment is necessary to identify any adverse effects such as oxidative stress, DNA damage, mutagenesis, or apoptosis. By evaluating the biosafety of gold nanoparticles, we can ensure their safe and responsible use in agriculture and other industries [26]. The synthesis of gold nanoparticles can be achieved through various methods, allowing for control over their size, shape, and properties. Common techniques include chemical reduction, photochemical synthesis, and biological synthesis. In chemical reduction, parameters such as pH, concentration, and temperature are carefully controlled to manipulate the particle size and formation rate. Photochemical synthesis involves the use of light to initiate and control the nanoparticle synthesis process. Biological synthesis utilizes living organisms or their extracts to produce gold nanoparticles [15]. By understanding and utilizing these different synthesis methods, researchers can tailor the characteristics of gold nanoparticles to suit specific applications. The Mi theory, also known as the Mie theory, is a fundamental principle in nanotechnology that explains the relationship between the shape of gold nanoparticles and their physical properties. It states that the plasmon band frequency, the specific point at which nanoparticles absorb and emit light, is determined by their shape. This theory is crucial for designing and engineering gold nanoparticles with desired properties for various applications. Understanding the relationship between nanoparticle shape and physical properties is essential for tailoring their characteristics for specific applications [32] (Table 1).

Table 1 Green synthesis of AuNPs using microorganisms, fungi, algae, and plants

Organism	Shape	References
Pseudomonas putida and *Pseudomonas fluorescence*	Spherical	[19]
Deinococcus radiodurans	Spherical, pseudo-spherical, truncated triangular and irregular	[30]
Bacillus cereus	Spherical, hexagonal, and octagonal with irregular contours	[47]
Pycnoporus sanguineus	Spherical, pseudo-spherical, triangular, truncated triangular, pentagonal, and hexagonal	[56]
Magnusiomyces ingens	Spherical, triangular, hexagonal	[72]
Trichoderma hamatum	Spherical, pentagonal and hexagonal	[1]
Stoechospermum marginatum	Spherical, hexagonal and triangle	[6]
Galaxaura elongata	Rod, triangular, truncated triangular and hexagonal	[2]
Turbinaria conoides (brown algae)	Small spherical, triangle and pseudo-spherical	[3]
Sargassum polycystum (brown algae)	Spherical	[9]
Mango peel extract	Spherical	[68]
Mentha piperita leaf extract	Hexagonal	[3]
Coleus aromaticus leaf extract	Spherical, rod, and triangular	[9]
Papaver somniferum seed pulp extract	Spherical	[63]
Punica Granatum fruit extract	Triangular and spherical	[34]
Capsicum annuum var. grossum pulp extract	Triangle, hexagonal, and quasi-spherical	[70]
Plumeria alba flower extract	Spherical	[40]
Platycodon grandiflorum leaf extract	Spherical	[5]

3.3 Genotoxic Effects of Au-NPs on Plants

Au-NPs are the class of nanoscale materials with unique physicochemical properties that have gained significant attention due to their broad range of applications. With dimensions less than 100 nm, Au-NPs exhibit different behaviours compared to their bulk forms, making them highly desirable for various industries. However, their increasing use and consumption have raised concerns about their potential environmental impact and accumulation in ecosystems [71]. Plants can be exposed to Au-NPs through various pathways, including the use of nanomaterial-based fertilizers and pesticides. Industrial activities and wastewater discharge can cause soil and water contamination, further exposing plants to these nanoparticles. Once in the soil, plants can uptake Au-NPs through their roots, affecting growth, development, and overall health [66]. Studies show that Au-NPs can interfere with nutrient uptake, leading to nutrient deficiencies and stunted growth. They can also disrupt photosynthesis, reducing plants' ability to convert sunlight into energy. Exposure to Au-NPs can inhibit root development, limiting the plant's ability to anchor itself in the soil and access water and nutrients [18]. When taken up by plants, Au-NPs can accumulate in edible plant parts, potentially transferring to the food web. The transfer of Au-NPs through the food chain raises ecological concerns, as they can accumulate in tissues, affect metabolism, and impact the entire food web. The presence of Au-NPs in decomposer organisms can affect nutrient cycling and decomposition rates, further influencing ecosystem processes [65]. To mitigate risks associated with nanomaterial applications, it is essential to consider these risks and develop strategies, such as implementing regulations and environmental assessments, before widespread adoption. By doing so, we can harness the benefits of NPs while minimizing their potential negative consequences [14]. Plants absorb Au-NPs through various pathways, such as cracks during lateral root formation and stomata on plant leaves. Stomata, with larger apertures than biological barriers like cell walls, provide more opportunities for nanomaterial to enter the plant system [49]. Au-NPs, due to their small size and high surface area-to-volume ratio, can negatively impact plants by generating reactive oxygen species (ROS), causing oxidative stress and cell damage. They can also interfere with nutrient uptake and transport, leading to nutrient deficiencies and impaired growth. Developing safer and more sustainable nanotechnology applications in agriculture can help mitigate these risks [60].

Gold nanoparticles (Au-NPs) have gained attention in fields like medicine, electronics, and environmental remediation. However, the potential risks associated with their accumulation in plants and their effects on plant growth and development are not well understood [54]. A study using inductively coupled plasma mass spectrometry (ICP-MS) quantified the accumulation of Au-NPs in plants using two exposure methods, drop-cast, and aerosol. The study found that the highest accumulation of Au-NPs was observed in roots, indicating that they can be transported through the vascular system [51]. The absorption of Au-NPs by plants resulted in cytotoxic and genotoxic effects, including oxidative stress, altered antioxidant enzyme activity, and enhanced micronucleus and chromosome aberration indices. These effects affected plant seed germination and root elongation [50]. The accumulation of Au-NPs in plants not only affects plant development and nutrient uptake but also poses potential risks to human health through high-level enrichment and migration in the food web [33]. Studies show that these Au-NPs can enter plant roots, stems, and leaves, potentially causing harm to human health if they reach edible parts. For example, mature peanuts can contain 20 nm Ag-NPs, while onion cells can absorb and accumulate Ag-NPs. 3.5-nm Au-NPs can travel through roots and into plant vascular systems. Phloem transport is the mechanism behind Au-NPs movement and is taken up by plants, with the proportion of absorption varying depending on the species [69]. Au-NPs have gained significant attention in environmental toxicology due to their potential to cause plant genotoxicity. Understanding the mechanisms through which Au-NPs exert their genotoxic effects on plants is crucial for assessing their environmental impact [16]. Plant genotoxicity is caused by three main mechanisms: increased reactive oxygen species (ROS), transcriptional response, and interaction with DNA or organelles. High concentrations of ROS can aggravate damage and lead to oxidative stress, causing damage to cellular components like lipids, proteins, and DNA [48]. Exposure to Au-NPs triggers transcriptional changes to activate defense mechanisms and mitigate genotoxicity, enabling plants to scavenge ROS, repair damaged biomolecules, and protect cellular structures. However, direct interaction with DNA and organelles can lead to DNA damage and genomic instability, affecting plant health and vitality, such as decreased ATP production and impaired metabolic processes [23].

4 Mechanism of Genotoxicity of Gold Nanoparticle in Plants

4.1 Cellular Uptake of Au-NPs

Gold nanoparticles (Au-NPs) enter cells through three mechanisms: direct diffusion, a process where small Au-NPs move across the membrane based on various factors such as size, charge, hydrophobicity, composition, and shape. Another mechanism is endocytosis, which involves the cell engulfing the Au-NPs. The third mechanism involves channels and membrane transporter proteins facilitating the entry of Au-NPs into the cell. Factors such as lipid composition, fluidity, and bounded molecular species also affect their passage [8]. Cells can take in Au-NPs by engulfing them through a process called endocytosis. This can happen through different mechanisms, like clathrin-mediated endocytosis, caveolae-mediated endocytosis, and macropinocytosis. Endocytosis is the process of bringing molecules into cells. It can happen through fluid-phase mechanisms, where molecules diffuse into vesicles and enter the cell, or receptor-mediated mechanisms, where molecules bind to specific components on the cell surface and are then taken in Sousa de Almeida et al. [61]. Channels and membrane transporter proteins are key transmission mechanisms for Au-NPs into cells, but their high selectivity, low open probability, and small pore size limit their ability to enter cells. Therefore, fluid-phase and receptor-mediated endocytosis are the primary methods for Au-NPs to enter cells. These mechanisms allow for efficient uptake of Au-NPs and play a crucial role in their cellular internalization [64].

4.2 Interaction with Cellular Components

Au-NPs can be taken up by plant cells through various routes, such as root uptake and foliar application. The size, shape, and surface charge of Au-NPs influence their uptake efficiency and distribution within plant tissues. Smaller Au-NPs have higher uptake efficiency and spherical-shaped Au-NPs are more readily taken up by plant cells. The surface charge also affects their interaction with cell membranes and penetration into plant tissues. Despite their potential benefits in agriculture, such as improving crop growth, enhancing nutrient uptake, and increasing plant resistance to stressors, there are limitations and challenges to their application. These include concerns about potential toxicity, the cost of large-scale production, and their stability in different environmental conditions. Therefore, understanding these factors is essential for optimizing Au-NPs use in agriculture [35]. Au-NPs accumulate in plant cells, especially cell walls, cytoplasm, and organelles. The cell wall acts as a protective barrier, facilitating the uptake and signaling of nutrients. Accumulation of Au-NP in the plant cell wall can affect nutrient uptake and signaling processes by creating a protective barrier, potentially limiting the entry of essential nutrients and potentially causing nutrient depletion. In addition, the presence of Au-NP in cell matter and organelles can interfere with cellular processes such as photosynthesis or respiration, which are essential for plant growth and development [42]. Negatively charged groups in cell wall components, such as pectin and cellulose, help bind Au-NPs through electrostatic forces. Additionally, studies have shown that transport proteins help Au-NPs move across cell membranes [10]. For example, when plants are exposed to Au-NPs, it has been observed that the expression of genes related to oxidative stress and defence mechanisms becomes dysregulated [28]. The activation of antioxidant systems such as superoxide dismutase (SOD) and catalase reflects the plant's attempt to reduce oxidative stress induced by Au-NP [14]. The production of reactive oxygen species (ROS), which are highly reactive molecules, increases when plants are exposed to Au-NP. These ROS molecules then participate in signaling pathways that are involved in stress responses. The subcellular localization of Au-NP within plant cells extends to specific organs, further affecting cellular functions [32]. For example, when Au-NP accumulates in mitochondria, it can impact mitochondrial function, potentially leading to changes in cellular energy dynamics and redox homeostasis. Similarly, when Au-NP interacts with chloroplasts, the sites of photosynthesis, it can cause changes in chloroplast structure and function, which may have implications for photosynthetic efficiency [37]. Au-NPs have intricate effects on plant function that demand in-depth knowledge. To maximize the potential advantages of Au-NPs and minimize any negative consequences, it is essential to understand how they interact with various plant cell components. Studying how Au-NPs accumulate and distribute within plants is likewise crucial, as is how they may affect human health by way of the food chain [29].

4.3 Oxidative Stress and DNA Damage

When plants are exposed to Au-NPs, they can take in these nanoparticles through their roots, leaves, and other parts. When Au-NPs enter plant cells, they cause a series of reactions that affect the plant's physiology and molecular processes. The increased production of ROS, which includes superoxide radicals ($O_2^{\cdot -}$), hydrogen peroxide (H_2O_2), and hydroxyl radicals ($\cdot OH$), can overpower the cellular antioxidant systems [44]. Oxidative stress is a condition where there is an imbalance between the production of reactive oxygen species (ROS) and the body's ability to defend against them. This imbalance can result in the accumulation of ROS, leading to damage to cellular components such as lipids, proteins, and DNA. Additionally, the oxidative stress caused by Au-NPs can trigger a cascade of signaling pathways that further exacerbate cellular damage and impair plant growth and development. According to studies, Au-NPs may prevent plant cells' mitochondria and chloroplasts from operating normally. These organelles are essential for both energy production and photosynthesis. When these mechanisms are disturbed, ROS generation may increase and oxidative stress may be exacerbated [53]. Plants have a strong antioxidant defense system against oxidative stress, which is counteracted by enzymes like superoxide dismutase (SOD), catalase (CAT), and peroxidase (POD). However, prolonged exposure to Au-NPs can override these defenses, leading to oxidative damage in plant cells. DNA damage is a significant consequence of oxidative stress, causing strand breaks, cross-linking, and base modifications. Genotoxicity studies show that Au-NPs can induce mutations and chromosomal aberrations, potentially impacting plant growth and viability. The unique physicochemical properties of Au-NPs contribute to their bioavailability and internalization, disrupting cellular processes. Further research is needed to fully understand the implications of Au-NP exposure on plant health and ecosystem dynamics [62] (Fig. 2).

4.4 Future Directions and Challenges

The genotoxicity of Au-NPs on plants poses significant ecological and human health risks due to their transfer and accumulation in various parts of plants, affecting growth, development, and reproduction. Exposure to Au-NPs can result in stunted growth, decreased seed production, and inhibited root development, and reduced fruit yield in tomato plants [38]. One potential mechanism for these effects is the generation of reactive oxygen species (ROS), which can cause oxidative stress, DNA damage, and disruptions in cellular processes. Another mechanism is the interference with gene expression and protein synthesis, altering developmental pathways in plants [57]. The development of engineered nanomaterials (ENMs), including gold nanoparticles (Au-NPs), has led to their potential applications in agriculture, such as enhancing plant growth and increasing crop yields. However, their potential environmental and health impacts raise concerns. Understanding the behavior

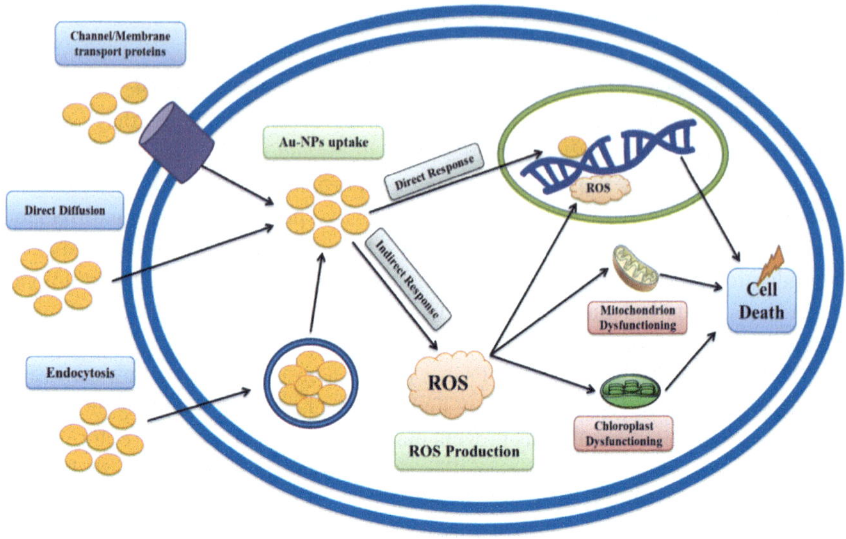

Fig. 2 Mechanism of uptake and action of Au-NPs on plant cells

and fate of Au-NPs in plants and the environment is crucial for ensuring their safe and sustainable use. The genotoxic effects of Au-NPs in plants can have long-term consequences for plant growth and development. Additionally, the potential transfer of genotoxicity from plants to higher trophic levels in the food chain highlights the need for thorough risk assessments before widespread use of Au-NPs in agriculture or other applications [13]. When consumed by animals higher in the food chain, Au-NPs can accumulate in their tissues, leading to potential health risks. Studies have shown that Au-NPs can cause organ damage and immune system dysfunction in animals, which can have cascading effects on ecosystem dynamics. Additionally, the transfer of Au-NPs to humans through the consumption of plant products can pose risks to human health, including potential toxicity and long-term health consequences [4]. In conclusion, comprehensive studies are essential to understand the underlying mechanisms and address the associated challenges to ensure the safe and sustainable use of Au-NPs in agriculture and other fields.

5 Conclusion

Gold nanoparticles (Au-NPs) are widely used due to their high surface area, biocompatibility, and catalytic activity. However, their increasing release into the environment may pose potential risks to living organisms, particularly plants, which are the primary producers and consumers in the food chain. Understanding the toxicity mechanisms and effects of Au-NPs on plants is crucial for developing strategies to mitigate their adverse effects on plant health and ecosystem stability. Factors such as size, shape, surface charge, coating, concentration, exposure time, species, genotype, and physiological state of plants also influence the toxicity of Au-NPs. Au-NPs, which can enter plant cells through various pathways, can cause oxidative stress, alter gene expression, disrupt nutrient metabolism, inhibit photosynthesis, and cause genotoxic damage. The genotoxicity of Au-NPs can be assessed using methods like comet assay, micronucleus test, chromosome aberration test, and DNA damage repair assay. The effects may result from direct or indirect mechanisms, such as DNA intercalation, strand breakage, base modification, cross-linking, or interference with DNA replication and repair. The increasing use of Au-NPs in various fields raises concerns about their potentially harmful effects on plant health. Understanding their genotoxicity is crucial for assessing their impact on plant ecosystems and ensuring safe utilization. The genotoxicity of Au-NPs in plants is primarily due to the production of reactive oxygen species (ROS), which can cause damage to cellular components, including DNA. The genotoxic effects of Au-NPs are still under investigation, with challenges in understanding their bioavailability, toxicity in soil, long-term effects on plant health and productivity, determining nanotoxicity minimization thresholds, and interactions with plant genetic material at genomic and organellar DNA levels. Au-NPs, despite their potential applications, have potential genotoxic effects on plants. Understanding these mechanisms is crucial for safe and sustainable use. Future research should focus on genotoxicity, developing strategies to mitigate these effects, and investigating long-term effects on plant populations and ecosystems. This will help make informed decisions about the use of Au-NPs and protect human health and the environment.

References

1. Abdel-Kareem MM, Zohri AA (2018) Extracellular mycosynthesis of gold nanoparticles using *Trichoderma hamatum*: optimization, characterization and antimicrobial activity. Lett Appl Microbiol 67(5):465–475
2. Abdel-Raouf N, Al-Enazi NM, Ibraheem IBM (2017) Green biosynthesis of gold nanoparticles using *Galaxaura elongata* and characterization of their antibacterial activity. Arab J Chem 10:3029–3039
3. Ahmad N, Bhatnagar S, Saxena R, Iqbal D, Ghosh AK, Dutta R (2017) Biosynthesis and characterization of gold nanoparticles: Kinetics, in vitro and in vivo study. Mater Sci Eng C Mater Biol Appl 78:553–564
4. Ali S, Mehmood A, Khan N (2021) Uptake, translocation, and consequences of nanomaterials on plant growth and stress adaptation. J Nanomater 1–17
5. Anbu P, Gopinath SCB, Jayanthi S (2020) Synthesis of gold nanoparticles using *Platycodon grandiflorum* extract and its antipathogenic activity under optimal conditions. Nanomater Nanotechnol 10:1–9
6. Arockiya Aarthi Rajathi F, Parthiban C, Ganesh Kumar V, Anantharaman P (2012) Biosynthesis of antibacterial gold nanoparticles using brown alga, *Stoechospermum marginatum* (kützing). Spectrochim Acta A Mol Biomol Spectrosc 99:166–173

7. Badeggi UM, Ismail E, Adeloye AO, Botha S, Badmus JA, Marnewick JL, Cupido CN, Hussein AA (2020) Green synthesis of gold nanoparticles capped with procyanidins from *Leucosidea sericea* as potential antidiabetic and antioxidant agents. Biomolecules 10:452
8. Behzadi S, Serpooshan V, Tao W, Hamaly MA, Alkawareek MY, Dreaden EC, Brown D, Alkilany AM, Farokhzad OC, Mahmoudi M (2017) Cellular uptake of nanoparticles: journey inside the cell. Chem Soc Rev 46(14):4218–4244
9. Boomi P, Ganesan RM, Poorani G, Gurumallesh Prabu H, Ravikumar S, Jeyakanthan J (2019) Biological synergy of greener gold nanoparticles by using *Coleus aromaticus* leaf extract. Mater Sci Eng C Mater Biol Appl 99:202–210
10. DeRosa MC, Monreal C, Schnitzer M, Walsh R, Sultan Y (2010) Nanotechnology in fertilizers. Nat Nanotechnol 5(2):91
11. Elespuru RK, Doak SH, Collins AR, Dusinska M, Pfuhler S, Manjanatha M, Cardoso R, Chen CL (2022) Common considerations for genotoxicity assessment of nanomaterials. Front Toxicol. 4:859122
12. Fan J, Cheng Y, Sun M (2020) Functionalized gold nanoparticles: synthesis, properties and biomedical applications. Chem Rec 20(12):1474–1504
13. Gao M, Chang J, Wang Z, Zhang H, Wang T (2023) Advances in transport and toxicity of nanoparticles in plants. J Nanobiotechnology. 21(1):75
14. Gardea-Torresdey JL, Rico CM, White JC (2014) Trophic transfer, transformation, and impact of engineered nanomaterials in terrestrial environments. Environ Sci Technol 48(5):2526–2540
15. Gericke M, Pinches A (2006) Biological synthesis of metal nanoparticles. Hydrometallurgy 83:132–140
16. Ghosh M, Ghosh I, Godderis L, Hoet P, Mukherjee A (2019) Genotoxicity of engineered nanoparticles in higher plants. Mutat Res Genet Toxicol Environ Mutagen 842:132–145
17. Grandviewresearch (2019). Gold nanoparticles market worth $6.33. Billion by 2025. Grand View Research Accessed (15/03/2021), Available at: https://www.grandviewresearch.com/press-release/global-gold-nanoparticles-market
18. Ha N, Seo E, Kim S, Lee SJ (2021) Adsorption of nanoparticles suspended in a drop on a leaf surface of *Perilla frutescens* and their infiltration through stomatal pathway. Sci Rep 11(1):11556
19. Hosseini M, Mashreghi M, Eshghi H (2016) Biosynthesis and antibacterial activity of gold nanoparticles coated with reductase enzymes. Micro Nano Lett 11:484–489
20. Husen A (2017) Gold nanoparticles from plant system: synthesis, characterization and their application. In: Ghorbanpourn M, Manika K, Varma A (Eds.) Nanoscience and plant–soil systems Vol.–48. Springer International Publishing AG, Gewerbestrasse 11, 6330 Cham, pp 455–479
21. Husen A (2023) Nanomaterials and Nanocomposites Exposures to Plants (Response, Interaction, Phytotoxicity and Defense Mechanisms). Springer Nature Singapore Pte Ltd., 152 Beach Road, #21-01/04 Gateway East, Singapore 189721, Singapore. https://doi.org/10.1007/978-981-99-2419-6
22. Husen A, Rahman QI, Iqbal M, Yassin MO (2019) Plant-mediated fabrication of gold nanoparticles and their applications. In: Husen A, Iqbal M (Eds.) N*anomaterials* and plant potential. Springer International Publishing AG, Gewerbestrasse 11, 6330 Cham, pp 71–110 https://doi.org/10.1007/978-3-030-05569-1_3
23. Jogaiah S, Paidi MK, Venugopal K, Geetha N, Mujtaba M, Udikeri SS, Govarthanan M (2021) Phytotoxicological efects of engineered nanoparticles: an emerging nanotoxicology. Sci Total Environ 801:149809
24. Kirkland D, Reeve L, Gatehouse D, Vanparys P (2011) A core in vitro genotoxicity battery comprising the Ames test plus the in vitro micronucleus test is sufficient to detect rodent carcinogens and in vivo genotoxins. Mutat Res 721(1):27–73
25. Kirsch-Volders M, Sofuni T, Aardema M, Albertini S, Eastmond D, Fenech M, Ishidate M Jr, Kirchner S, Lorge E, Morita T, Norppa H, Surrallés J, Vanhauwaert A, Wakata A (2003) Report from the in vitro micronucleus assay working group. Mutat Res 540(2):153–163

26. Kumawat M, Madhyastha H, Umapathi A, Singh M, Revaprasadu N, Daima HK (2022) Surface engineered peroxidase-mimicking gold nanoparticles to subside cell inflammation. Langmuir 38(5):1877–1887
27. Lee KX, Shameli K, Yew YP, Teow SY, Jahangirian H, Rafiee-Moghaddam R, Webster TJ (2020) Recent developments in the facile bio-synthesis of gold nanoparticles (AuNPs) and their biomedical applications. Int J Nanomed 15:275–300
28. Lee WM, An YJ, Yoon H, Kweon HS (2008) Toxicity and bioavailability of copper nanoparticles to the terrestrial plants mung bean (*Phaseolus radiatus*) and wheat (*Triticum aestivum*): plant agar test for water-insoluble nanoparticles. Environ Toxicol Chem 27(9):1915–1921
29. Lee WM, Kwak JI, An YJ (2012) Effect of silver nanoparticles in crop plants *Phaseolus radiatus* and *Sorghum bicolor*: media effect on phytotoxicity. Chemosphere 86(5):491–499
30. Li J, Li Q, Ma X, Tian B, Li T, Yu J, Dai S, Weng Y, Hu Y (2016) Biosynthesis of gold nanoparticles by the extreme bacterium *Deinococcus radiodurans* and an evaluation of their antibacterial properties. Int J Nanomed 11:5931–5944
31. Li T, Albee B, Alemayehu M, Diaz R, Ingham L, Kamal S, Rodriguez M, Bishnoi SW (2010) Comparative toxicity study of Ag, Au, and Ag-Au bimetallic nanoparticles on Daphnia magna. Anal Bioanal Chem 398(2):689–700
32. Li X, Xu H, Chen ZS, Chen G (2011) Biosynthesis of nanoparticles by microorganisms and their applications. J Nanomater 8:270974
33. Lian J, Liu W, Meng L, Wu J, Chao L, Zeb A, Sun Y (2021) Foliar-applied polystyrene nanoplastics (PSNPs) reduce the growth and nutritional quality of lettuce (*Lactuca sativa* L.). Environ Pollut 280:116978
34. Lokina S, Suresh R, Giribabu K, Stephen A, Lakshmi Sundaram R, Narayanan V (2014) Spectroscopic investigations, antimicrobial, and cytotoxic activity of green synthesized gold nanoparticles. Spectrochim Acta A Mol Biomol Spectrosc 129:484–490
35. Ma X, Geisler-Lee J, Deng Y, Kolmakov A (2010) Interactions between engineered nanoparticles (ENPs) and plants: phytotoxicity, uptake and accumulation. Sci Total Environ 408(16):3053–3061
36. Madhusudhan A, Bandi R, Sharma G, Dadigala R, Husen A, Kim JC (2022) Current trends in engineered gold nanoparticles for cancer therapy. In: Kim JC, Alle M, Husen A (eds) Smart nanomaterials in biomedical applications. Nanotechnology in the Life Sciences. Springer, Cham, pp 1–40 https://doi.org/10.1007/978-3-030-84262-8_1
37. Majumdar S, Peralta-Videa JR, Bandyopadhyay S, Castillo-Michel H, Hernandez-Viezcas JA, Sahi S, Gardea-Torresdey JL (2014) Exposure of cerium oxide nanoparticles to kidney bean shows disturbance in the plant defense mechanisms. J Hazard Mater 278:279–287
38. Marmiroli M, Marmiroli N, Pagano L (2022) Nanomaterials induced genotoxicity in plant: methods and strategies. Nanomaterials (Basel) 12(10):1658
39. Martínez G, Merinero M, Pérez-Aranda M, Pérez-Soriano EM, Ortiz T, Begines B, Alcudia A (2020) Environmental impact of nanoparticles' application as an emerging technology: a review. Materials (Basel) 14(1):166. Published 2020 Dec 31. https://doi.org/10.3390/ma1401 0166
40. Mata R, Bhaskaran A, Sadras SR (2016) Green-synthesized gold nanoparticles from *Plumeria alba* flower extract to augment catalytic degradation of organic dyes and inhibit bacterial growth. Particuology 24:78–86
41. Menon S, Rajeshkumar S, VenkatKumar S (2017) A review on biogenic synthesis of gold nanoparticles, characterization, and its applications. Resour-Effic Technol 3:516–527
42. Mirshafiee V, Mahmoudi M, Lou K, Cheng J, Kraft ML (2013) Protein corona significantly reduces active targeting yield. Chem Commun (Camb) 49(25):2557–2559
43. Mosquera J, García I, Henriksen-Lacey M, Martínez-Calvo M, Dhanjani M, Mascareñas JL, Liz-Marzán LM (2020) Reversible control of protein corona formation on gold nanoparticles using host-guest interactions. ACS Nano 14(5):5382–5391
44. Nel A, Xia T, Mädler L, Li N (2006) Toxic potential of materials at the nanolevel. Science 311(5761):622–627

45. Oh KH, Soshnikova V, Markus J, Kim YJ, Lee SC, Singh P, Castro-Aceituno V, Ahn S, Kim DH, Shim YJ, Kim YJ, Yang DC (2018) Biosynthesized gold and silver nanoparticles by aqueous fruit extract of *Chaenomeles sinensis* and screening of their biomedical activities. Artif Cells Nanomed Biotechnol 46(3):599–606
46. Okamoto T, Nakamura T, Sakota K, Yatsuhashi T (2019) Synthesis of single-nanometer-sized gold nanoparticles in liquid–liquid dispersion system by femtosecond laser irradiation. Langmuir 35(37):12123–12129
47. Pourali P, Badiee PSH, Manafi S, Noorani T, Rezaei A, Yahyaei B (2017) Biosynthesis of gold nanoparticles by two bacterial and fungal strains, *Bacillus cereus* and *Fusarium oxysporum*, and assessment and comparison of their nanotoxicity in vitro by direct and indirect assays. Electron J Biotechnol 29:86–93
48. Qamer Z, Chaudhary MT, Du X, Hinze L, Azhar MT (2021) Review of oxidative stress and antioxidative defense mechanisms in *Gossypium hirsutum L* in response to extreme abiotic conditions. J Cotton Res. 4:160–168
49. Rajput V, Minkina T, Ahmed B, Sushkova S, Singh R, Soldatov M, Laratte B, Fedorenko A, Mandzhieva S, Blicharska E, Musarrat J, Saquib Q, Flieger J, Gorovtsov A (2020) Interaction of copper-based nanoparticles to soil, terrestrial, and aquatic systems: critical review of the state of the science and future perspectives. Rev Environ Contam Toxicol 252:51–96
50. Rajput V, Minkina T, Mazarji M, Shende S, Sushkova S, Mandzhieva S, Burachevskaya M, Chaplygin V, Singh A, Jatav H (2020) Accumulation of nanoparticles in the soil-plant systems and their effects on human health. AOAS 2(65):137–143
51. Raliya R, Franke C, Chavalmane S, Nair R, Reed N, Biswas P (2016) Quantitative understanding of nanoparticle uptake in watermelon plants. Front Plant Sci 7:1288
52. Rawat P, Rajput YS, Bharti MK, Sharma R (2016) A method for synthesis of gold nanoparticles using 1-amino-2-naphthol-4- sulphonic acid as reducing agent. Curr Sci 110:2297–2300
53. Sani A, Cao C, Cui D (2021) Toxicity of gold nanoparticles (AuNPs): A review. Biochem Biophys Rep 26:100991
54. Scherer MD, Sposito JCV, Falco WF, Grisolia AB, Andrade LHC, Lima SM, Machado G, Nascimento VA, Gonçalves DA, Wender H, Oliveira SL, Caires ARL (2019) Cytotoxic and genotoxic efects of silver nanoparticles on meristematic cells of *Allium cepa* roots: a close analysis of particle size dependence. Sci Total Environ 660:459–467
55. Sharma V, Park K, Srinivasarao M (2009) Colloidal dispersion of gold nanorods: Historical background, optical properties, seed mediated synthesis, shape separation and self-assembly. Mater Sci Eng R Rep 65:1–38
56. Shi C, Zhu N, Cao Y, Wu P (2015) Biosynthesis of gold nanoparticles assisted by the intracellular protein extract of *Pycnoporus sanguineus* and its catalysis in degradation of 4-nitroaniline. Nanoscale Res Lett 10:147
57. Shukla RK, Badiye A, Vajpayee K, Kapoor N (2021) Genotoxic potential of nanoparticles: structural and functional modifications in DNA. Front Genet 12:728250
58. Siddiqi KS, Husen A (2016) Engineered gold nanoparticles and plant adaptation potential. Nanoscale Res Lett 11(400):1–10
59. Siddiqi KS, Husen A (2017) Recent advances in plant-mediated engineered gold nanoparticles and their application in biological system. J Trace Elem Med Biol 40:10–23
60. Singla R, Kumari A, Yadav SK (2019) Impact of nanomaterials on plant physiology and functions. In: Husen A, Iqbal M (eds) Nanomaterials and plant potential. Springer, Cham
61. Sousa de Almeida M, Taladriz-Blanco P, Drasler B, Balog S, Yajan P, Petri-Fink A, Rothen-Rutishauser B (2022) Cellular uptake of silica and gold nanoparticles induces early activation of nuclear receptor NR4A1. Nanomaterials 12(4):690
62. Tripathi DK, Tripathi A, Shweta Singh S, Singh Y, Vishwakarma K, Yadav G, Sharma S, Singh VK, Mishra RK, Upadhyay RG, Dubey NK, Lee Y, Chauhan DK (2017) Uptake, accumulation and toxicity of silver nanoparticle in autotrophic plants, and heterotrophic microbes: a concentric review. Front Microbiol 8:07
63. Wali M, Sajjad AS, Sumaira S, Muhammad N, Safia H, Muhammad J (2017) Green synthesis of gold nanoparticles and their characterizations using plant extract of *Papaver somniferum*. Nano Sci Nano Technol 11:118

64. Wang H, Chen B, He M, Li X, Chen P, Hu B (2019) Study on uptake of gold nanoparticles by single cells using droplet microfluidic chip-inductively coupled plasma mass spectrometry. Talanta 200:398–407
65. Wang P, Lombi E, Zhao FJ, Kopittke PM (2016) Nanotechnology: a new opportunity in plant sciences. Trends Plant Sci 21(8):699–712
66. Xu L, Wang YY, Huang J, Chen CY, Wang ZX, Xie H (2020) Silver nanoparticles: synthesis, medical applications and biosafety. Theranostics 10(20):8996–9031
67. Yan A, Chen Z (2019) Impacts of silver nanoparticles on plants: a focus on the phytotoxicity and underlying mechanism. Int J Mol Sci 20(5):1003. Published 2019 Feb 26. https://doi.org/10.3390/ijms20051003
68. Yang N, Lib WH, Hao L (2014) Biosynthesis of Au nanoparticles using agricultural waste mango peel extract and its in vitro cytotoxic effect on two normal cells. Mater Lett 134:67–70
69. Youssef MS, Elamawi RM (2020) Evaluation of phytotoxicity, cytotoxicity, and genotoxicity of ZnO nanoparticles in *Vicia faba*. Environ Sci Pollut Res 16(27):18972–18984
70. Yuan CG, Huo C, Yu S, Gui B (2017) Biosynthesis of gold nanoparticles using *Capsicum annuum* var. grossum pulp extract and its catalytic activity. Phys E 85:9–26
71. Zhang P, Ma Y, Zhang Z (2015) Interactions between engineered nanomaterials and plants: phytotoxicity, uptake, translocation, and biotransformation. In: Siddiqui M, Al-Whaibi M, Mohammad F (eds) Nanotechnology and plant sciences. Springer, Cham
72. Zhang X, Qu Y, Shen W, Wang J, Li H, Zhang Z, Li S, Zhou J (2016) Biogenic synthesis of gold nanoparticles by yeast *Magnusiomyces ingens* LH-F1 for catalytic reduction of nitrophenols. Colloids Surf A Phys Eng 497:280–285

Changes in Plant Gene Expression in Response to Gold Nanoparticles and Gold Ions

Kaan Hürkan and Yasemin Kemeç Hürkan

Abstract Plants have been struggled by changing environmental conditions for millions of years. Fortunately, plants have the ability to accommodate it by fine-tuning itself. Genes, which are commonly known as functional unit of heredity in books, are secret weapons of plants that are always ready to be used for any changing circumstances. These DNA-level units open, close, and change their expressions to finalise a protein that survives the plant. Recently, nanoparticle (NP) synthesis and application to various fields such as agriculture, biomedicine, and industry, is a new avenue for science and a pioneering topic for researchers. Although there are industrial ways to synthsise NPs, biological synthesis, which is done by bacteria, viruses, fungi and some algae, is the the most eco-friendly and reasonably priced way to obtain. The synthesised NPs are employed in numerous fields including medicine (metal–organic cages), pharmacology (nanocarriers), agriculture (remediation of soil), food industry (preservatives and packaging), heavy industry (automotive and petrol), cosmetics, chemical industry (coatings, paints, water purification, energy source in batteries and supercapacitors). The list is being extended by each scientific article published. However, each NP used whether produced by chemical or biomaterial-based, has an affect on organisms. The present chapter focuses on how plants respond to Au-NPs on gene level by criticising recent studies and discussing prospects of how can human benefit NPs to "design" better plants to increase yield and be tolerant to harsh environmental conditions.

Keywords Nanomaterials · Au nanoparticles · Gene regulation · Plant physiology

K. Hürkan (✉)
Department of Agricultural Biotechnology, Faculty of Agriculture, Iğdır University, 76000 Iğdır, Türkiye
e-mail: kaanhurkan@hotmail.com

Y. K. Hürkan
Independent Scholar, 76000 Iğdır, Türkiye

A. Husen (ed.), *Plant Response to Gold Nanoparticles*, Smart Nanomaterials Technology, https://doi.org/10.1007/978-981-97-7174-5_5

1 Introduction

The history of plants is a fascinating journey spanning millions of years, from the emergence of simple, non-vascular plants to the diverse array of flora we observe today. Fossil records suggest that plants began to colonize terrestrial environments around 500 million years ago, evolving from aquatic ancestors. Terrestrial plants played a significant role in shaping Earth's ecosystems, increasing oxygen levels and providing habitats for various organisms. The transition from spore-bearing to seed-producing plants caused a significant evolutionary jump by enhancing their reproductive success. There were many milestones including the emergence of angiosperms approximately 140 million years ago, leading to the remarkable biodiversity seen in current ecosystems. Throughout the dynamic history of our planet, plants have showcased a remarkable ability to adjust to evolving ecosystems, employing a diverse range of evolutionary tactics. Extensive research into the evolution of plants and their responses to environmental challenges has yielded valuable insights into the intricate mechanisms that underpin adaptation. Scientists have delved deeply into the historical context of plant evolution, illuminating key adaptations crucial for plant success in varied ecological settings [50]. A critical adaptation within the plant kingdom involves the development of vascular tissues, a topic thoroughly discussed using transcriptome-scale analysis [37]. Vascular tissues, encompassing xylem and phloem, have facilitated the effective transport of water, nutrients, and sugars, enabling plants to navigate the shift from aquatic to terrestrial environments.

Another pivotal adaptation in the evolutionary journey of plants is the development of seeds [51]. This innovation provides plants with a protective mechanism for embryos, ensuring their survival and enabling dispersal to new locations, thereby contributing significantly to the success of seed-producing plants. Plants' adaptive responses to changing climates and environmental stresses are evident through various mechanisms. The exploration of water conservation strategies by researchers [3] has revealed that many plants have evolved features such as succulent leaves and the ability to close stomata, effectively reducing transpiration and enhancing resilience in arid conditions. The coevolutionary dance between plants and other organisms has resulted in the development of intricate floral structures, colours, and scents designed to attract specific pollinators [14]. Similarly, the evolution of fruits with adaptations for dispersal highlights the collaborative relationship between plants and various dispersal agents [39].

In the context of human-influenced ecosystems, the emphasis is on understanding how certain species thrive in urban environments by adapting to pollutants and disturbances associated with human activities. The ongoing coevolution between plants and human-altered environments serves as a testament to the resilience of plant life [8]. As we confront contemporary challenges such as climate change and habitat degradation, recognising and understanding plant adaptation becomes imperative for effective conservation. The referenced literature offers a glimpse into the wealth of knowledge accumulated by researchers, providing insights into the captivating ways in which plants have evolved and persistently adapted to Earth's ever-changing ecosystems.

Nanomaterials have emerged as promising tools to alleviate plant stress and enhance crop resilience in challenging environmental conditions [16, 43, 44]. The distinctive features of nanomaterials, specifically their elevated surface area and reactivity, enable precise interactions with plants at the molecular level. One area of investigation explores the potential of nanomaterials as stress alleviators, with a focus on mitigating the impact of abiotic stressors such as drought, salinity, and extreme temperatures. For instance, the application of nano-sized water-retention polymers can augment soil water-holding capacity, ensuring a consistent water supply to plants during dry periods and ameliorating the effects of drought stress [36]. Additionally, nanomaterials can play a crucial role in counteracting oxidative stress, a common consequence of various environmental stressors. Nanoparticles exhibiting antioxidant properties, including specific metal and metal oxide nanoparticles, have illustrated their capability to scavenge reactive oxygen species (ROS) and safeguard plant cells from oxidative damage [16, 46]. This versatile approach of employing nanomaterials as stress alleviators holds substantial potential for sustaining agricultural productivity amidst shifting climates and environmental uncertainties. In this chapter, we stressed on how gold nanoparticles (Au-NPs) affects gene expression of plants and how we beneficially use them for better and more resistant plants against a rapidly changing environment and feeding the record-breaking growing human population.

2 Plants and Stress

In the realm of plant biology, physiological stress refers to the intricate and dynamic responses that plants exhibit when faced with adverse environmental conditions. Abiotic stress factors, such as drought, salinity, extreme temperatures, and nutrient deficiencies, trigger a cascade of physiological changes aimed at enhancing the plant's adaptive capacity. Mittler [32], elucidates the molecular and biochemical mechanisms underlying plant responses to abiotic stress, emphasising the role of ROS, heat shock proteins, and various signalling pathways. Additionally, Choudhury et al. [7], delve into the intricate crosstalk between different stress signalling networks, providing insights into how plants integrate information from various stressors to orchestrate an optimized response. The concept of hormesis, adds nuance to our understanding, suggesting that sub-lethal stress exposure can induce beneficial responses, contributing to enhanced stress tolerance [21]. Overall, the scientific literature on physiological stress in plants not only unravels the intricacies of their adaptive mechanisms but also informs strategies for improving crop resilience in the face of challenging environmental conditions [17, 18].

2.1 How Are Plants Affected Stress?

Physiological stress in plants manifests morphologically and can have significant economic implications. Abiotic factors like drought, salinity, and extreme temperatures prompt morphological adjustments vital for a plant's adaptive responses. When confronted with water deficiency, plants may undergo changes such as reducing its leaf size, modifying its root architecture, and triggering senescence [47]. These alterations are adaptive mechanisms to minimize water loss and enhance resource use efficiency. The economic drawbacks of physiological stress on plants are significant, particularly within agriculture. The potential for diminished crop yield and quality due to stressors can lead to economic loss for farmers and pose risks to food security. Lobell et al. [29], stress on the economic consequences of climate-based yield fluctuations in major crops, emphasising the susceptibility of agricultural systems to physiological stress. A comprehensive look at the morphological changes linked to physiological stress and their economic repercussions is indispensable for the development of robust crop varieties and sustainable agricultural practices. This knowledge is essential for challenging against the adverse impacts of stressors on plant morphology and productivity.

The economic impact of physiological stress on crop plants is various, exerting considerable repercussions on both agricultural productivity and global food security. Abiotic stressors, capable of inducing yield losses and compromising crop quality have negative effect on farmers' income and the broader economy. Lobell et al. [29] emphasise the economic vulnerability of agriculture to climate-induced variations in crop yields, stressing the need to address changing environmental conditions. The economic challenges extend beyond agriculture, impact downstream industries and potentially contributing to fluctuations in food prices. Thus, we need to develop strategies to enhance crop resilience to physiological stress emerges as a critical imperative to alleviate these economic challenges and secure a stable global food supply.

The applications of physiological stress factors to crop plants are crucial for advancing sustainable agriculture. By subjecting plants to controlled stress conditions, researchers can acquire profound insights into the intricate mechanisms governing plant responses to environmental challenges. This systematic exploration helps to unravel the molecular and physiological pathways activated during stress, providing a foundation for the development of crop varieties with enhanced stress tolerance. Controlled stress challenges replicate real-world conditions, enabling scientists to pinpoint and comprehend key adaptive traits contributing to plant resilience. Furthermore, purposeful exposure of crops to physiological stress becomes a strategic tool in breeding programs, streamlining the selection of individuals with superior stress tolerance. This approach facilitates the development of robust crops capable of withstanding adverse conditions, ultimately fortifying global food security amidst climate uncertainties and environmental fluctuations. In summary, the deliberate application of physiological stress factors is indispensable for fostering innovation and sustainability in agriculture.

2.2 *Why Do We Need More Food?*

The need to increase food production to accommodate a rising global population within the confines of limited arable land is a significant challenge, deeply rooted in scientific disciplines like agronomy, genetics, molecular biology, and environmental science. Addressing the surging demand for food requires a scientific perspective, recognising that conventional agricultural methods may fall short of meeting the demands of a growing population. Scientific advancements in plant genetics and breeding are crucial for developing crop varieties that can thrive in diverse environmental conditions and exhibit heightened yields. The application of genetic modification technologies allows the creation of crops with increased resistance to pests, diseases, and environmental stressors, thereby contributing to elevated agricultural productivity on a per-land-unit basis.

Precision agriculture, an emerging scientific frontier, leverages technology such as remote sensing, GPS-guided machinery, and data analytics to optimize resource usage and amplify crop yields. This approach precisely tailors inputs like water, fertilisers, and pesticides to the specific needs of different areas within a field, minimising waste and environmental impact while maximising overall efficiency. Additionally, interdisciplinary research in agroecology underscores the importance of sustainable farming practices that prioritize ecosystem health, soil fertility, and biodiversity. A nuanced understanding of the intricate relationships between plants, soils, and the environment guides scientific efforts to sustain or boost productivity on limited land resources without compromising long-term ecological well-being. In summary, the scientific pursuit of innovative agricultural methods is paramount for meeting the increasing demand for food within the confines of diminishing arable land.

3 Genes: The Hidden Weapons of Organisms

Genes represent the fundamental units of heredity, encapsulating the genetic instructions essential for the development, function, and regulation of organisms. Defined by specific sequences of DNA, genes govern the synthesis of proteins through the intricate processes of transcription and translation. The intricacies of gene expression play a pivotal role in determining an organism's traits and characteristics. The groundbreaking revelation by Watson and Crick in 1953, uncovering the double-helical structure of DNA, laid the groundwork for comprehending the molecular essence of genes and underscored their significance in heredity [49]. The genome, encapsulating an organism's entire genetic material, extends beyond individual genes to encompass non-coding segments and regulatory elements. Genomic research, exemplified notably by the exhaustive mapping of the human genome conducted by the Human Genome Project Consortium, has brought about a paradigm shift in our comprehension of genetic architecture [19]. Genomics entails a methodical exploration of genes within chromosomes, casting light on genetic variability, evolutionary dynamics, and the intricate molecular processes steering complex traits. The continual evolution of DNA sequencing technologies has hastened genomic analyses, enabling a more profound grasp of genetic diversity and the functional constituents embedded within genomes. The literature, enriched by these seminal contributions, stands as an invaluable wellspring, melding the contemporary comprehension of genes and genomes, thus establishing the foundation for ongoing genetic exploration and its transformative potential across diverse scientific domains.

3.1 Gene Expression: The Silent Symphony of Plant Genetics

Let's unravel the fascinating journey of how plants express their genes, a crucial dance for their growth, development, and response to the world around them. Picture this: at the heart of this intricate process is transcription. Here, the genetic information packed in DNA undergoes transcription into RNA. Now, a bunch of insightful studies have delved deep into this molecular magic of gene expression in plants, uncovering the involvement of diverse transcription factors, regulatory elements, and even epigenetic tweaks.

In a study led by Yadav and colleagues [52], they take a closer look at these rockstar transcription factors, highlighting how they pull the strings in plant development. These factors are like conductors, shaping when and where genes play their tune, influencing the plant's look and feel. Now, picture another gang led by Sequeira-Mendes et al. [40], digging into epigenetic tweaks, like DNA methylation and histone modifications. These tweaks play a pivotal role, tweaking the structure of the genetic material and deciding which genes get the spotlight and which ones take a backseat.

When the environment throws a curveball, plants jam out a dynamic symphony of gene expression to adapt and thrive. The study by Lämke and Bäurle [26] provides insights into the molecular mechanisms underlying environmental sensing and subsequent modulation of gene expression. The authors indicated the role of chromatin modifications and small RNAs in plant responses to environmental stresses, offering a comprehensive view of the interconnected regulatory networks involved.

Plants fine-tune gene expression through intricate molecular mechanisms, allowing them to respond to environmental stimuli, developmental cues, and internal signals. Transcriptional regulation plays the main role, with genes selectively activated or repressed through the action of transcription factors and regulatory elements. For example, Raffaele et al. [34] investigate the role of WRKY (pronounced worky) transcription factors in plant defence responses, showcasing their involvement in regulating the expression of genes associated with pathogen resistance.

DNA methylation and histone modifications significantly contribute to gene regulation in plants. Zhang et al. [54] explored the role of histone modifications in regulating flowering time in Arabidopsis, emphasising the dynamic interplay between chromatin modifications and gene expression during different developmental processes.

Post-transcriptional and post-translational mechanisms also contribute to fine-tuning gene expression in plants. Small RNAs, including microRNAs (miRNAs) and small interfering RNAs (siRNAs), play a crucial role in post-transcriptional gene silencing. Mallory et al. [31] investigated the role of miRNAs in Arabidopsis development, providing insights into their regulatory functions.

Plant hormones, such as auxins, gibberellins, and abscisic acid, act as signalling molecules influencing gene expression in response to various environmental and developmental cues. Santner and Estelle [38] comprehensively discussed the role of auxin in plant growth and development, shedding light on intricate signalling networks modulating gene expression.

Advancements in high-throughput technologies, such as RNA sequencing, have revolutionized the study of gene expression in plants. The work of Oono et al. [33] utilizes RNA-seq to unravel the transcriptomic landscape of plants under stress conditions, providing a genome-wide perspective on dynamic changes in gene expression. This technological leap has empowered researchers to comprehensively analyse the plant transcriptome, uncovering novel genes and pathways involved in various biological processes. In summary, gene expression in plants is a multifaceted process regulated by transcription factors, epigenetic modifications, and environmental stimuli. These insights, derived from a rich body of literature, contribute to our understanding of the molecular intricacies governing plant development and responses to the ever-changing environment.

4 Nanotechnology

Advancements in nanotechnology, a swiftly evolving domain, have impacted diverse sectors by introducing innovative materials at a nanoscale level [53]. The resulting nanomaterials find application across a broad spectrum of industries including engineering, agriculture, medicine, cosmetics, and electronics [13].

The integration of nanotechnology and utilisation of nanomaterials have triggered notable progress and enhancements across these industries. Incorporating nanotechnology and employing nanomaterials has led to noteworthy breakthroughs. For instance, the application of nanomaterials has contributed to the creation of materials that are both stronger and lighter, consequently advancing construction methodologies within the field of engineering. Additionally, nanomaterials have transformed agriculture by elevating crop production through enhanced nutrient delivery systems for plants.

In the realm of medicine, nanotechnology plays a pivotal role in shaping targeted drug delivery systems and diagnostic tools. The unique properties of nanomaterials contribute significantly to the cosmetics industry, where they are frequently employed for attributes such as UV protection and antiaging properties [4]. These multifaceted applications highlight the far-reaching impact of nanotechnology and underscore its role in propelling innovation across various sectors.

4.1 Nanomaterials in Agriculture: A Rapid Growing Frontier

The integration of nanotechnology into agricultural practices has facilitated progress in crop production, pest management, and nutrient regulation. Nanomaterials, functioning at the nanoscale, possess distinctive characteristics that can potentially revolutionize traditional farming techniques. This segment explores the diverse applications of nanomaterials in agriculture, highlighting their potential benefits and addressing associated considerations. A particularly encouraging application involves the use of nanomaterials as inventive fertilizers. Nanoparticles, encompassing nano-sized nutrients and metal oxides, provide an improved mechanism for delivering nutrients to plants, enhancing absorption efficiency, and reducing environmental impact [11]. This innovative approach shows potential for mitigating nutrient runoff and enhancing overall crop yields. Furthermore, nanoparticles play a role in strategies for safeguarding crops. Nanoencapsulation of pesticides and herbicides improves targeted delivery, diminishing the required chemical quantities and lessening environmental impact [23]. This precise delivery not only amplifies the efficacy of pest control but also minimizes detrimental effects on non-target organisms.

4.2 Nanomaterials in Agriculture

Nanomaterials have become valuable tools in the agricultural sector, offering inventive solutions to improve crop productivity and support sustainable farming practices. Within the array of nanomaterials explored in agriculture, metal and metal oxide nanoparticles stand out due to their distinctive properties. Notably, silver nanoparticles showcase antimicrobial characteristics, positioning them as potential candidates for nano-pesticides against various plant pathogens [30]. Research has demonstrated the efficacy of silver nanoparticles in managing fungal and bacterial infections in crops, presenting a hopeful avenue for eco-friendly disease control.

Zinc oxide nanoparticles have gained attention for their role in enhancing plant growth and nutrient uptake. Tripathi et al. [45, 46] indicated that zinc oxide nanoparticles can improve nutrient absorption efficiency, leading to increased crop yields. The use of these nanoparticles contributes to sustainable agricultural practices by optimising essential nutrient utilisation and reducing environmental impact.

Silica nanoparticles constitute another category of nanomaterials with significant applications in agriculture. These nanoparticles have shown the ability to enhance nutrient delivery to plants, thereby improving nutrient absorption efficiency. Studies, such as those conducted by Raliya et al. [35], highlight the potential of silica nanoparticles in optimising crop nutrient management, providing a route to more efficient and sustainable agricultural practices.

In the context of crop protection, nanoencapsulation has emerged as a promising strategy. Khodakovskaya et al. [22] investigated the nanoencapsulation of pesticides and herbicides, illustrating that this approach can enhance the targeted delivery of agrochemicals. Controlled release mechanisms associated with nanoencapsulation not only improve the efficacy of pest control but also reduce the overall environmental impact by minimising the amounts of chemicals required. These instances underscore the varied applications of nanomaterials in agriculture, representing an innovative frontier for enhancing crop yields and addressing environmental concerns.

5 Nanoscale Marvels: Gold Nanoparticles (Au-NPs)

Gold nanoparticles (Au-NPs) have become captivating entities within the field of nanotechnology, capturing the interest of researchers owing to their distinct properties and versatile applications. These nanoscale structures, composed of gold atoms typically ranging from 1 to 100 nM in size, exhibit remarkable physicochemical characteristics that set them apart from their bulk counterparts. The distinctive optical, electronic, and catalytic attributes of Au-NPs have sparked widespread exploration of their potential applications across diverse fields, encompassing medicine, electronics, and environmental science.

The captivating characteristics of Au-NPs find their roots in surface plasmon resonance (SPR), a phenomenon where the collective oscillation of free electrons on the nanoparticle's surface leads to heightened absorbance and light scattering. This optical behaviour is highly adaptable, contingent on factors like size, shape, and surface chemistry, providing a foundation for customising their applications to specific requirements.

In the realm of medicine, gold nanoparticles have gained recognition for their compatibility with biological systems and their potential roles in imaging, drug delivery, and therapy. The surface functionalisation of Au-NPs facilitates the targeted delivery of therapeutic agents to specific cells or tissues, mitigating side effects and enhancing treatment efficacy [15]. Additionally, the capacity of gold nanoparticles to amplify contrast in imaging modalities, such as computed tomography (CT) and photoacoustic imaging, has unveiled new possibilities for early disease diagnosis and monitoring [2].

In catalysis, Au-NPs have demonstrated outstanding performance as catalysts in various reactions owing to their distinctive electronic structure and high surface area [12]. Their catalytic prowess has been explored in fields ranging from organic synthesis to environmental remediation, emphasising their potential for sustainable and efficient processes.

As the exploration of Au-NPs continues, researchers are consistently uncovering novel applications and refining their understanding of the underlying principles governing their behaviour. This dynamic and multidisciplinary field holds promise for addressing complex challenges across scientific and technological domains.

Some examples of where Au-NPs are used are listed.

Biomedical Imaging and Diagnostics:

Au-NPs have been employed for enhancing contrast in various imaging techniques. For instance, in the field of photoacoustic imaging, gold nanoparticles have shown promise for improving imaging depth and resolution. Research by Jokerst et al. utilized gold nanorods as contrast agents for in vivo photoacoustic imaging, demonstrating their potential for non-invasive imaging in biological tissues [20].

Drug Delivery and Therapeutics:

Au-NPs play a crucial role in drug delivery systems due to their biocompatibility and surface functionalization capabilities. In a study by Dreaden et al., Au-NPs were utilized as carriers for the targeted delivery of chemotherapeutic drugs, demonstrating enhanced therapeutic efficacy with reduced side effects [9].

Catalysis and Green Chemistry:

Au-NPs exhibit noteworthy catalytic prowess in diverse reactions, making significant contributions to the domain of sustainable chemistry. The efficacy of gold nanoparticles as catalysts in the low-temperature oxidation of carbon monoxide was illustrated by Haruta et al., underscoring their promise for environmentally conscious catalytic processes [12].

The generation of gold nanoparticles encompasses several methodologies, each possessing distinct advantages and limitations. Here are a few common approaches.

Chemical Reduction

Within this approach, the synthesis of Au-NPs transpires through the reduction of gold ions within a solution, giving rise to nanoparticles. Typically, a gold salt, like gold chloride, undergoes reduction facilitated by a reducing agent. An illustrious example is citrate reduction, where trisodium citrate acts to reduce gold ions, culminating in the creation of stable Au-NPs. The manipulation of reaction parameters, such as the ratio of reactants, temperature, and reaction time, offers control over the size and shape of the resulting nanoparticles.

6 Green Synthesis

The green synthesis methodology employs natural extracts, including plant extracts or microorganisms, as reducing agents for the production of Au-NPs. This approach is deemed environmentally preferable compared to conventional chemical methods. Numerous plant extracts encompass biomolecules proficient in reducing gold ions and stabilising the resultant nanoparticles Fig. 1). The prominence of green synthesis lies in its capacity to diminish the reliance on toxic chemicals in the production of nanoparticles.

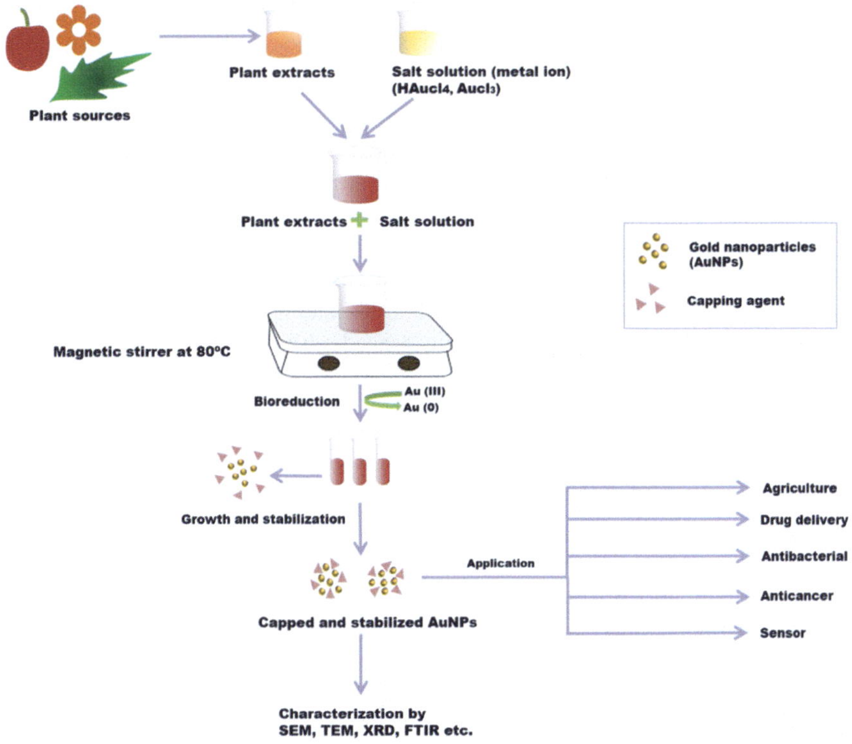

Fig. 1 Scheme of green synthesis of AuNPs from a plant [5]

7 Microemulsion

Microemulsion entails utilising surfactants to generate nanoscale compartments where the reduction and stabilisation of Au-NPs transpire. This technique enables enhanced control over the size and shape of the nanoparticles. The microemulsion stabilized by surfactants furnishes a constrained environment for the reduction reaction, leading to the creation of precisely defined Au-NPs.

8 Electrochemical Synthesis

In simple terms, electrochemical methods work by reducing gold ions at the surface of an electrode. When you apply a voltage difference across the electrodes in a solution with gold ions, it triggers the reduction process, and you end up with gold nanoparticles deposited on the electrode. The cool thing is that you can tweak the size and shape of these tiny gold bits by playing around with the electrochemical settings. It's like customising the outcome to get exactly what you want.

9 Template-Assisted Synthesis

So, picture this: template-assisted methods are like using melds or templates to give shape to the nanoparticles. The usual drill is to lay down some gold onto a template surface and then whisk away the template, leaving you with the exact gold nanoparticle structure you had in mind. It's like crafting these tiny gold wonders with the shapes and sizes you want.

Now, here's the scoop: when it comes to choosing how to whip up these gold nanoparticles, it all boils down to what you're aiming for and where you plan to use them. Each method has its own perks and hurdles, and researchers usually pick the one that fits the bill for their study's unique needs. It's all about finding the right dance partner for the nanoparticle tango.

9.1 How Au-NPs Effect Plants' Physiology?

The exploration of how Au-NPs influence plant physiology has become an increasingly captivating area of scientific inquiry, driven by the potential applications of these nanoparticles in agriculture and environmental science. Renowned for their distinctive physicochemical attributes—like small size, expansive surface area and easy surface functionalisation—Au-NPs emerge as promising candidates for diverse biological interactions. Recent scientific inquiries have delved into unravelling the mechanisms through which Au-NPs impact various facets of plant physiology, spanning growth, development and biochemical responses.

In a study by Li et al. [28], exposure to low concentrations of Au-NPs was found to significantly boost the germination and growth of Arabidopsis thaliana seeds. This effect was linked to the modulation of key physiological processes, including heightened water uptake and enhanced nutrient absorption facilitated by the presence of Au-NPs. Furthermore, Sharma et al. [41] demonstrated that Au-NPs could positively influence photosynthetic activity in plants, resulting in improved carbon assimilation and overall biomass production. The authors proposed that the enhanced photosynthetic performance might be attributed to the ability of Au-NPs to serve as electron shuttles within the photosynthetic apparatus.

Conversely, alternative investigations have probed the potential adverse effects of Au-NPs on plants. Wang et al. [48] noted that exposure to high concentrations of Au-NPs induced oxidative stress in rice plants, evidenced by increased levels of reactive oxygen species (ROS) and lipid peroxidation. The study posited that the generation of ROS could disrupt cellular homeostasis and adversely impact plant physiology. Grasping the dualistic nature of Au-NPs in plants, acting both as growth promoters and stress inducers, is pivotal for harnessing their potential in agriculture and environmental management.

The interaction between Au-NPs and plant physiology unfolds as a complex phenomenon, with studies revealing both positive and negative outcomes. Further research is imperative to unveil the intricate molecular mechanisms underpinning these effects and establish guidelines for the safe and effective utilisation of Au-NPs in sustainable agriculture. The referenced studies contribute significantly to the expanding knowledge on this subject, offering valuable insights into the intricate interplay between nanomaterials and plant biology.

9.2 Understanding the Impact of Gold Nanoparticles on Plant Gene Expression

Au-NPs emerge as promising contenders for diverse applications, notably in agriculture, owing to their distinct physicochemical attributes. As the integration of Au-NPs into agriculture gains traction, it becomes imperative to explore their potential impact on fundamental biological processes in plants. This review centres on elucidating the influence of Au-NPs on plant gene expression, synthesising insights from recent studies to provide a glimpse into the involved molecular mechanisms.

Cutting-edge transcriptomic analyses have been employed in recent research to untangle the intricate relationship between Au-NPs and plant gene expression. In a research endeavour led by Li et al. [28], *Arabidopsis thaliana* subjected to Au-NPs exhibited significant alterations in the expression of genes associated with stress response and defence mechanisms. Comprehensive transcriptomic analyses, including RNA sequencing and microarray studies, provided a comprehensive perspective on shifts in gene expression patterns, unveiling a nuanced interplay between Au-NPs and the plant's molecular machinery (Fig. 2).

The findings proposed by Li et al. [28] suggest that Au-NPs may serve as elicitors, instigating the upregulation of stress-responsive genes in plants. The activation of these genes can be perceived as a defence mechanism, empowering plants to cope with the stress induced by nanoparticle exposure. Unravelling the specific pathways and regulatory elements involved in this response becomes paramount for comprehending the adaptive strategies employed by plants in the presence of Au-NPs. Singh et al. [42] delved into the impact of Au-NPs on rice plants, uncovering shifts in the expression of genes associated with oxidative stress responses and antioxidant defence mechanisms. This observation implies that exposure to Au-NPs triggers a molecular response geared toward mitigating oxidative stress, potentially serving as a protective mechanism against cellular damage. Deciphering the intricate network of genes participating in these pathways contributes to a more comprehensive understanding of how plants respond to Au-NP-induced oxidative stress.

While the mentioned studies yield valuable insights, acknowledging the challenges in studying nanoparticle-plant interactions is crucial. Variability in experimental conditions, nanoparticle characteristics, and plant species may contribute to divergent findings. Future research should strive to standardise methodologies and explore the long-term effects of Au-NPs on plant gene expression to ensure the reliability and applicability of results.

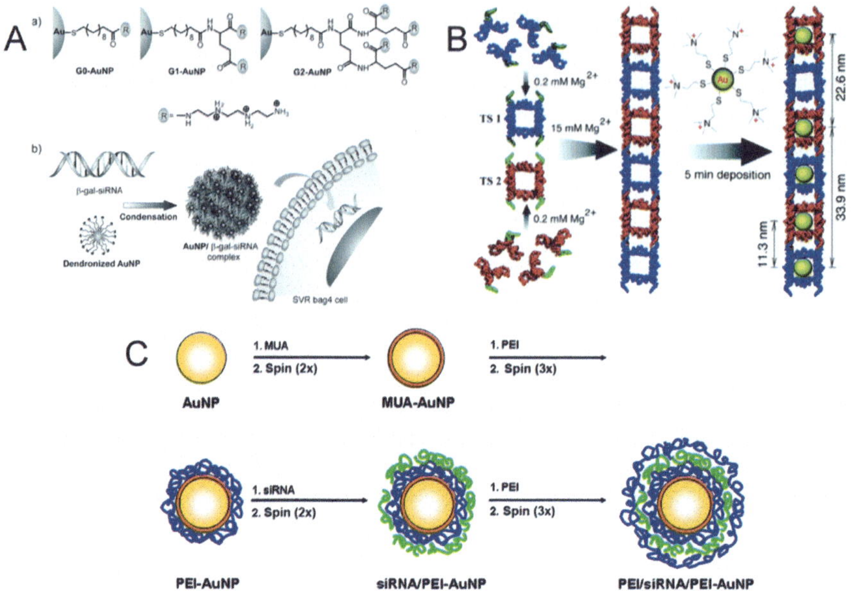

Fig. 2 Instances of interactions between RNA and gold nanoparticles through electrostatic forces are illustrated in various scenarios. These include RNA trapped within a dendronised polymer layer (depicted in A) [24], a hierarchical supramolecular assembly of TS ladder adorned with cationic AuNPs (depicted in B) [25], and a layer-by-layer deposition of siRNA and poly(ethylene imine) (PEI) on the AuNPs surface (depicted in C) [10]. The color scheme in Figure C represents PEI in blue and siRNA in green

The interplay between Au-NPs and plant gene expression unfolds as a multifaceted and dynamic process. Studies by Li et al. [28] and Singh et al. [42] underscore the complexity of these interactions, indicating both the upregulation of stress-responsive genes and the activation of antioxidant defence mechanisms. Continued research in this field remains essential to unveil the specific molecular mechanisms dictating Au-NP-induced changes in gene expression, paving the way for informed and sustainable applications of nanotechnology in agriculture.

References

1. Alberts B, Johnson A, Lewis J, Raff M, Roberts K, Walter P (2002) Molecular Biology of the Cell (4th ed.). Garland Science
2. Alkilany AM, Murphy CJ (2010) Toxicity and cellular uptake of gold nanoparticles: what we have learned so far? J Nanopart Res 12(7):2313–2333
3. Austin AL, Pockman WT (2013) Dynamics of water uptake and belowground interactions in a Semiarid Grassland. J Ecol 101(2):489–500

4. Badhusha AMI, Khan MS, Prabhu S, Kumar P, Karthiga R, Odeibat HA, Naz H, Buvaneswari K, Islam MR (2022) Solar power light-driven improved photocatalytic action of Mg-doped CuO nanomaterial modified with polyvinylalcohol. https://doi.org/10.1155/2022/2430840
5. Bharadwaj KK, Rabha B, Pati S, Sarkar T, Choudhury BK, Barman A, Mohd Noor NH (2021) Green synthesis of gold nanoparticles using plant extracts as beneficial prospect for cancer theranostics. Molecules 26(21):6389
6. Brown TA (2002) Genomes, 2nd edn. Wiley-Liss, Oxford
7. Choudhury FK, Rivero RM, Blumwald E, Mittler R (2017) Reactive oxygen species, abiotic stress and stress combination. Plant J 90(5):856–867
8. Corlett RT (2015) The anthropocene concept in ecology and conservation. Trends Ecol Evol 30(1):36–41
9. Dreaden EC, Alkilany AM, Huang X, Murphy CJ, El-Sayed MA (2012) The golden age: gold nanoparticles for biomedicine. Chem Soc Rev 41(7):2740–2779
10. Elbakry A, Zaky A, Liebl R, Rachel R, Goepferich A, Breunig M (2009) Layer-by-layer assembled gold nanoparticles for siRNA delivery. Nano Lett 9(5):2059–2064
11. Giraldo JP, Landry MP, Faltermeier SM, McNicholas TP, Iverson NM, Boghossian AA, Strano MS (2019) Plant nanobionics approach to augment photosynthesis and biochemical sensing. Nat Mater 13(4):400–408
12. Haruta M, Kobayashi T, Sano H, Yamada N (2007) Novel gold catalysts for the oxidation of carbon monoxide at a temperature far below 0°C. Chem Lett 36(2):220–221
13. Hegazy AA, Ahmed MM, Shehata M, Abdelfattah M (2018) Changes in rats' liver structure induced by zinc oxide nanoparticles and the possible protective role of Vitamin E. https://doi.org/10.14302/issn.2577-2279.ijha-18-2384
14. Hodges SA (2013) Reproductive isolation as a consequence of adaptive divergence in plants. Am Nat 183(S1):S104–S115
15. Huang X, El-Sayed IH, Qian W, El-Sayed MA (2007) Cancer cell imaging and photothermal therapy in the near-infrared region by using gold nanorods. J Am Chem Soc 128(6):2115–2120
16. Husen A (2023) Nanomaterials and nanocomposites exposures to plants (response, interaction, phytotoxicity and defense mechanisms). Springer Nature Singapore Pte Ltd., 152 Beach Road, #21–01/04 Gateway East, Singapore 189721, Singapore. https://doi.org/10.1007/978-981-99-2419-6
17. Husen A (2021a) Harsh environment and plant resilience (*Molecular and Functional Aspects*). Springer Nature Switzerland AG, Gewerbestrasse 11, 6330 Cham, Switzerland. https://doi.org/10.1007/978-3-030-65912-7
18. Husen A (2021b) Plant performance under environmental stress (*Hormones, Biostimulants and Sustainable Plant Growth Management*). Springer Nature Switzerland AG, Gewerbestrasse 11, 6330 Cham, Switzerland. https://doi.org/10.1007/978-3-030-78521-5
19. Human Genome Project Consortium (2001) Initial sequencing and analysis of the human genome. Nature 409(6822):860–921
20. Jokerst JV, Thangaraj M, Kempen PJ, Sinclair R, Gambhir SS (2011) Photoacoustic imaging of mesenchymal stem cells in living mice via silica-coated gold nanorods. ACS Nano 6(7):5920–5930
21. Khan MI, Fatma M, Per TS, Anjum NA, Khan NA (2015) Salicylic acid-induced abiotic stress tolerance and underlying mechanisms in plants. Front Plant Sci 6:462
22. Khodakovskaya M, Dervishi E, Mahmood M, Xu Y, Li Z, Watanabe F, Biris AS (2012) Carbon nanotubes are able to penetrate plant seed coat and dramatically affect seed germination and plant growth. ACS Nano 6(5):4093–4103
23. Khot LR, Sankaran S, Maja JM, Ehsani R, Schuster EW (2020) Applications of nanomaterials in agricultural production and crop protection: a review. Crop Prot 35:64–70
24. Kim ST, Chompoosor A, Yeh YC, Agasti SS, Solfiell DJ, Rotello VM (2012) Dendronized gold nanoparticles for siRNA delivery. Small 8(21):3253
25. Koyfman AY, Braun G, Magonov S, Chworos A, Reich NO, Jaeger L (2005) Controlled spacing of cationic gold nanoparticles by nanocrown RNA. J Am Chem Soc 127(34):11886–11887

26. Lämke J, Bäurle I (2017) Epigenetic and chromatin-based mechanisms in environmental stress adaptation and stress memory in plants. Genome Biol 18(1):124
27. Lander ES, Linton LM, Birren B, Nusbaum C, Zody MC, Baldwin J, International Human Genome Sequencing Consortium (2001) Initial sequencing and analysis of the human genome. Nature 409(6822):860–921
28. Li X, Xu H, Chen Z-S (2018) Effects of gold nanoparticles on stress-responsive gene expression in *Arabidopsis thaliana.* J Nanopart Res 20(3):65. https://doi.org/10.1007/s11051-018-4150-1
29. Lobell DB, Schlenker W, Costa-Roberts J (2011) Climate trends and global crop production since 1980. Science 333(6042):616–620
30. Maksoud MI, Agrawal N, Patel MK (2020) Antimicrobial properties of silver nanoparticles: potential candidates for nanopesticides in agriculture. J Nanosci Nanotechnol 20(7):4210–4218
31. Mallory AC, Vaucheret H (2008) Functions of microRNAs and related small RNAs in plants. Nat Genet 38(Suppl):S31–S36
32. Mittler R (2006) Abiotic stress, the field environment and stress combination. Trends Plant Sci 11(1):15–19
33. Oono Y, Yazawa T, Kawahara Y, Kanamori H, Kobayashi F, Sasaki H, Mori S (2011) Genome-wide transcriptome analysis reveals that cadmium stress signaling controls the expression of genes in drought stress signal pathways in rice. PLoS ONE 6(7):e21945
34. Raffaele S, Vailleau F, Léger A, Joubès J, Miersch O, Huard C, Mongrand S (2008) A MYB transcription factor regulates very-long-chain fatty acid biosynthesis for activation of the hypersensitive cell death response in Arabidopsis. Plant Cell 20(3):752–767
35. Raliya R, Biswas P, Tarafdar JC (2015) Improving nutrient use efficiency of crops with silica nanoparticles. Front Plant Sci 6:121
36. Raliya R, Tarafdar JC, Biswas P (2016) Enhancing the mobilization of native phosphorus in the mung bean rhizosphere using ZnO nanoparticles synthesized by soil fungi. J Agric Food Chem 64(15):3111–3118
37. Rothfels CJ et al (2018) Transcriptome-scale homology search reveals vascular plant-specific gene determinants. Nature Plants 4(7):498–509
38. Santner A, Estelle M (2009) The ubiquitin-proteasome system regulates plant hormone signaling. Plant J 61(6):1029–1040
39. Schemske DW, Lande R (1985) The evolution of self-fertilization and inbreeding depression in Plants II. Empirical observations. Evolution 39(1):41–52
40. Sequeira-Mendes J, Aragüez I, Peiró R, Mendez-Giraldez R, Zhang X, Jacobsen SE, Gutierrez C (2014) The functional topography of the Arabidopsis genome is organized in a reduced number of linear motifs of chromatin states. Plant Cell 26(6):2351–2366
41. Sharma P, Jha AB, Dubey RS, Pessarakli M (2020) Reactive oxygen species, oxidative damage, and antioxidative defense mechanism in plants under stressful conditions. J Bot 2020:1–26. https://doi.org/10.1155/2020/4674526
42. Singh R, Sharma A, Patel S (2019) Impact of gold nanoparticles on oxidative stress-related gene expression in rice (*Oryza sativa*). Environ Nanotechnol 7(2):123–134. https://doi.org/10.1080/17435390.2018.1511035
43. Singh S, Husen A (2020) Behavior of agricultural crops in relation to nanomaterials under adverse environmental conditions. In: Husen, A, Jawaid M (Eds.) Nanomaterials for agriculture and forestry applications. Elsevier Inc. 50 Hampshire St., 5th Floor, Cambridge, MA 02139, USA, pp 219–256 https://doi.org/10.1016/B978-0-12-817852-2.00009-3
44. Singh S, Husen A (2019) Role of nanomaterials in the mitigation of abiotic stress in plants. In: Husen A, Iqbal M (Eds.) Nanomaterials and plant potential. Springer International Publishing AG, Gewerbestrasse 11, 6330 Cham, pp 441–471. https://doi.org/10.1007/978-3-030-05569-1_18
45. Tripathi DK, Singh S, Singh S, Pandey R, Singh VP, Sharma NC, Chauhan DK (2017) An overview on manufactured nanoparticles in plants: uptake, translocation, accumulation and phytotoxicity. Plant Physiol Biochem 110:2–12
46. Tripathi S, Mishra G, Mishra A, Srivastava S (2017) Zinc oxide nanoparticles: a boon for enhancing plant growth and nutrient uptake. J Nanopart Res 19(12):360

47. Verslues PE, Agarwal M, Katiyar-Agarwal S, Zhu J, Zhu JK (2006) Methods and concepts in quantifying resistance to drought, salt and freezing, abiotic stresses that affect plant water status. Plant J 45(4):523–539
48. Wang C, Wang Q, Li J (2019). Gold nanoparticles induce oxidative stress and apoptosis in rice (*Oryza sativa* L.) roots. Environ Sci Pollut Res 26(10):10515–10525. https://doi.org/10.1007/s11356-019-04388-6
49. Watson JD, Crick FH (1953) Molecular structure of nucleic acids: a structure for deoxyribose nucleic acid. Nature 171(4356):737–738
50. Willis K, McElwain J (2002) The Evolution of Plants. Oxford University Press, Oxford
51. Wolff K, Agren J (2008) The evolution of seed dormancy: environmental cues, evolutionary costs, and benefits. Q R Biol 83(2):149–175
52. Yadav RK, Girke T, Pasala S, Xie M, Reddy GV (2019) Gene expression map of the Arabidopsis shoot apical meristem stem cell niche. Proc Natl Acad Sci 116(34):16800–16805
53. Yassin MT, Mostafa AA, Al-Askar AA, Al-Otibi FO (2022) Facile green synthesis of zinc oxide nanoparticles with potential synergistic activity with common antifungal agents against multidrug-resistant Candidal strains. https://doi.org/10.3390/cryst12060774
54. Zhang X, Clarenz O, Cokus S, Bernatavichute YV, Pellegrini M, Goodrich J, Jacobsen SE (2018) Whole-genome analysis of histone H3 lysine 27 trimethylation in arabidopsis. PLoS Biol 6(5):e143
55. Zhu JK (2016) Abiotic stress signaling and responses in plants. Cell 167(2):313–324

Proteomic Insights into the Impact of Au-NPs on Plants Under Abiotic Stress

Serdar Altintaş, Sedriye Çatkin, and Behcet İnal

Abstract Proteomics, a prominent investigative field in the past decade, has garnered substantial attention for comprehending cellular functions, fostering potential advancements in disease diagnosis, prevention, and targeted therapeutic strategies. To attain a profound understanding, an exhaustive examination of all organism-expressed proteins under specific conditions is imperative. This involves a comprehensive analysis encompassing protein structure, localization, modifications, functions, and interactions with other proteins. Post-translation modifications (PTMs), including glycosylation, S-nitrosylation, and phosphorylation, bear critical importance, emphasizing the intricate nature of proteomic studies. The complexity of the proteome, illustrated by limited knowledge of a significant portion of protein-coding human genes, poses a formidable challenge. Recent studies involving Arabidopsis thaliana exposed to Au NPs unveil the induction of glutathione S-transferases in the shoot, suggesting potential applications for AuNPs in proteomic techniques. The anticipated continuous growth in AuNPs potentialities over the next five years positions them as valuable tools for protein enrichment, structural characterization, and analyte detection due to their distinctive characteristics. Transcriptomics efforts have been made to comprehend metal-related regulatory processes in plants, yet challenges persist in translating gene expressions to protein synthesis. Quantitative proteomic analysis emerges as a crucial improvement, aiding in molecular event comprehension and identification of key regulators. Comparative proteome analysis, particularly in response to various stimuli, demonstrates Au's impact on proteins, revealing alterations in cellular metabolism and redox-maintaining proteins. Notably, AuNPs induce oxidative stress, affecting mitochondrial proteins, ROS scavengers, and tonoplast-associated proteins. This study delves into the potential role of AuNPs in regulating defense systems, ion balance, and stomatal dynamics in wheat plants under salt stress. The synthesized AuNPs exhibit efficacy in mitigating salt stress responses, highlighting their significant role in enhancing wheat plant productivity. The study scrutinizes stress markers, such as H_2O_2 and LOX, elucidating the positive

S. Altintaş · S. Çatkin · B. İnal (✉)
Department of Agricultural Biotechnology Faculty of Agriculture, Siirt University, Siirt 56100, Türkiye
e-mail: behcetinal@siirt.edu.tr

A. Husen (ed.), *Plant Response to Gold Nanoparticles*, Smart Nanomaterials Technology, https://doi.org/10.1007/978-981-97-7174-5_6

interaction between AuNPs and antioxidants under salt stress conditions. Furthermore, AuNPs-mediated activation of the glyoxalase system demonstrates its potential in mitigating oxidative stress-induced responses, emphasizing the importance of this approach in adjusting osmotic balance under salt stress.

Keywords AuNP · Protein · Stress · Gene

1 Abiotic Stress in Plants

Abiotic stress represents a crucial issue currently challenging agriculture, resulting in significant global crop production losses and a reduction in cultivated land. According to reports, abiotic stress factors can lead to an annual reduction in agricultural yield ranging from 51% to 82% [109, 115]. Abiotic stressors that can influence agricultural productivity on a global scale comprise high or reduced temperatures, waterlogging, drought, salinity, heavy metals (HMs), and ultraviolet (UV) radiation. Abiotic stress adversely affects economically important crop plants by impacting crucial processes like plant photosynthesis, respiration, growth, and development. As sessile organisms, plants have developed the capacity to adjust to and exploit shifts in climate and environment [74]. Various environmental challenges frequently trigger signals and pathways involved in cellular responses including an increase in antioxidants, the accumulation of solutes, and alterations in protein trafficking.

The global challenge of soil salinity significantly diminishes crop productivity. More than 6% of the world's land is impacted by salinity, according to the FAO Land and Plant Nutrition Management Service. 45 million hectares (19.5%) of the 230 million hectares of irrigated land currently under cultivation are impacted by salt, while 32 million hectares (2.1%) of the 1,500 million hectares under dry land agriculture are affected by salt to varied degrees. Salinity adversely affects plant growth and development by causing water stress, cytotoxicity from the excessive absorption of ions like sodium ($Na+$) and chloride ($Cl-$), and an imbalance in nutrition. Furthermore, salinity is commonly associated with oxidative stress due to the generation of reactive oxygen species (ROS) [57, 148]. The harmful impact is noticed on the entire plant, leading to either plant death or a decrease in productivity. Salt stress influences fundamental processes like germination, growth, levels of photosynthetic pigments and photosynthesis, water balance, nutrient imbalance, oxidative stress, and crop yield [118].

Drought is another significant factor that limits agricultural production, leading to a substantial impact on crop yield. Additionally, serving as a primary constraint in the plant growth process, drought can impede plant respiration, photosynthesis, and stomatal movement, thereby influencing overall plant growth and physiological metabolism. In reaction to drought stress, plants initiate mechanisms in response, including morphological and structural adjustments, the expression of drought-resistant genes, the synthesis of hormones, and the production of osmotic regulatory substances to mitigate the adverse effects of drought stress. The reduction in water availability negatively affects both the yield and quality of plants [20]. Plant responses to drought are influenced by factors such as growth stages, age, plant species, and the severity and duration of the drought [49]. Additionally, the resistance mechanism to drought varies among different plant species. Consequently, plants possess the capability to minimize their use of resources and adapt their growth to withstand adverse environmental conditions, such as drought.

While all plants need water for survival, excessive water, waterlogging, or flooding can induce stress and impede gas exchanges between the soil and atmosphere. As a result, prolonged waterlogging adversely affects nearly every developmental stage in a plant's lifecycle, ultimately leading to a reduction in yield.

Heavy metals (HMs) comprise a collection of 52 metallic elements, such as lead (Pb), manganese (Mn), copper (Cu), nickel (Ni), cobalt (Co), cadmium (Cd), mercury (Hg), and arsenic (As). These elements have a direct impact on the performance of plants, and their effects vary depending on the concentration [24, 112]. Certain metals like zinc, copper, and nickel serve as vital micronutrients, necessary in minimal quantities, as they function as cofactors for numerous enzymes. Conversely, metals like cadmium and lead found in pesticides lack any beneficial role and turn toxic when their concentration surpasses a specific limit [5]. Heavy metals (HMs) disturb food chains, water sources, the atmosphere, and soil. They also have mutagenic, genotoxic, and cytotoxic effects on humans, animals, and plants [39]. High concentrations of heavy metals (HMs) damage physiological and biochemical systems like photosynthesis, respiration, water relations, organic matter transport, stress tolerance, and mineral feeding, which negatively impact plant survival, growth, and development.

The growth and development of plants are significantly influenced by temperature, which is a crucial environmental factor. Plant exposure to low temperatures can be divided into two categories: chilling stress (0–20 °C) and freezing stress (<0 °C). These categories are determined by specific temperature ranges and the distinct physiological mechanisms that operate at different temperatures [146, 149]. Nevertheless, the majority of temperate plants possess the capacity to adjust to low temperatures through a process known as cold acclimation. Following exposure to nonfreezing low temperatures, these plants exhibit enhanced tolerance to freezing conditions [25]. Cold acclimation induces numerous physiological and molecular modifications, encompassing alterations in transcription, translation, and metabolism. These changes result in the elevation or reduction of specific proteins, metabolites, and phytohormones [97, 113]. Given the critical role of membrane stabilization in surviving freezing conditions, cold acclimation also triggers dynamic shifts in lipid composition [31, 149]. Due to the abundance of abiotic stresses and the increasing probability of global climate change, there is an urgent requirement for plants to adapt and mitigate these challenging conditions [160]. It is crucial to emphasize the significance of diverse plant pathways in addressing abiotic stresses and supporting continuous growth. The adoption of nanomaterials signifies an innovative trend among recognized plant strategies for addressing abiotic stresses [30].

A burgeoning area of the twenty-first century is nanotechnology, which significantly impacts the global way of life, economy, and industry by introducing nanomaterials, nanorods, nanotubes, nanosensors, nano drugs, and other innovations [17,

135]. In general, nanoparticles have the potential to serve as materials for delivering nucleotides, proteins, and chemicals to specific sites under in vitro conditions. This capability aims to achieve important objectives, including enhancing crop growth and yield, along with bolstering resistance against various stressful conditions [127]. Due to their small size, specific shape, diverse chemical characteristics, and increased surface area, nanoparticles (NPs) have demonstrated remarkable effectiveness in enhancing plant growth mechanisms and nutrient accessibility. However, careful consideration and expertise are essential in choosing an appropriate range of NPs, as they can induce oxidative stress depending on factors such as type, application ratio, surface size, and, most importantly, the targeted plant species [26]. Out of numerous applied methods, nanotechnology is recognized as a crucial application in agriculture and has been actively utilized for many years to enhance crop productivity, alleviate biotic and abiotic plant stress, and to maintain a sustainable agriculture approach [15, 136]. Furthermore, nanoparticles find application in agriculture for the effective and regulated release of fertilizers, herbicides, pesticides, and for enhancing the physico-chemical properties of soil, such as water retention capacity [129]. In response to environmental stress, plants have evolved diverse and effective molecular programs to quickly detect stressors and adapt accordingly [154]. The interaction of nanoparticles (NPs) with plants can enhance this response. Nanotechnology holds the potential to boost crop yield by enhancing plant tolerance mechanisms under abiotic stress conditions [13]. Numerous studies have indicated that NPs play a crucial role in enhancing plant tolerance to abiotic stresses by influencing various physiological, biochemical, and molecular processes.

The effects of water-deficit stress extend across the entire life cycle of a plant, from germination to maturation. It negatively impacts plant yield and productivity by disrupting a variety of morphological, physiological, and biochemical processes [114, 137]. Physiological changes induced by drought stress include altered stomatal closure, reduced relative water content, decreased photosynthetic activity, lower transpiration, and increased electrolyte leakage [11, 89]. Additionally, drought stress alters the plant's biochemical profile by decreasing chlorophyll content, promoting the accumulation of compatible osmolytes (such as proline), increasing the production of reactive oxygen species (ROS, e.g., hydrogen peroxide, superoxide radical), enhancing lipid peroxidation, and activating the antioxidant machinery [11, 166].

Nanoparticles (NPs) have the potential to mitigate the impact of abiotic stresses on diverse plant species by triggering physiological, biochemical, and molecular mechanisms (see Figure 1). Under conditions of drought stress, NPs can enhance morphological traits, stimulate the production of both enzymatic and non-enzymatic antioxidants, facilitate the biosynthesis of osmolytes, and promote the expression of genes associated with stress response in plants [56]. This includes improvements in the photosynthetic rate, stomatal conductance, relative water content, and mitigation of cell membrane damage by reducing the levels of stress-related metabolites and electrolyte leakage. Additionally, a general mechanism for overcoming and alleviating drought stress involves increases in osmolyte levels, carotenoids, chlorophyll, proteins, and phenolic substances (e.g., rosmarinic acid and chlorogenic acid). Moreover, there is an enhancement in the activities of antioxidant enzymes such as catalase (CAT), superoxide dismutase (SOD), and peroxidase (POX) [124].

Among the different abiotic stressors, salinity presents a considerable obstacle to the growth and productivity of plants, particularly in arid and semi-arid regions. The increased concentration of salt in the soil disturbs the ion balance in plants, obstructs water absorption, and triggers oxidative damage, ultimately leading to decreased crop yields and lower crop quality. Several methods can be employed to mitigate salt stress in plants. Utilizing nanoparticles represents a hopeful approach to alleviating salt stress in plants. Several research studies have established that the utilization of nanoparticles can boost plant growth, elevate crop yield, and enhance the quality of crops cultivated under conditions of salt stress [118, 167].

Nanoparticles can assist plants coping with salt stress by regulating ion balance, mitigating Na+ ion toxicity, increasing K+ uptake, activating the antioxidative defense system, enhancing pigment and compatible solute contents, and increasing stomatal conductance. In salt-stressed *Triticum aestivum* L., the application of magnetite nanoparticles led to improved chlorophyll levels and antioxidative

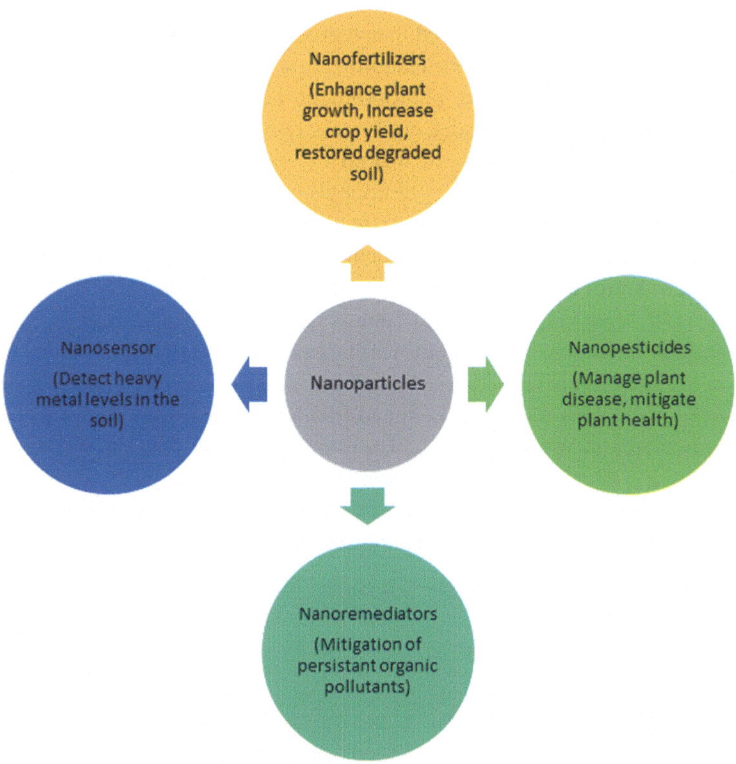

Fig. 1 Possible uses of nanoparticles in agriculture

enzymes, coupled with the amelioration of various polypeptide chains associated with salinity stress tolerance [38]. Nano-SiO_2 enhanced the growth of Glycine max under salt stress by elevating leaf K+ levels and biological antioxidant activities [40]. Similarly, in salt-stressed *Triticum aestivum* L. cultivars, nano-SiO_2 was observed to enhance seed germination and growth [110].

Heavy metal stress has been observed to impact the plant's morphology, physiology, and biochemistry. Recently, researchers have suggested the integration of metal-based nanoparticles (NPs) as a viable approach to mitigate environmental pollutants [37]. For instance, lowering the bioavailable concentration of heavy metals in the soil [103], controlling the activation of heavy metal transport genes in plants [27], augmenting the effectiveness of plant antioxidant systems, and enhancing physiological functions [159], and encouraging the generation of defensive substances like root exudates, phytochelatin, and organic acids [98]. Despite specific concerns related to NP synthesis methods, strategies for cellular penetration, and ecotoxicological effects, the overall durable structure, small size, extensive surface area, and significant adsorption potential make NPs promising candidates for the removal of heavy metals from plants [41].

Cold or low-temperature stress, which negatively impacts plant growth and development, is further exacerbated by global climate change. Chitosan nanoparticles and TiO_2 NPs have been widely employed in numerous studies due to their effectiveness in enhancing cold-stress tolerance. The utilization of Ti NPs has proven effective in enhancing electrolyte leakage, photosynthetic activity, and membrane damage regulation in chickpea plants under cold-stress conditions through transcriptional regulation [101]. The application of nano-TiO_2 elevates the expression levels of genes that encode for Rubisco- and chlorophyll-binding proteins. The alteration in gene expression linked to the mitigating effects of nanoparticles (NPs) on photosynthesis in cold conditions has been documented in two investigations by Hasanpour et al. [54] and Amini et al. [9]. Both studies concentrate on the role of TiO_2NPs in alleviating cold stress in two varieties of chickpea (*Cicer arietinum* L.): one resistant to cold and the other sensitive to cold. Hasanpour et al. [54] observed a significant reduction in H_2O_2 levels in cold-resistant plants treated with TiO_2NPs compared to the sensitive ones. This reduction is linked to a heightened capacity for photosynthesis, particularly in resistant plants. Additionally, the application of TiO_2NPs is associated with a decrease in H_2O_2 content and an improvement in photosynthetic activity. Singh and Lee [141] demonstrated that utilizing TiO_2 nanoparticles could diminish Cd toxicity and improve the photosynthetic rate in soybeans [141].

2 Nanoparticles as Potential Stress Alleviators and Using in Agriculture

Nanoparticles find wide-ranging applications in agricultural sectors (Fig. 1), encompassing nano biosensors, plant growth regulators, promoters of plant growth, nanofertilizers, nanopesticides, nutrient management, and defense against phytopathogens [62–66, 65].

2.1 *Nanofertilizer*

Nanofertilizers, as nanomaterials, are designed to supply one or more types of nutrients to promote the growth of plants and enhance overall production [91]. Nanofertilizers provide a unique opportunity to evolve into plant nutrients with a high absorption rate and enhanced utilization efficiency, resulting in minimal losses. This makes them a more effective option for facilitating nutrient uptake in plants. Nanofertilizers exhibit greater efficacy compared to conventional fertilizers in minimizing nitrogen loss caused by liberation, leaching, and sustained assimilation by soil microorganisms [88]. Moreover, Nanofertilizers offer greater advantages to plants compared to conventional fertilizers as they enhance developmental parameters, including plant growth, height, leaf surface area, dry mass, optimal chlorophyll production, and photosynthesis rate. This leads to elevated photosynthesis production and increased translocation to various plant parts [6, 132].

Nano-fertilizers can be categorized into three groups depending on the nutritional needs of plants: (1) macro-nutrient nano-fertilizers, (2) micro-nutrient nano-fertilizers, and (3) nanoparticulate nano-fertilizers [28]. Nano-fertilizers containing macronutrients consist of a blend of elements, including potassium (K), magnesium (Mg), nitrogen (N), calcium (Ca), and phosphorus (P). Nitrogen (N), phosphorus (P), potassium (K), magnesium (Mg), sulfur (S), and calcium (Ca) are categorized as macronutrients essential for plant growth. Urea-coated zeolite chips and urea-modified hydroxyapatite nanoparticles have been created as nitrogen sources, demonstrating their effectiveness in providing a slow and controlled release of nitrogen over an extended period [81, 99]. In a study conducted by Liu et al. [92], calcium nanoparticles were synthesized as a substitute for traditional calcium macronutrients, resulting in a 15% increase in the biomass of *Arachis hypogaea*. Micronutrients essential for plant growth encompass molybdenum (Mo), copper (Cu), iron (Fe), nickel (Ni), manganese (Mn), and zinc (Zn). In contrast to macronutrients, these elements are needed in significantly lesser quantities for the proper development of crops. A combination of three micronutrient nanoparticles (ZnO, CuO, and B_2O_3) has been successfully formulated for agricultural enhancement, with the aim of alleviating drought stress in soybean plants. Application of manganese nanoparticles (Mn NP) on *Vigna radiata* resulted in a 52% increase in root length, 38% in shoot length, 71% in rootlet growth, and a 38% improvement in biomass at a concentration of 0.05 ppm, as compared to using bulk manganese sulfate ($MnSO_4$). Nanoparticulate fertilizers include additional nanoparticles like CNTs, TiO_2, and SiO_2, which contribute to plant growth. The combination of TiO_2 and SiO_2 resulted in increased nitrogen fixation, enhanced growth, and improved seed germination in soybeans [94]. Positive nanoparticles also demonstrate their influence on the absorption of nutrients. The utilization of specific nanoparticles like ZnONPs, silicaNPs, titanium dioxide NPs, ZnS/ZnCdSe core–shell quantum dots (QDs), InP/ZnS core–shell QDs, Mn/ZnSe QDs, gold NPs, Al2O3 NPs, CeO2 NPs, and FeO NPs has been documented to enhance nutritional effectiveness. Applying nano-anatase treatment (TiO_2) at concentrations ranging from 10 to 40 mg/mL to parsley (Petroselinum crispum) results in increased uptake of nitrogen, calcium, potassium, and magnesium, thereby enhancing the nutritional content [32].

2.2 Nanopesticides

Nanopesticides have exhibited a beneficial effect on the regulation of plant pests and diseases. The development of pesticide nanoformulations using polymers or metal nanoparticles represents a significant and essential aspect of the pesticide industry. The process of nanoencapsulation offers benefits by enabling controlled and gradual release of the active ingredient. This is achieved by manipulating the outer shell of nanocapsules, resulting in the discharge of a minimal dose over an extended period. Consequently, this approach helps decrease the excess runoff of undesirable pesticides (Pérez-de-Luque and Rubiales 2009). Silver nanoparticles (Ag) serve as a broad-spectrum active agent against phytopathogens. Silver nanoparticles produced using the latex of the *Plumeria rubra* plant exhibit effective larvicidal activity against *Aedes aegypti* and Anopheles stephensi [120]. Jafir et al. [71] documented that silver nanoparticles derived from the leaf extract of *Ocimum basilicum* are a safe alternative for managing S. litura. They observed an elevated mortality rate in *S. litura* when using these nanoparticles compared to other synthetic insecticides.

2.3 Soil Texture

The fertility of the soil is primarily controlled by its texture, and for less fertile soils, nanomaterials offer an intelligent, innovative, eco-friendly, and sustainable solution for soil improvement. The utilization of nanoparticles in soil has enhanced various soil attributes, including color, texture, structure, porosity, density, consistency, aggregate stability, and temperature. These characteristics play a crucial role in influencing processes such as infiltration, erosion, nutrient cycling, and biological activity [35, 58, 90]. The primary nanoparticles employed in soil applications are those derived from silver and silica, which effectively regulate soil degradation. Alsharef et al. [7] demonstrated that incorporating approximately 0.2% of multiwall carbon nanotubes (MWCNTs) and carbon nanofibers (CNFs) into clayey sand soil (UKM soil) improves soil hydraulic conductivity and diminishes the occurrence of soil cracks. The application of metallic nanoparticles as soil additives has the potential to enhance soil characteristics. Iron oxide nanoparticles, in particular, can bring about alterations in both the physical and chemical properties of the soil [106].

2.4 Seed Dormancy

Currently, employing nanoparticles (NPs) in seed priming represents an innovative approach to enhancing agricultural productivity. Seed priming is a viable method for overcoming seed dormancy, enhancing seed viability, and optimizing germination potential [117]. Nanotechnology significantly contributes to seed science research, primarily by improving seed quality, germination, storage, development, and protection [143]. Mahakham et al. [96] have suggested that nanopriming with AgNPs not only promotes seed germination but also induces the formation of nanopores, leading to improved water uptake. This process revitalizes ROS/antioxidant systems, facilitates cell wall loosening through the generation of hydroxyl radicals, and expedites starch hydrolysis in seeds with the assistance of nanocatalysts. Treating seeds with nanoparticles can activate seed enzymes, leading to improved germination. For example, applying TiO_2 nanoparticles to seeds boosts the activity of antioxidant enzymes such as superoxide dismutase, catalase, and peroxidase, thereby enhancing seed vigor [59]. Seed pretreatment with nanoparticles not only stimulates seed germination but also influences the subsequent growth of seedlings and enhances plant resistance to both biotic and abiotic stresses.

2.5 Use of NPs in Water Treatment

Water pollution has consistently emerged as a persistent environmental problem and a worldwide challenge [3]. This is largely attributed to the escalating pace of human activities, including population growth, industrialization, and agricultural practices [68], which have raised concerns about water quantity, quality, and availability in numerous regions across the globe. Due to an increasing number of pollutants and a rapid pace of industrialization, numerous traditional approaches, such as coagulation [69], membrane processes [23], adsorption [2], dialysis [95], advanced oxidation processes [29], photocatalytic degradation [8], and biological methods [53], have been employed for the removal of hazardous pollutants from water and wastewater. Over the past decade, nanoparticles have become a promising choice for addressing the worldwide challenge of water pollution and its treatment. Recent publications detail various technological approaches utilizing carbon nanoparticles for the extraction of pathogenic microorganisms from wastewater [78, 79].

3 Significance of Proteomic and Proteomic Analysis Techniques

Proteomics, transcriptomics, genomics, and metabolomics are important omics technologies or approaches that allow analyzing molecular components in plant systems [72]. By integrating data obtained from omics techniques, scientists create a holistic perspective on the molecular interactions of plants against abiotic stress factors [130]. Considering the data obtained with omics techniques holistically provides important information for identifying genes and proteins in stress management and metabolic pathways in stress adaptation. Additionally, the obtained data are integrated with advanced computational tools to create predictive models and allow optimizing strategies to increase plant stress tolerance [72]. While transcriptome analyzes show what the cell is capable of doing at any given time, proteome analyzes provide information about what the cell is. Proteins, which are the functional machines of the cell, are key molecules that play a role in the organism's growth, development, reproduction, and response to stress [36]. Proteomics includes the application of technologies for the identification and quantification of general proteins found in a cell, tissue or organism. It is a supporting technology in determining the identity and structure function of a particular protein, together with the data provided by other omics approaches [16]. Thus, it is possible to reveal the functional information of genes. "Proteome" can be defined as the general protein content of a cell characterized in terms of their localization, interactions, post-translational modifications, and transformations at a specific time. The term "proteomics" was first used by Marc Wilkins in 1996 to refer to "the PROTein complement of a genOME" [161]. Proteomics technology is defined as the systematic analysis of the proteome, which is the protein complement of the genome [116]. This technology allows qualitative and quantitative measurements of a large number of proteins that directly affect cellular biochemistry, thereby enabling accurate analysis of cellular state or system changes during growth, development, and response to environmental factors. It is an important technology for the study of complex and dynamic biological systems. Although gene expression at the transcriptional level provides important information regarding early-stage transfer from the genome to the cellular machinery, mRNA levels are not always consistent with the abundance of cognate proteins (Gygi et al. 1999,[67]). Moreover, one gene can produce many different types of proteins due to various alternative splicing, mRNA processing, protein proteolysis, and PTM (post-translational modification) [116],Service 2001). Because proteomics can reveal chemical complexity and biological dynamics, it provides functional information about the biochemical processes underlying phenotypes that cannot be accessed or predicted by other means. Although advances in proteomics technology have helped answer complex biological problems, there are also some technical challenges for plant proteomics studies. The presence of secondary metabolites, proteases, and oxidative enzymes in plant cells and the low protein content in plant cells make it difficult to obtain higher quality protein extracts [46]. Additionally, the presence of the cell wall is another factor that complicates the situation. For this purpose, the use of TCA (Trichloroacetic acid), acetone and phenol has made it easier to overcome this difficulty [70]. Appropriate techniques and data sets must be used to accurately characterize proteins that are key to biological functions in the cell. The workflow of proteomic analysis techniques is generally applied by following the protein purification, analysis, characterization, sequence analysis, quantification, structural analysis, and bioinformatics steps (Aslam et al. 2006). Chromatography-based techniques such as ion exchange and affinity chromatography are used in the protein purification step. Ion exchange chromatography (IEC) is a useful technique due to its low cost and based on the anionic or cationic separation of proteins [73]. Western Blotting and Enzyme-Linked Immuno Sorbent Assay (ELISA techniques are used to analyze selective proteins. These techniques can analyze protein but may be insufficient to estimate expression level. Western Blotting is a powerful and useful technique used for the precise detection of low-density target proteins in the cell using the electrophoresis system (Leguin 2005,[100]). Gel-based approaches (Sodium dodecyl sulfatepoly-acrylamide gel electrophoresis (SDS-PAGE) and two-dimensional electrophoresis (2-DE)) (Wittmann et al. 2006,Dunn and Burkhes 1986) and mass spectroscopy (MS), especially liquid chromatography coupled with tandem mass spectrometry (LC-MS/MS) are techniques used in the separation and identification of complex proteins (Newton et al. 2004). 2-DE matrix-assisted laser desorption/ionization-time of flight (MALDI-TOF) mass spectrometry (MS) is a widely used technique in the analysis of DNA, protein, peptide biomolecules, and various organic substances. It is relatively less useful than other techniques because it has low efficiency in determining proteins with very high and very low pH values [4]. Gel-based methods are widely used because they are simple, reproducible, and detect post-translational changes [46]. However, LC-based methods are also preferred by researchers as they identify low amounts of proteins and cover a wide molecular weight scale [45]. High-throughput genome-based omics techniques have been used extensively to study plant responses to NPs [33]. Proteomics uses high-throughput approaches to uncover information about protein structure and conformation, as well as interactions between proteins and ligands. Compared to genome- or transcriptome-based technologies, proteomics has various advantages because it works with functional molecules directly instead of DNA or mRNA [19]. For the purpose of studying plant stress signaling pathways, protein biomarkers, protein chips/microarrays, and gel-based or gel-free proteomic approaches have all been extensively utilized for the accurate identification and quantification of stress-responsive proteins [1]. Proteomics is a fast, sensitive, and trustworthy method for identifying and characterizing differentially regulated proteins to assess the potential effects of NPs on crops, thanks to advancements in mass spectrometry and improved protein extraction protocols. As an alternative to the single-omics approach, multi-omics techniques such as the combination of transcriptomics, proteomics, and metabolomics offer further advantages in identifying the underlying response mechanisms of plants to environmental pollutants, including NPs [158]. The protein response of silver, aluminum oxide, and zinc oxide nanoparticles in the roots and leaves of the soybean plant was analyzed using gel-free techniques (LC MS/MS) [61, 111]. Protein responses occurring in roots and leaves of silver nanoparticle application in Eruca sativa, wheat, and tobacco plants were measured using 2-DE and MALDI-TOF gel-based proteomic techniques [121, 152, 153]. The Edman degradation method is used to determine the amino acid sequence of a certain protein [87]. Additionally, X-ray crystallography and nuclear magnetic resonance (NMR) spectroscopy are two important high-throughput techniques that provide the three-dimensional (3D) structure of the protein and can help understand the biological function of the protein. While X-ray crystallography is more sensitive to the general shape of the molecule, NMR technology has sensitivity at the atomic level (Schiro et al. 2020). Bioinformatics databases have been created to store and manage proteomic data obtained with advanced techniques. Using these bioinformatics tools, the three-dimensional structure of the protein can be predicted, protein-protein interaction can be detected, and sequencing can be performed to determine the relationship [123, 155]. To summarize the subject, proteome analysis helps to understand the structural and functional information of the cell, as well as the response mechanism of the cell to various types of stress, at the protein level, by using proteome techniques individually or in combination.

4 Effect of Au-NPs on Plant Physıology

Plant physiology is defined as the branch of science that studies plant functions, in short, including the dynamic processes of plant matter exchange, growth, development and movement, metabolism, and reproduction (Strasburger 2008) [144]. Plant physiology studies how plants function by focusing on their physics and chemistry. Plants capture light energy, produce sugars through the photosynthesis reaction, and break down these sugars through aerobic cellular respiration. They respond to various environmental conditions through changes in growth, vital events such as germination or flowering, or through movement. Plants need soil for mineral nutrients and water, and biogeochemical cycles replenish soil with these nutrients. Once water and minerals are taken up from the soil, they must be transported through the xylem, and this movement is driven by the loss of water vapor from the leaves (transpiration) and the cohesive property of water. Similarly, sugar-rich assimilate must also be transported or translocated through the phloem. All of these processes occurring in plants are examined within the scope of plant physiology.

Plants can be exposed to many environmental stress factors that will affect crop yield in their natural environment. It is estimated that there may be a decrease in crop production in the future due to climate changes, decrease in agricultural lands due to increasing population, decrease in water resources, and global changes [14, 150]. Therefore, there is a need to develop new technologies and sustainable techniques in agriculture. Although genetic modifications, plant breeding and the use of synthetic chemicals have increased yields, they have some limitations. Therefore, it is thought that the use of nanotechnology, which is more environmentally friendly and more beneficial for health, will meet this need to meet the increasing food need [10, 44, 128]. Nanotechnology, accepted as "Key Enabling Technology" by the European Commission, is one of the promising and innovative tools [119].

Nanoparticles used in nanotechnology show their effects by overcoming cellular barriers in living organisms, thanks to their physico-chemical properties [76, 125]. NPs that can be applied to soil and leaves enter the root epidermis through apoplastic and symplastic pathways [12, 165]. While the uptake of NPs from soil occurs mainly through physiologically active lateral roots and root hairs, in the leaf NPs enter largely through stomata but also through trichomes [80]. Once in the roots and leaves, NPs spread from one cell to another and move up and down the xylem or phloem tissues, respectively [55]. In many physiological studies conducted with NPs, it has been shown that the plant increases macro and micro element uptake [43], photosynthesis [83], plant growth and development [164], and total biomass [131]. *Capsicum annuum* L. [163], *Cucumis sativus* L. [93], *Zea mays* [151], *Triticum aestivum* [22] and Hordeum vulgare under different stress conditions [21] has been shown that manganese (Mn), manganese oxide (Mn_3O_4), copper (Cu) and chitosane nanoparticles increase the seed germination, photosynthetic pigment content, total photosynthesis level and relative water content (RWC) of the plants, respectively. Au-NPs are used in various sectors, including biology, medicine, chemistry, cosmetics, and electronics [86, 126]. Gold nanoparticles (Au-NPs, which are in the group of beneficial nanoparticles, are metal-based nanoparticles and increase plant tolerance to environmental stresses (Ficheira et al. 2020) [140]). In a study conducted on the Solanum lycopersicum plant, it was shown that Au-NPs increased seed germination, shoot elongation, chlorophyll content and photosynthesis level. [145]. Beneficial NPs also have an effect on the intake of macro and micronutrients. As Au NPs, ZnO NPs, TiO_2 NPs, Mn/ZnSe, Al2O3 NPs, CeO2 NPs, and FeO beneficial NPs have been shown to increase nutritional efficiency. However, the effect of NP use varies from one plant species to another. The physical and chemical properties of NPs may also modify the effect of NPs on nutritional parameters. NPs facilitate the slow release of nutrients, which reduces leakage. NPs, easily absorbed by porous soil, enrich soil nutrients. Moreover, these nutrients are absorbed and accumulated by plants, which helps them to compensate for nutrient deficiency and overall increase in crop production [140]. Feichtmeier et al [42] showed that gold nanoparticles (Au-NPs) in barley plants caused concentration-dependent changes in the plant's macro and micronutrient uptake, biomass production, and especially root growth. Application of NPs increases the rate of photosynthesis, total chlorophyll content, oxygen evolution, chlorophyll formation and protects the chloroplast. In a study conducted with Au NPs, it was reported that seed germination and electron transfer efficiency increased in application groups [18]. Additionally, another study showed that gold nanoparticles (10 ppm/1000 μM increased product yield by causing an increase in sugar and chlorophyll levels [14, 48]. Beneficial NPs have an effect on seed germination percentage, germination rate, seed viability index. It also has an effect on physiological properties such as osmolyte accumulation (soluble sugars, proline, glycine betaine, and free amino acids), balance in photosynthetic pigments, transpiration, and stomatal conductance, which help plants combat abiotic stresses [52, 75]. Studies have shown that Au NPs increase seed germination at concentrations of 5–15 mg/L and 10–80 μg/ml [18, 82]. Gold nanoparticle studies are still limited. Considering the limited literature studies, the results obtained show that Au-NPs can have effective results on the physiological parameters that are necessary for the plant to survive. The use of gold, which is known to be a useful nanoparticle, may enable the dissemination of an alternative technology in order to increase world food security and product yield.

5 Proteome Modulation in Response to Au Treatment Under the Abiotic Stresses

In the past ten years, the topic of proteomics has received a lot of interest as a means of improving our knowledge of how cells function. Gaining insight into the basic role of a cell can lead to many opportunities for diagnosis, prevention, and the creation of targeted therapy plans for diseases for which there is no known cure or where existing therapies are failing. Studying every protein that an organism expresses under specific circumstances is required to reach this deeper level of consciousness. Understanding their structure, localization, variations, function, and interactions with other proteins is essential to navigating this. Moreover comprehensive details regarding post-translational modifications (PTMs), including glycosylation, S-nitrosylation, and phosphorylation [139]. Mutations in proteins associated with conditions and relations between proteins and medicines are of extreme significance. With this quantum of information needed, and due to the complexity of the proteome, it is apparent that this task has not been easy over times. Until lately, proteomics exploration refocused on 20,300 protein-rendering mortal genes expressed in eukaryotic cells, of which there is limited knowledge of only two- thirds of the proteins [84]. It is known that NPs are associated with proteins that play a role in many physiological and biochemical mechanisms at the cellular level in plants (Fig. 2). Thus, the complexity of the proteome poses the first challenge. After exposing *Arabidopsis thaliana* to Au NPs, a proteome analysis revealed the induction of glutathione S-transferases in the shoot. Also, experimenters have also suggested that auxin, the factory hormone, plays a significant part in shaping Au-convinced [108]. In a five-time perspective, the nonstop growth of AuNPs possibilities is anticipated, as an advantage to enhance several proteomic ways formerly in use to study the proteome, its proteins, and protein–protein relations. The three topmost characteristics of AuNPs size and morphology, face functionalization according to the operation, and unique optic and electrochemical features, make them decreasingly seductive to be enforced in laboratory routines as tools for protein enrichment, structural characterization and analytic discovery. Toward understanding the essence-related nonsupervisory processes in shops, several transcriptomics grounded sweats have been made in recent times. Still, it is inversely accepted that gene expressions at a recap could not always transduce up to the protein conflation [51]. Accordingly, a quantitative proteomic analysis improves our understanding of the molecular events and helps us identify the key regulators [156]. Relative proteome analysis is currently widely used to identify the probable role of particular proteins in reaction to vibrant biotic and abiotic stimulants seen in retail settings. A relative proteome study of Au-treated root and shoot was carried out in order to determine the impact of Au on proteins and to establish a correlation between transcriptome and proteome. Certain proteins involved in cellular metabolism, such as fructokinase-1, 40S ribosomal protein S, glycerophosphodiester phosphodiesterase (GDPD3), NADH dehydrogenase 1, and glyceraldehyde-3-phosphate dehydrogenase-C subunit (GAPC-1) were shown to be persuasive [147]. Generally speaking, GAPC-1 is recognized for its involvement in the glycolytic pathway; nevertheless, it can also interact with H_2O_2 to obtain a portion of the ROS50 signaling cascade. The alteration of multiple major metabolism-related proteins implies that exposure to Au has a relative impact on carbon metabolism. These include the overexpression of fructokinase-1 in root tissues and GAPC1, 3-hydroxyisobutyrate dehydrogenase-like 3 (from pentose phosphate shunt) [147]. The ROS-scavenging protein Mn superoxide dismutase-1 (MSD1) was downregulated in the root. According to [147], MSD1 exertion is essential for preserving ROS localization, which is crucial for the development of the Arabidopsis embryo sac. This finding provided more evidence that Au used cellular proteins and enzymes to counteract oxidative stress, which in turn produced free radicals and reactive oxygen species (ROS). One likely pathway for ROS production inside cells is the dislocation or failure of the electron transport system caused by Au when it moves inside the mitochondria and chloroplast.

The proteome analysis demonstrated the overexpression of key components of the electron transport chain, such as the main subunits of the cytochrome c oxidase subunit 5b-1 and beta subunit 9 of NADH dehydrogenase (ubiquinone)-1 in shoots. The expression of lactoylglutathione lyase (glyoxalase-I), two thioredoxin-like proteins, and several other oxidative stress-regulated proteins was elevated. As predicted, cellular proteins that maintain redox, such as cytochrome c oxidase and 3-hydroxyisobutyrate dehydrogenase, were activated following exposure to aurum [147]. One interesting consequence of the proteome analysis was the upregulation of HMA3 (7.98 pack). According to [105], AtHMA3 is found on tonoplast and contributes to the detoxification of zinc and copper by vacuolar insulation. Serine racemase was the only one of the 15 related proteins in the shoot that had down-regulation (−3.44 fold). According to [47], serine racemase bears magnesium as an asco-factor and catalyzes the conversion of D-serine to L-serine enantiomer. Indole-3-acetic acid inducible 5 was the other elevated protein associated, it is involved in the initiation of side roots in Arabidopsis [107]. The proteome analysis has connected the four GSTs. One member of the GST family, GST F6, was eliminated in the root in accordance with the transcriptome data, while GST F2, GST U15, and GST U21 significantly increased in the shoot. Given the advanced Au buildup in the shoot, the identification of significant numbers of GSTs in the proteome of the shoot is especially significant. It is commonly recognized that GSTs have a role in xenobiotic detoxification, redox conservation, and subcellular transport of flavonoids. The biotransformation of Au ions into AuNPs to protect cells against related toxicity is probably linked to the stimulation of numerous GSTs. Less research has been done on the defensive mechanisms that AuNPs regulate, such as the glyoxalase and AsA-GSH pathways, proline and nitrogen absorption, ionic balance, and stomatal

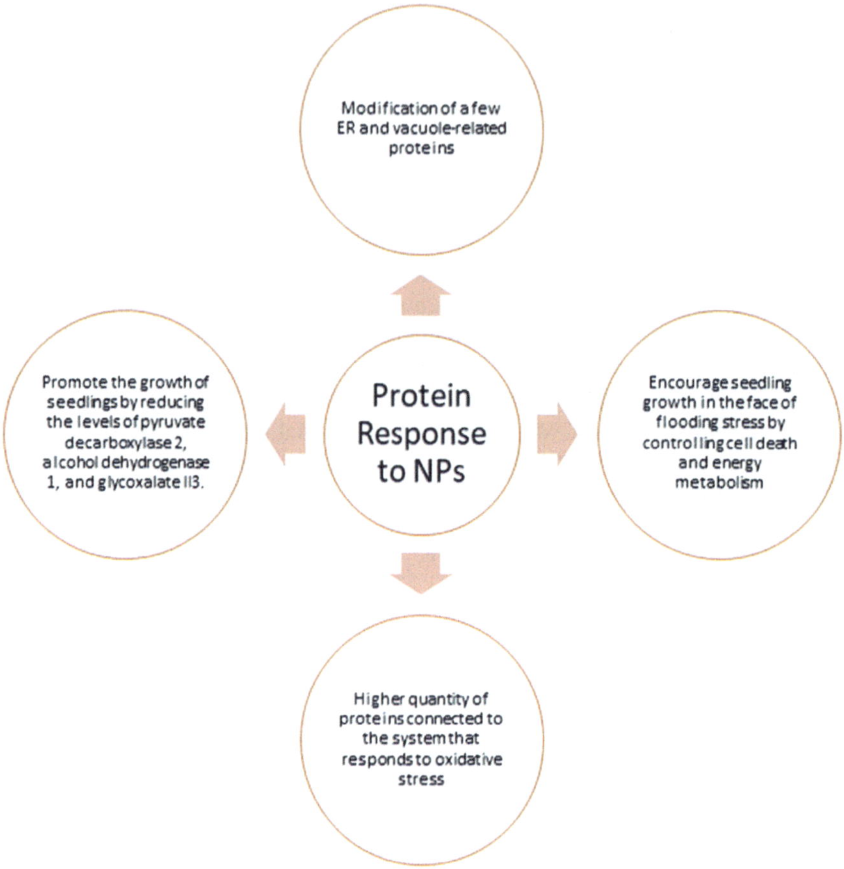

Fig. 2 Diagrammatic depiction of the transcriptome and proteome interface between plants and NPs

dynamics of wheat shops under swab stress. Here, wheat leaf extract-derived AuNPs were produced and described in vibrant ways, which made them useful for attenuating swab stress-induced reactions.

Hence, the aforementioned studies showed a significant part of green synthesized AuNPs in modulating the swab stress-responsive traits in wheat shops, and farther aid in adding the factory productivity. Swab stress induces accumulation of stress labels including H_2O_2, LOX, and MDA in shops. Among these, H_2O_2 is considered a largely diffusible and stable stress patch, which generally causes oxidative stress. High attention of H_2O_2 negatively alters macromolecules and membrane structures resulting in inhibited factory growth and development [60]. Piecemeal from ROS including H_2O_2 and O2. -, lipid peroxidation is also touched off by increased exertion of LOX enzyme under stress conditions [104]. AuNPs intermediated proliferation in AsA and GSH content suggest the positive commerce between AuNPs and non-enzymatic antioxidants under swab stress. In addition, the forenamed antioxidants help in maintaining redox eventuality in AuNPs-treated swab-treated wheat shops. Another potent oxidative stress marker, MG, generated via the fragmentation of glyceraldehydes-3-phosphate (G-3-P) and dihydroxyacetone phosphate (DHAP), oxidative declination of glucose and emerges as the by-product of lipid peroxidation [77]. Thus, in this study, we have also concentrated on this system to estimate and give an understanding of AuNPs-intermediated activation of glyoxalase system in response to swab stress. Herein, AuNPs operation significantly increased the conditioning of GLY- I(glyoxalase-I and GLY-II(glyoxalase-II which in turn facilitates the detoxification of MG(Methylglyoxal in swab stressed shops. In the former study, AuNPs specially reduced Spell-intermediated proline catabolism, further, the increased GK exertion enhanced proline content in swab-stressed shops. This ultimately redounded in proline assimilation in AuNPs treated shops and farther backed in conforming the bibulous balance under swab stress [157].

6 Conclusion

In the prospective outlook for proteomics over the next five years, there is anticipated sustained growth in the potential applications of gold nanoparticles (AuNPs). This development is expected to provide a strategic advantage for enhancing existing proteomic techniques utilized in the investigation of the proteome, its constituent proteins, and protein-protein interactions. The three predominant attributes of AuNPs—namely, size and morphology, surface functionalization tailored to specific applications, and distinctive optical and electrochemical features—contribute to their increasing attractiveness for integration into laboratory procedures as valuable tools for protein enrichment, structural characterization, and analytic detection. Despite the rapid progression of proteomic methodologies, limitations persist. However, the incorporation of AuNPs is anticipated to lead to the exclusive adoption of these nanoparticles in some conventional techniques, enhancing their overall performance in the future. AuNPs have demonstrated considerable success in proteomic applications, and their potential applications are deemed limitless. It is noteworthy that despite substantial advancements facilitated by nanoparticle introduction in proteomic techniques, challenges persist, particularly in distinguishing and identifying proteins and small molecules. The complexity of proteins, necessitating a combination of techniques and substantial data for accurate identification, contrasts with the relatively straightforward spectra of small molecules. The use of nanoparticles as matrices has significantly aided in mitigating issues such as ion interference. The consistent application of mass spectrometry (MS) in conjunction with kits containing AuNPs for protein enrichment and quantitative measurements is expected to evolve into user-friendly solutions. These solutions are anticipated to surpass the sensitivity levels of currently available kits designed for similar purposes. This transition is projected to simplify the processes of protein identification and quantity measurement, contributing to the ongoing advancement of proteomic methodologies.

References

1. Abdelhamid HN, Wu HF (2015) Proteomics analysis of the mode of antibacterial action of nanoparticles and their interactions with proteins. TrAC, Trends Anal Chem 65:30–46
2. Adeniyi AG, Ighalo JO (2019) Biosorption of pollutants by plant leaves: an empirical review. J Environ Chem Eng 7(3):103100
3. Ajala O, Ighalo J, Adeniyi A, Samuel O, Comfort A (2020) Contamination issues in sachet and bottled water in Nigeria: a mini-review. Sustain Water Resour Manage 6:112
4. Alaoui-Jamali MA, Xu YJ (2006) Proteomic technology for biomarker profiling in cancer: an update. J Zhejiang Univ Sci B 7:411–420
5. Ali I, Alothman ZA, Alwarthan A (2017) Supra molecular mechanism of the removal of 17-β-estradiol endocrine disturbing pollutant from water on functionalized iron nano particles. J Mol Liq 241:123–129
6. Ali N, Al-Juthery H (2017) The application of nanotechnology for micronutrient in agricultural production. Iraqi J Agric Sci 48(4):984
7. Alsharef J, Taha MR, Firoozi AA, Govindasamy P (2016) Potential of using nanocarbons to stabilize weak soils. Appl Environ Soil Sci
8. Amar I, Harara H, Baqul Q, Qadir MA, Altohami F, Ahwidi M, Abdalsamed I, Saleh F (2020) Photocatalytic degradation of malachite green dye under UV light irradiation using calcium-doped ceria nanoparticles. Asian J Nanosci Mater 3(1):1–14
9. Amini S, Maali-Amiri R, Mohammadi R, Kazemi-Shahandashti S-S (2017) CDNA-AFLP analysis of transcripts induced in chickpea plants by TiO_2 nanoparticles during cold stress. Plant Physiol Biochem 111:39–49
10. Amjad S, Sharma AK, Serajuddin M (2018) Toxicity assessment of cypermethrin nanoparticles in *Channa punctatus*: behavioural response, micronuclei induction and enzyme alteration. Regul Toxicol Pharmacol 100:127–133
11. Ansari WA, Atri N, Pandey M, Singh AK, Singh B, Pandey S (2019) Influence of drought stress on morphological, physiological and biochemical attributes of plants: a review. Biosci Biotechnol Res Asia 16(4):697–709
12. Arif N, Yadav V, Singh S, Tripathi DK, Dubey NK, Chauhan DK, & Giorgetti L (2018) Interaction of copper oxide nanoparticles with plants: uptake, accumulation, and toxicity. In Nanomaterials in plants, algae, and microorganisms; Academic Press, pp 297–310
13. Arif Y, Singh P, Siddiqui H, Bajguz A, Hayat S (2020) Salinity induced physiological and biochemical changes in plants: an omic approach towards salt stress tolerance. Plant Physiol Biochem 156:64–77
14. Arora S, Sharma P, Kumar S, Nayan R, Khanna PK, Zaidi MGH (2012) Gold-nanoparticle induced enhancement in growth and seed yield of *Brassica juncea*. Plant Growth Regul 66:303–310
15. Ashraf SA, Siddiqui AJ, Abd Elmoneim OE, Khan MI, Patel M, Alreshidi M, Moin A, Singh R, Snoussi M, Adnan M (2021) Innovations in nanoscience for the sustainable development of food and agriculture with implications on health and environment. Sci Total Environ 768:144990
16. Aslam B, Basit M, Nisar MA, Khurshid M, Rasool MH (2016) Proteomics: technologies and their applications. J Chromatogr Sci 1:15
17. Aziz N, Faraz M, Pandey R, Shakir M, Fatma T, Varma A, Barman I, Prasad R (2015) Facile algae-derived route to biogenic silver nanoparticles: synthesis, antibacterial, and photocatalytic properties. Langmuir 31(42):11605–11612
18. Barazzouk S, Kamat PV, Hotchandani S (2005) Photoinduced electron transfer between chlorophyll a and gold nanoparticles. J Phys Chem B 109(2):716–723
19. Barkla BJ, Vera-Estrella R, Pantoja O (2013) Progress and challenges for abiotic stress proteomics of crop plants. Proteomics 13(12–13):1801–1815
20. Battaglia ML, Lee C, Thomason W (2018) Corn yield components and yield responses to defoliation at different row widths. Agron J 110(1):210–225

21. Behboudi F, Tahmasebi Sarvestani Z, Kassaee MZ, Modares Sanavi SAM, Sorooshzadeh A, Ahmadi SB (2018) Evaluation of chitosan nanoparticles effects on yield and yield components of barley (*Hordeum vulgare* L.) under late season drought stress. J Water Environ Nanotechnol 3(1): 22–39
22. Behboudi F, Tahmasebi-Sarvestani Z, Kassaee MZ, Modarres-Sanavy SAM, Sorooshzadeh A, Mokhtassi-Bidgoli A (2019) Evaluation of chitosan nanoparticles effects with two application methods on wheat under drought stress. J Plant Nutr 42(13):1439–1451
23. Boussouga Y-A, Richards BS, Schäfer AI (2021) Renewable energy powered membrane technology: system resilience under solar irradiance fluctuations during the treatment of fluoride-rich natural waters by different nanofiltration/reverse osmosis membranes. J Membr Sci 617:118452
24. Boyd RS, Rajakaruna N (2013) Heavy metal tolerance. Oxford University Press, Oxford, UK
25. Browse J, Xin Z (2001) Temperature sensing and cold acclimation. Curr Opin Plant Biol 4(3):241–246
26. Burman U, Saini M, Kumar P (2013) Effect of zinc oxide nanoparticles on growth and antioxidant system of chickpea seedlings. Toxicol Environ Chem 95(4):605–612
27. Cao F, Dai H, Hao P-F, Wu F (2020) Silicon regulates the expression of vacuolar H+-pyrophosphatase 1 and decreases cadmium accumulation in rice (*Oryza sativa* L.). Chemosphere 240:124907
28. Chhipa H (2017) Nanofertilizers and nanopesticides for agriculture. Environ Chem Lett 15:15–22
29. Cuerda-Correa EM, Alexandre-Franco MF, Fernández-González C (2019) Advanced oxidation processes for the removal of antibiotics from water. An overview. Water 12(1):102
30. de la Rosa G, García-Castañeda C, Vázquez-Núñez E, Alonso-Castro ÁJ, Basurto-Islas G, Mendoza Á, Cruz-Jiménez G, Molina C (2017) Physiological and biochemical response of plants to engineered NMs: Implications on future design. Plant Physiol Biochem 110:226–235
31. Degenkolbe T, Giavalisco P, Zuther E, Seiwert B, Hincha DK, Willmitzer L (2012) Differential remodeling of the lipidome during cold acclimation in natural accessions of Arabidopsis thaliana. Plant J 72(6):972–982
32. Dehkourdi EH, Mosavi M (2013) Effect of anatase nanoparticles (TiO_2) on parsley seed germination (*Petroselinum crispum*) in vitro. Biol Trace Elem Res 155:283–286
33. Dietz KJ, Herth S (2011) Plant nanotoxicology. Trends Plant Sci 16(11):582–589
34. Dunn MJ, Burghes AH (1986) High resolution two-dimensional polyacrylamide-gel electrophoresis. In Gel Electrophoresis of Proteins. Butterworth-Heinemann, pp 203–261
35. Elemike EE, Uzoh IM, Onwudiwe DC, Babalola OO (2019) The role of nanotechnology in the fortification of plant nutrients and improvement of crop production. Appl Sci 9(3):499
36. Elmore JM, Griffin BD, Walley JW (2021) Advances in functional proteomics to study plant-pathogen interactions. Curr Opin Plant Biol 63:102061
37. El-Saadony MT, Saad AM, Najjar AA, Alzahrani SO, Alkhatib FM, Shafi ME, Selem E, Desoky E-SM, Fouda SE, El-Tahan AM (2021) The use of biological selenium nanoparticles to suppress *Triticum aestivum* L. crown and root rot diseases induced by Fusarium species and improve yield under drought and heat stress. Saudi J Biol Sci 28(8):4461–4471
38. El-Saber MM, Mahdi AA, Hassan AH, Farroh KY, Osman A (2021) Effects of magnetite nanoparticles on physiological processes to alleviate salinity induced oxidative damage in wheat. J Sci Food Agric 101(13):5550–5562
39. Emamverdian A, Ding Y, Mokhberdoran F, Xie Y (2015) Heavy metal stress and some mechanisms of plant defense response. Scientif World J
40. Farhangi-Abriz S, Torabian S (2018) Nano-silicon alters antioxidant activities of soybean seedlings under salt toxicity. Protoplasma 255:953–962
41. Fatemi H, Esmaiel Pour B, Rizwan M (2021) Foliar application of silicon nanoparticles affected the growth, vitamin C, flavonoid, and antioxidant enzyme activities of coriander (*Coriandrum sativum* L.) plants grown in lead (Pb)-spiked soil. Environ Sci Pollut Res 28:1417–1425

42. Feichtmeier NS, Walther P, Leopold K (2015) Uptake, effects, and regeneration of barley plants exposed to gold nanoparticles. Environ Sci Pollut Res 22:8549–8558
43. Feizi H, Rezvani Moghaddam P, Shahtahmassebi N, Fotovat A (2012) Impact of bulk and nanosized titanium dioxide (TiO_2) on wheat seed germination and seedling growth. Biol Trace Elem Res 146:101–106
44. Fincheira P, Tortella G, Duran N, Seabra AB, Rubilar O (2020) Current applications of nanotechnology to develop plant growth inducer agents as an innovation strategy. Crit Rev Biotechnol 40(1):15–30
45. Fournier ML, Gilmore JM, Martin-Brown SA, Washburn MP (2007) Multidimensional separations-based shotgun proteomics. Chem Rev 107(8):3654–3686
46. Ghosh D, Xu J (2014) Abiotic stress responses in plant roots: a proteomics perspective. Front Plant Sci 5:6
47. Gogami Y, Ito K, Kamitani Y, Matsushima Y, Oikawa T (2009) Occurrence of D-serine in rice and characterization of rice serine racemase. Phytochemistry 70(3):380–387
48. Gopinath K, Gowri S, Karthika V, Arumugam A (2014) Green synthesis of gold nanoparticles from fruit extract of *Terminalia arjuna*, for the enhanced seed germination activity of *Gloriosa superba*. J Nanostruct Chem 4:1–11
49. Gray SB, Brady SM (2016) Plant developmental responses to climate change. Dev Biol 419(1):64–77
50. Gygi SP, Rochon Y, Franza BR, Aebersold R (1999) Correlation between protein and mRNA abundance in yeast. Mol Cell Biol
51. Gygi SP, Rochon Y, Franza BR, Aebersold R (1999) Correlation between protein and mRNA abundance in yeast. Mol Cell Biol
52. Haghighi M, Afifipour Z, Mozafarian M (2012) The effect of N-Si on tomato seed germination under salinity levels. J Biol Environ Sci 6(16)
53. Hamza RA, Iorhemen OT, Tay JH (2016) Advances in biological systems for the treatment of high-strength wastewater. J Water Process Eng 10:128–142
54. Hasanpour H, Maali-Amir R, Zeinali H (2015) Effect of TiO_2 nanoparticles on metabolic limitations to photosynthesis under cold in chickpea. Russ J Plant Physiol 62:779–787
55. Hauck TS, Ghazani AA, Chan WC (2008) Assessing the effect of surface chemistry on gold nanorod uptake, toxicity, and gene expression in mammalian cells. Small 4(1):153–159
56. Heikal YM, El-Esawi MA, El-Ballat EM, Abdel-Aziz HM (2023) Applications of nanoparticles for mitigating salinity and drought stress in plants: an overview on the physiological, biochemical and molecular genetic aspects. N Z J Crop Hortic Sci 51(3):297–327
57. Hernández JA, Ferrer MA, Jiménez A, Barceló AR, Sevilla F (2001) Antioxidant systems and $O_2.-/H_2O_2$ production in the apoplast of pea leaves. Its relation with salt-induced necrotic lesions in minor veins. Plant Physiol 127(3):817–831
58. Hess FD (2018) Herbicide effects on plant structure, physiology, and biochemistry. In: Pesticide interactions in crop production. CRC Press, pp 13–34
59. Hong-Bo S, Zong-Suo L, Ming-An S (2005) LEA proteins in higher plants: structure, function, gene expression and regulation. Colloids Surf B 45(3–4):131–135
60. Hossain MA, Bhattacharjee S, Armin S-M, Qian P, Xin W, Li H-Y, Burritt DJ, Fujita M, Tran L-SP (2015) Hydrogen peroxide priming modulates abiotic oxidative stress tolerance: insights from ROS detoxification and scavenging. Front Plant Sci 6:420
61. Hossain Z, Mustafa G, Sakata K, Komatsu S (2016) Insights into the proteomic response of soybean towards Al2O3, ZnO, and Ag nanoparticles stress. J Hazard Mater 304:291–305
62. Husen A (2023) Nanomaterials and nanocomposites exposures to plants (response, interaction, phytotoxicity and defense mechanisms). Springer Nature Singapore Pte Ltd., 152 Beach Road, #21-01/04 Gateway East, Singapore 189721, Singapore https://doi.org/10.1007/978-981-99-2419-6
63. Husen A, Bachheti RK, Bachheti A (2023) Current trends in green nano-emulsions (food, agriculture and biomedical sectors). Springer Nature Singapore Pte Ltd., 152 Beach Road, #21-01/04 Gateway East, Singapore 189721, Singapore

64. Husen A, Iqbal M (2019) Nanomaterials and plant potential. Springer International Publishing AG, Gewerbestrasse 11, 6330 Cham, Switzerland. https://doi.org/10.1007/978-3-030-05569-1
65. Husen A, Jawaid M (2020) Nanomaterials for Agriculture and Forestry Applications. Elsevier Inc., 50 Hampshire St., 5th Floor, Cambridge, MA 02139, USA. https://doi.org/10.1016/C2018-0-02349-X
66. Bachheti RK, Bachheti A, Husen A (2024) Metal and metal-oxide based nanomaterials (*Synthesis, Agricultural, Biomedical and Environmental Interventions*). Springer Nature Singapore Pte Ltd., 152 Beach Road, #21-01/04 Gateway East, Singapore 189721, Singapore
67. Ideker T, Galitski T, Hood L (2001) A new approach to decoding life: systems biology. Ann Rev Genomics Hum Genet 2(1):343–372
68. Ighalo JO, Adeniyi AG, Marques G (2021) Internet of things for water quality monitoring and assessment: a comprehensive review. Artificial intelligence for sustainable development: theory, practice and future applications 245–259
69. Igwegbe CA, Onukwuli OD (2019) Removal of total dissolved solids (TDS) from aquaculture wastewater by coagulation-flocculation process using *Sesamum indicum* extract: effect of operating parameters and coagulation-flocculation kinetics. Pharmaceut Chem J 6(4):32–45
70. Isaacson T, Damasceno CM, Saravanan RS, He Y, Catalá C, Saladié M, Rose JK (2006) Sample extraction techniques for enhanced proteomic analysis of plant tissues. Nat Protoc 1(2):769–774
71. Jafir M, Ahmad JN, Arif MJ, Ali S, Ahmad SJN (2021) Characterization of *Ocimum basilicum* synthesized silver nanoparticles and its relative toxicity to some insecticides against tobacco cutworm, *Spodoptera litura* Feb.(Lepidoptera; Noctuidae). Ecotoxicol Environ Safety 218:112278
72. Joshi S, Patil S, Shaikh A, Jamla M, Kumar V (2023) Modern omics toolbox for producing combined and multifactorial abiotic stress tolerant plants. Plant Stress 100301
73. Jungbauer A, Hahn R (2009) Ion-exchange chromatography. Methods Enzymol 463:349–371
74. Kalinowska K, Isono E (2018) All roads lead to the vacuole—autophagic transport as part of the endomembrane trafficking network in plants. J Exp Bot 69(6):1313–1324
75. Kalteh M, Alipour ZT, Ashraf S, Marashi Aliabadi M, Falah Nosratabadi A (2018) Effect of silica nanoparticles on basil (*Ocimum basilicum*) under salinity stress. J Chem Health Risks 4(3)
76. Kaphle A, Navya PN, Umapathi A, Daima HK (2018) Nanomaterials for agriculture, food and environment: applications, toxicity and regulation. Environ Chem Lett 16:43–58
77. Kaur C, Sharma S, Singla-Pareek SL, Sopory SK (2016) Methylglyoxal detoxification in plants: role of glyoxalase pathway. Indian J Plant Physiol 21:377–390
78. Khan FSA, Mubarak NM, Khalid M, Tan YH, Abdullah EC, Rahman ME, Karri RR (2021) A comprehensive review on micropollutants removal using carbon nanotubes-based adsorbents and membranes. J Environ Chem Eng 9(6):106647
79. Khan M, Khan MSA, Borah KK, Goswami Y, Hakeem KR, Chakrabartty I (2021) The potential exposure and hazards of metal-based nanoparticles on plants and environment, with special emphasis on ZnO NPs, TiO_2 NPs, and AgNPs: a review. Environ Adv 6:100128
80. Khan MR, Rizvi TF (2014) Nanotechnology: scope and application in plant disease management. Plant Pathol J 13(3):214–231
81. Kottegoda N, Munaweera I, Madusanka N, Karunaratne V (2011) A green slow-release fertilizer composition based on urea-modified hydroxyapatite nanoparticles encapsulated wood. Curr Sci 73–78
82. Kumar V, Guleria P, Kumar V, Yadav SK (2013) Gold nanoparticle exposure induces growth and yield enhancement in *Arabidopsis thaliana*. Sci Total Environ 461:462–468
83. Lahiani MH, Dervishi E, Chen J, Nima Z, Gaume A, Biris AS, Khodakovskaya MV (2013) Impact of carbon nanotube exposure to seeds of valuable crops. ACS Appl Mater Interfaces 5(16):7965–7973
84. Legrain P, Aebersold R, Archakov A, Bairoch A, Bala K, Beretta L, Bergeron J, Borchers CH, Corthals GL, Costello CE (2011) The human proteome project: current state and future direction. Mol Cell Proteom 10(7)

85. Lequin RM (2005) Enzyme immunoassay (EIA)/enzyme-linked immunosorbent assay (ELISA). Clin Chem 51(12):2415–2418
86. Leso V, Fontana L, Iavicoli I (2019) Biomedical nanotechnology: occupational views. Nano Today 24:10–14
87. Lin GD, Li RW (2022) Analytical characterization of herbal biomolecules. In: Herbal Biomolecules in Healthcare Applications. Academic Press, pp 121–144
88. Liu F, Wen L-X, Li Z-Z, Yu W, Sun H-Y, Chen J-F (2006) Porous hollow silica nanoparticles as controlled delivery system for water-soluble pesticide. Mater Res Bull 41(12):2268–2275
89. Liu J, Guo Y, Bai Y, Camberato J, Xue J, Zhang R (2018) Effects of drought stress on the photosynthesis in maize. Russ J Plant Physiol 65:849–856
90. Liu M, Feng S, Ma Y, Xie C, He X, Ding Y, Zhang J, Luo W, Zheng L, Chen D (2019) Influence of surface charge on the phytotoxicity, transformation, and translocation of CeO_2 nanoparticles in cucumber plants. ACS Appl Mater Interfaces 11(18):16905–16913
91. Liu R, Lal R (2015) Potentials of engineered nanoparticles as fertilizers for increasing agronomic productions. Sci Total Environ 514:131–139
92. X-m LIU, F-d ZHANG, S-q ZHANG, X-s HE, R-f WANG, Z-b FENG, Y-j WANG (2005) Responses of peanut to nano-calcium carbonate. J Plant Nutrit Fertil 11(3):385–389
93. Lu L, Huang M, Huang Y, Corvini PFX, Ji R, Zhao L (2020) Mn 3 O 4 nanozymes boost endogenous antioxidant metabolites in cucumber (*Cucumis sativus*) plant and enhance resistance to salinity stress. Environ Sci Nano 7(6):1692–1703
94. Luyckx M, Hausman J-F, Lutts S, Guerriero G (2017) Silicon and plants: current knowledge and technological perspectives. Front Plant Sci 8:411
95. Mafukidze N, Majozi T (2016) Synthesis and optimisation of an integrated water and membrane network framework with multiple electrodialysis regenerators. Comput Chem Eng 85:151–161
96. Mahakham W, Sarmah AK, Maensiri S, Theerakulpisut P (2017) Nanopriming technology for enhancing germination and starch metabolism of aged rice seeds using phytosynthesized silver nanoparticles. Sci Rep 7(1):8263
97. Maruyama K, Todaka D, Mizoi J, Yoshida T, Kidokoro S, Matsukura S, Takasaki H, Sakurai T, Yamamoto YY, Yoshiwara K (2012) Identification of cis-acting promoter elements in cold- and dehydration-induced transcriptional pathways in Arabidopsis, rice, and soybean. DNA Res 19(1):37–49
98. Mengmeng S, Jie X, Lingling Z, Dejian S, Xitong Z (2019) Research progress on effects of silicon on plant growth under cadmium stress. Guangzhou Chem Ind 47:41–43
99. Millán G, Agosto F, Vázquez M, Botto L, Lombardi L, Juan L (2008) Use of clinoptilolite as a carrier for nitrogen fertilizers in soils of the Pampean regions of Argentina. Cien Inv Agr 35(3):293–302
100. Mishra M, Tiwari S, Gomes AV (2017) Protein purification and analysis: next generation Western blotting techniques. Expert Rev Proteomics 14(11):1037–1053
101. Mohammadi R, Maali-Amiri R, Mantri N (2014) Effect of TiO_2 nanoparticles on oxidative damage and antioxidant defense systems in chickpea seedlings during cold stress. Russ J Plant Physiol 61:768–775
102. Mohana NC, Mithun P, Rao HY, Mahendra C, Satish S (2020) Nanoparticle applications in sustainable agriculture, poultry, and food: trends and perspective. In: Nanotoxicity. Elsevier, pp 341–353
103. Moharem M, Elkhatib E, Mesalem M (2019) Remediation of chromium and mercury polluted calcareous soils using nanoparticles: sorption–desorption kinetics, speciation and fractionation. Environ Res 170:366–373
104. Morales M, Munné-Bosch S (2019) Malondialdehyde: facts and artifacts. Plant Physiol 180(3):1246–1250
105. Morel M, Crouzet J, Gravot A, Auroy P, Leonhardt N, Vavasseur A, Richaud P (2009) AtHMA3, a P1B-ATPase allowing Cd/Zn/co/Pb vacuolar storage in Arabidopsis. Plant Physiol 149(2):894–904

106. Mukhopadhyay SS (2014) Nanotechnology in agriculture: prospects and constraints. Nanotechnol Sci Appl 63–71
107. Müller K, Linkies A, Leubner-Metzger G, Kermode AR (2012) Role of a respiratory burst oxidase of *Lepidium sativum* (cress) seedlings in root development and auxin signalling. J Exp Bot 63(18):6325–6334
108. Munir N, Gulzar W, Abideen Z, Hasanuzzaman M, El-Keblawy A, Zhao F (2023) Plant-nanoparticle interactions: transcriptomic and proteomic insights. Agronomy 13(8):2112
109. Muntha St, Faizan M, Mehreen S, Shareen (2022) Exploring nanotechnology to reduce stress: mechanism of nanomaterial-mediated alleviation. In: Sustainable agriculture reviews 53: Nanoparticles: A New Tool to Enhance Stress Tolerance. Springer, pp 93–113
110. Mushtaq A, Rizwan S, Jamil N, Ishtiaq T, Irfan S, Ismail T, Malghani MN, Shahwani MN (2019) Influence of silicon sources and controlled release fertilizer on the growth of wheat cultivars of Balochistan under salt stress. Pak J Bot 51(5):1561–1567
111. Mustafa G, Sakata K, Komatsu S (2016) Proteomic analysis of soybean root exposed to varying sizes of silver nanoparticles under flooding stress. J Proteomics 148:113–125
112. Nagajyoti PC, Lee KD, Sreekanth T (2010) Heavy metals, occurrence and toxicity for plants: a review. Environ Chem Lett 8:199–216
113. Nakaminami K, Matsui A, Nakagami H, Minami A, Nomura Y, Tanaka M, Morosawa T, Ishida J, Takahashi S, Uemura M (2014) Analysis of differential expression patterns of mRNA and protein during cold-acclimation and de-acclimation in Arabidopsis. Mol Cell Proteomics 13(12):3602–3611
114. Osakabe Y, Osakabe K, Shinozaki K, Tran L-SP (2014) Response of plants to water stress. Front Plant Sci 5:86
115. Oshunsanya SO, Nwosu NJ, Li Y (2019) Abiotic stress in agricultural crops under climatic conditions. Sustain Agricul Forest Environ Manage 71–100
116. Pandey A, Mann M (2000) Proteomics to study genes and genomes. Nature 405(6788):837–846
117. Paparella S, Araújo S, Rossi G, Wijayasinghe M, Carbonera D, Balestrazzi A (2015) Seed priming: state of the art and new perspectives. Plant Cell Rep 34:1281–1293
118. Parihar P, Singh S, Singh R, Singh VP, Prasad SM (2015) Effect of salinity stress on plants and its tolerance strategies: a review. Environ Sci Pollut Res 22:4056–4075
119. Parisi C, Vigani M, Rodríguez-Cerezo E (2015) Agricultural nanotechnologies: what are the current possibilities? Nano Today 10(2):124–127
120. Patil CD, Patil SV, Borase HP, Salunke BK, Salunkhe RB (2012) Larvicidal activity of silver nanoparticles synthesized using *Plumeria rubra* plant latex against *Aedes aegypti* and *Anopheles stephensi*. Parasitol Res 110:1815–1822
121. Peharec Štefanić P, Jarnević M, Cvjetko P, Biba R, Šikić S, Tkalec M, Balen B (2019) Comparative proteomic study of phytotoxic effects of silver nanoparticles and silver ions on tobacco plants. Environ Sci Pollut Res 26:22529–22550
122. Pérez-de-Luque A, Rubiales D (2009) Nanotechnology for parasitic plant control. Pest Manag Sci: Former Pestic Sci 65(5):540–545
123. Perez-Riverol Y, Alpi E, Wang R, Hermjakob H, Vizcaíno JA (2015) Making proteomics data accessible and reusable: current state of proteomics databases and repositories. Proteomics 15(5–6):930–950
124. Rajput VD, Minkina T, Kumari A, Harish SVK, Verma KK, Mandzhieva S, Sushkova S, Srivastava S, Keswani C (2021) Coping with the challenges of abiotic stress in plants: New dimensions in the field application of nanoparticles. Plants 10(6):1221
125. Rajput VD, Minkina TM, Behal A, Sushkova SN, Mandzhieva S, Singh R, Movsesyan HS (2018) Effects of zinc-oxide nanoparticles on soil, plants, animals and soil organisms: A review. Environmental Nanotechnology, Monitoring & Management 9:76–84
126. Ramalingam V (2019) Multifunctionality of gold nanoparticles: plausible and convincing properties. Adv Coll Interface Sci 271:101989
127. Rastogi A, Tripathi DK, Yadav S, Chauhan DK, Živčák M, Ghorbanpour M, El-Sheery NI, Brestic M (2019a) Application of silicon nanoparticles in agriculture. 3 Biotech 9:1–11

128. Rastogi A, Tripathi DK, Yadav S, Chauhan DK, Živčák M, Ghorbanpour M, Brestic M (2019) Application of silicon nanoparticles in agriculture. 3 Biotech 9:1–11
129. Rastogi A, Zivcak M, Tripathi D, Yadav S, Kalaji H (2019b) Phytotoxic effect of silver nanoparticles in Triticum aestivum: improper regulation of photosystem I activity as the reason for oxidative damage in the chloroplast. Photosynthetica 57 (1)
130. Raza A, Bashir S, Salehi H, Jamla M, Charagh S, Rad AC, & Hossain MA (2023) Advanced techniques in omics research in relation to heavy metal/metalloid toxicity and tolerance in plants. Heavy Metal Toxicity and Tolerance in Plants: A Biological, Omics, and Genetic Engineering Approach 35–57
131. Rizwan M, Ali S, Ali B, Adrees M, Arshad M, Hussain A, Waris AA (2019) Zinc and iron oxide nanoparticles improved the plant growth and reduced the oxidative stress and cadmium concentration in wheat. Chemosphere 214:269–277
132. Salama DM, Abd El-Aziz M, Rizk FA, Abd Elwahed M (2021) Applications of nanotechnology on vegetable crops. Chemosphere 266:129026
133. Salısbury FB, ROSS CW (1992) Plant Physiology. Wadsworth Publishing Company, Belmont - California
134. Schirò A, Carlon A, Parigi G, Murshudov G, Calderone V, Ravera E, Luchinat C (2020) On the complementarity of X-ray and NMR data. J Struct Biology X 4:100019
135. Scott N, Chen H (2012) Nanoscale science and engineering for agriculture and food systems. Ind Biotechnol 8(6):340–343
136. Seleiman MF, Almutairi KF, Alotaibi M, Shami A, Alhammad BA, Battaglia ML (2020) Nano-fertilization as an emerging fertilization technique: Why can modern agriculture benefit from its use? Plants 10(1):2
137. Seleiman MF, Al-Suhaibani N, Ali N, Akmal M, Alotaibi M, Refay Y, Dindaroglu T, Abdul-Wajid HH, Battaglia ML (2021) Drought stress impacts on plants and different approaches to alleviate its adverse effects. Plants 10(2):259
138. Service RF (2001) High-speed biologists search for gold in proteins
139. Silva AM, Vitorino R, Domingues MRM, Spickett CM, Domingues P (2013) Post-translational modifications and mass spectrometry detection. Free Radical Biol Med 65:925–941
140. Singh A, Tiwari S, Pandey J, Lata C, Singh IK (2021) Role of nanoparticles in crop improvement and abiotic stress management. J Biotechnol 337:57–70
141. Singh J, Lee B-K (2016) Influence of nano-TiO_2 particles on the bioaccumulation of Cd in soybean plants (Glycine max): A possible mechanism for the removal of Cd from the contaminated soil. J Environ Manage 170:88–96
142. Sundararajan B, Kumari BR (2017) Novel synthesis of gold nanoparticles using *Artemisia vulgaris* L. leaf extract and their efficacy of larvicidal activity against dengue fever vector *Aedes aegypti* L. J Trace Elem Med Biol 43:187–196
143. Surabhi V, Rame Gowda NN (2021) Influence of seed treatment with nanoparticles on seed quality and storability of pigeonpea cv. BRG-2. Intl J Chem Stud 9 (1):3645–3651
144. Taiz L, Zeiger E (2002) And 2010) Plant Physiology, 3rd and, 5th edn. The Benjamin Cummings Publishing Company; Redwood City, California
145. Thakur RK, Dhirta B, Shirkot P (2018) Studies on effect of gold nanoparticles on Meloidogyne incognita and tomato plants growth and development. BioRxiv 428144
146. Thomashow MF (1999) Plant cold acclimation: freezing tolerance genes and regulatory mechanisms. Annu Rev Plant Biol 50(1):571–599
147. Tiwari M, Krishnamurthy S, Shukla D, Kiiskila J, Jain A, Datta R, Sharma N, Sahi SV (2016) Comparative transcriptome and proteome analysis to reveal the biosynthesis of gold nanoparticles in Arabidopsis. Sci Rep 6(1):21733
148. Tsugane K, Kobayashi K, Niwa Y, Ohba Y, Wada K, Kobayashi H (1999) A recessive Arabidopsis mutant that grows photoautotrophically under salt stress shows enhanced active oxygen detoxification. Plant Cell 11(7):1195–1206
149. Uemura M, Joseph RA, Steponkus PL (1995) Cold acclimation of Arabidopsis thaliana (effect on plasma membrane lipid composition and freeze-induced lesions). Plant Physiol 109(1):15–30

150. Usman M, Farooq M, Wakeel A, Nawaz A, Cheema SA, ur RehmanH, Sanaullah M (2020) Nanotechnology in agriculture: current status, challenges and future opportunities. Sci Total Environ 721:137778
151. Van Nguyen D, Nguyen HM, Le NT, Nguyen KH, Nguyen HT, Le HM, Van Ha C (2021) Copper nanoparticle application enhances plant growth and grain yield in maize under drought stress conditions. J Plant Growth Regulation 1–12
152. Vannini C, Domingo G, Onelli E, De Mattia F, Bruni I, Marsoni M, Bracale M (2014) Phytotoxic and genotoxic effects of silver nanoparticles exposure on germinating wheat seedlings. J Plant Physiol 171(13):1142–1148
153. Vannini C, Domingo G, Onelli E, Prinsi B, Marsoni M, Espen L, Bracale M (2013) Morphological and proteomic responses of *Eruca sativa* exposed to silver nanoparticles or silver nitrate. PLoS ONE 8(7):e68752
154. VanWallendael A, Soltani A, Emery NC, Peixoto MM, Olsen J, Lowry DB (2019) A molecular view of plant local adaptation: incorporating stress-response networks. Annu Rev Plant Biol 70:559–583
155. Vihinen M (2001) Bioinformatics in proteomics. Biomol Eng 18(5):241–248
156. Villiers F, Ducruix C, Hugouvieux V, Jarno N, Ezan E, Garin J, Junot C, Bourguignon J (2011) Investigating the plant response to cadmium exposure by proteomic and metabolomic approaches. Proteomics 11(9):1650–1663
157. Wahid I, Rani P, Kumari S, Ahmad R, Hussain SJ, Alamri S, Tripathy N, Khan MIR (2022) Biosynthesized gold nanoparticles maintained nitrogen metabolism, nitric oxide synthesis, ions balance, and stabilizes the defense systems to improve salt stress tolerance in wheat. Chemosphere 287:132142
158. Wang W, Luo M, Fu Y, Wang S, Efferth T, & Zu Y (2013) Glycyrrhizic acid nanoparticles inhibit LPS-induced inflammatory mediators in 264.7 mouse macrophages compared with unprocessed glycyrrhizic acid. Int J Nanomed 1377–1383
159. Wang Z, Yue L, Dhankher OP, Xing B (2020) Nano-enabled improvements of growth and nutritional quality in food plants driven by rhizosphere processes. Environ Int 142:105831
160. Wani SH, Kumar V, Shriram V, Sah SK (2016) Phytohormones and their metabolic engineering for abiotic stress tolerance in crop plants. Crop J 4(3):162–176
161. Wilkins MR, Sanchez JC, Gooley AA, Appel RD, Humphery-Smith I, Hochstrasser DF, Williams KL (1996) Progress with proteome projects: why all proteins expressed by a genome should be identified and how to do it. Biotechnol Genet Eng Rev 13(1):19–50
162. Wittmann-Liebold B, Graack HR, Pohl T (2006) Two-dimensional gel electrophoresis as tool for proteomics studies in combination with protein identification by mass spectrometry. Proteomics 6(17):4688–4703
163. Ye Y, Cota-Ruiz K, Hernández-Viezcas JA, Valdes C, Medina-Velo IA, Turley RS, Gardea-Torresdey JL (2020) Manganese nanoparticles control salinity-modulated molecular responses in *Capsicum annuum* L. through priming: a sustainable approach for agriculture. ACS Sustain Chem Eng 8(3):1427–1436
164. Yin L, Colman BP, McGill BM, Wright JP, & Bernhardt ES (2012) Effects of silver nanoparticle exposure on germination and early growth of eleven wetland plants
165. Zahedi SM, Karimi M, Teixeira da Silva JA (2020) The use of nanotechnology to increase quality and yield of fruit crops. J Sci Food Agric 100(1):25–31
166. Zhang L, Peng J, Chen T, Zhao X, Zhang S, Liu S, Dong H, Feng L, Yu S (2014) Effect of drought stress on lipid peroxidation and proline content in cotton roots. JAPS: J Animal Plant Sci 24 (6)
167. Zulfiqar F, Ashraf M (2021) Nanoparticles potentially mediate salt stress tolerance in plants. Plant Physiol Biochem 160:257–268

Gold Nanocomposites for Improved Plant-Soil System and Underlying Mechanism

Bhagya Ekanayaka, Nadun H. Madanayake, and Nadeesh M. Adassooriya

Abstract Nanotechnology has become a driving force in numerous industrial and technological fields to improve their efficiency and productivity. Gold nanomaterials (Au NMs) are one such category of metallic nanoparticles (NPs) that have been extensively utilized due to their unique physical and chemical properties. The incorporation of Au NPs into agricultural operations holds a promise for sustainable and efficient crop production as nanotechnology continues to change modern agriculture. Gold NPs, owing to their unique physicochemical properties, have shown great potential in plant and soil health conditioning. Recently, nanocomposites composed of Au NPs have been shown to improve plant and soil systems. Monitoring plant health, plant disease management, plant stress tolerance, plant growth promotion, and seed germination have been reported as some of the promising applications under different platforms. Hence, this book chapter elaborates on the application of gold nanocomposites in improved plant-soil systems.

Keywords Gold nanocomposites · Plant stress tolerance · Plant disease management · Plant growth improvement

1 Introduction

The introduction of nanotechnology to the science arena has widely expanded its applications in every field to uplift the efficiency of processes. Nanotechnology simply deals with nanomaterials (NMs), which are particles confined at least in one of its dimensions within 1–100 nm. NMs are synthesized in different forms depending on their size, shape, chemical composition, etc. The material world is undergoing a

B. Ekanayaka · N. M. Adassooriya (✉)
Department of Chemical and Process Engineering, Faculty of Engineering, University of Peradeniya, Peradeniya 20400, Sri Lanka
e-mail: nadeeshm@eng.pdn.ac.lk

N. H. Madanayake
Department of Botany, Faculty of Science, University of Peradeniya, Peradeniya 20400, Sri Lanka

A. Husen (ed.), *Plant Response to Gold Nanoparticles*, Smart Nanomaterials Technology, https://doi.org/10.1007/978-981-97-7174-5_7

revolution with the development of nanocomposites, a new class of composite materials composed of nanoparticles (NPs) embedded in a matrix. Composite materials fabricated using NMs are also in play at present where nanocomposites are widely applied in fields such as medicine, energy, electronics, environment, waste management, food, and agriculture [2–21, 27, 32, 39, 41, 46, 54]. These materials offer superior mechanical, electrical, magnetic, optical, thermal, biological, and transport properties, making them increasingly used in modern technologies. A fascinating and unique occurrence occurs as a NP decreases in size; its chemical and physical characteristics undergo significant changes, and the extent of these changes is dependent upon the particle size, or the number of atoms. In general, nanocomposites are composed of numerous NPs enmeshed within a bulk cargo. Nanocomposites may consist of different combinations of soft and hard NMs [66].

Composites have emerged as distinct materials in the mid-twentieth century, originating from the fabricating multiphase composites like fiberglass-reinforced polymers. Composite materials open interesting possibilities to create a wide range of materials with combined and inherent properties [26]. These are manmade constituents consisting of two or more constituent elements with vastly different physical, chemical, and mechanical properties. The concept of combined action states that the careful blending of two or more different materials which results in better property combinations [36]. Inherent properties of constituent materials including their relative volume fractions and configurations determine their typical attributes. Composite materials are utilized in various sectors including construction, automotive, aeronautics, housing, industrial parts, and medical fields for their high strength and low-density properties [25].

Gold-based NMs are the most famous metallic NPs used in various applications, including electronics, textiles, sensors, therapeutic agents, photodynamic therapy, electronics, and environmental monitoring and biomedicine. Gold's high free electron density, malleability, and conductivity make it ideal for producing stable NPs. Au NPs have several unique properties depending on their size, morphology, and surface charge, inspiring a vast range of applications [56]. The size and shape-dependent optical and electrical properties including the possible modifications to enhance biocompatibility allow people to use Au NMs in numerous fields [37]. Numerous gold-based nanocomposites have been developed in recent years, including gold-carbon nanotube nanocomposites, polymer-gold nanocomposites, graphene-gold nanocomposites, biomolecule-gold nanocomposites, and metal oxide-gold nanocomposites [60]. Agriculture is one of the fields where gold NMs and gold nanocomposites have been utilized. Sustainable agriculture is also a demanding approach where it always tries to enhance the crop productivity and yield by reducing crop damages due to biotic and abiotic stress, plant soil improvement and conditioning and plant growth promotion. Therefore, this chapter mainly focuses on gold nanocomposites for the improvement of plant-soil interaction.

2 Gold Nanocomposites for Improved Plant System

Plant health is utterly important in sustainable agricultural practices. Anthropogenic activities have resulted in several environmental consequences leading to significant effect on plant health. Therefore, improving plant health is important to enhance crop productivity by minimizing nutrient and resource wastage. Au nanocomposites have been reported to utilizing in several applications in plant sciences, including plant disease management, plant stress tolerance, plant growth, and seed germination.

2.1 Plant Disease Management

Food supply has a significant impact on plant diseases. Pathogens like viruses, bacteria, and fungi alongside with environmental pollutants lead to severely impact plant health. Plant disease management has become a dire requirement to obtain the maximum efficiency from agricultural practices. One of the approaches of plant disease management is to detect and diagnose the causative agents as early as possible and treat the relevant threat to control its dominance. Nanosensing technology is one of the highly highlighted strategies used in smart and sustainable agriculture for plant disease management. Simply, a nanosensor is an instrument fabricated to transmit information concerning changes in properties and actions of NPs at the nanoscale and around the immediate environment. Typically, sensors are developed to detect and diagnose biological elements composed of (bio)receptor, a transducer, and a detector system. Au NPs had raised much interest during recent years due to their optical properties like surface plasmon resonance and the surface immobilization to bio-recognition elements [53].

Ebrahimi et al. [30] fabricated a label-free electrochemical immunosensor to detect *Candidatus Phytoplasma Aurantifolia* which is the causative agent for witches' broom diseases. It was highlighted that the detection was carried out using antibodies specific to *Candidatus Phytoplasma Aurantifolia.* These were immobilized on a thiol terminated surface of a gold electrode modified by Au NPs. The authors have mentioned that the synthesized immunosensor with Au nanocomposites capable of detecting *Candidatus Phytoplasma Aurantifolia* with a minimum level of detection of 1.5 ng ml^{-1} which is lower than the conventional immune assays such as ELISA (enzyme-linked immunosorbent assay). Jian et al. [42] synthesized an anodic Aluminum oxide (AAO) barrier layer Au NPs nanosensor for the direct detection of *Odontoglossum ringspot virus* (ORSV) which is a orchid virus. Bioreceptor made using ORSV antibodies changes the interfacial charge transfer resistance of the electrode system, enabling to differentiate between healthy and infected orchids. It was highlighted that the detection limit of the developed sensor was 0.345 $ngmL^{-1}$. Razo et al. [61] developed a lateral flow immunoassay to perceive *Ralstonia solanacearum* using Au nanocomposites. Antibody modified Au NPs (17.4 ± 1.0 nm) thus incorporated have a detection limit of 3 $x10^4$ cells mL^{-1} in potato tuber extract. More importantly, this technique can detect the presence of *R. solanacearum* in infected potatoes within smaller duration (3 min). Luna-Moreno et al. [50] fabricated a surface plasmon resonance (SPR) active immunosensor for in-field detection of black Sigatoka disease caused by *Pseudocercospora fijiensis.* Immunosensor thus developed using anti-HF1 polyclonal antibodies immobilized on Au NPs showed a minimum level of detection of 11.7 μg mL^{-1} with a sensitivity of 0.0021 units of reflectance per ng mL^{-1}.

2.2 Plant Stress Tolerance

Abiotic stressors are significant issues for agricultural productivity and extension. Drought, salinity, alkalinity, submergence, mineral and metal toxicity/deficiencies, and a variety of other factors all limit crop growth and yield. To mitigate varying challenges, plants adapt and alleviate abiotic challenges through morphological, physiological, biochemical, and molecular changes [63]. Researchers discovered that Au-based nanocomposites assist plants in resisting abiotic challenges through their concentration-dependent effects on plant growth and development.

Among the food crops cultivated throughout the world rice is considered as the one of the most common and largely consumed food crop. Rice cultivation has a bad impact on quality and yield due to soil salinity. Song et al. [65] synthesized astaxanthin Au NPs (Ast-AuNPs)-based nanocomposites using astaxanthin and chloroauric acid to alleviate salinity stress. It was reported that the Au nanocomposites at 100 mgL^{-1} significantly reduced the stresses caused by high salinity in hydroponically grown rice seedlings. Furthermore, Ast-Au nanocomposites dramatically improved rice tolerance by restoring tetrapyrrole production, enhancing antioxidant enzyme activity, and decreasing reactive oxygen species (ROS) levels. Additionally, antioxidant enzyme activities increased by 1.85–3.85-fold under salt stress. Moreover, these nanocomposites have upregulated the expression of ROS-scavenging genes in leaves. Cadmium (Cd) is a major heavy metal found in rice. Cd uptake by cereal crops from soil causes plant toxicity and poses substantial health risks owing to human body accumulation through the food chain. Dai et al. [29] applied astaxanthin Au NPs (Ast-Au NPs) to alleviate the Cd toxicity in rice. It was reported that Au nanocomposites at 100 μg/mL have substantially reduced the Cd level of rice by 26.2% and 85.9% in roots and leaves, respectively. More importantly, this has restored chlorophyll biosynthesis and mitigated Cd-induced oxidative stresses in rice seedlings. Ferrari et al. [33] demonstrated that the Au NPs stabilized with sodium citrate and tannic acid positively affect the *Arabidopsis* growth and partially protect against oxidative stress. Transcriptomic and proteomic studies showed Au nanocomposites can downregulate the expression of oxidative stress and immune responses while upregulating the expression of growth-promoting genes.

Asefpour Vakilian [23] developed an optical biosensor using Au nanocomposites to determine microRNAs (miRNA-1886) (indicator of drought stress) in tomato roots. It was reported that miRNA-1886 concentration was independent of the plant's conventional morpho-physiological and biochemical characteristics, by other stresses, i.e., salinity, cold, and heat during the growth period. Furthermore, the analytical properties of the synthesized biosensors, such as linear range, limit of detection, accuracy, sensitivity, and specificity, were adequate to determine miRNAs molecules of small length. Wahid et al. [67] assessed the impact of green synthesized Au NPs on *Triticum aestivum* under salt stress conditions. It was reported that the Au NPs capable of moderating the K^+/Na^+ ratio, chlorophyll content, plant defense systems, nitrogen acclimatization, stomatal dynamics, and growth traits under salt stress condition in a positive manner. In addition, Au NPs are capable of alleviating the extreme accumulation of reactive oxygen/nitrogen species in wheat.

2.3 Plant Growth and Seed Germination

Mahakham et al. [51] reported that phytochemicals capped Au NPs synthesized using rhizome extract of Galanga plants were used as nanopriming agents to activate the germination and seedling growth of maize aged seeds. It was reported that the phytochemical capped Au nanocomposite at 5 ppm concentration promoted the seed emergence by 83% in comparison to unprimed seeds and hydroprimed seeds. In addition, this has improved the seedling vigor index by 3 folds. Au NPs green synthesized using *Tiliacora triandra* were used to treat rice (*Oryza sativa*) grains at different concentrations (0, 10, 100, 500, 1000, 2000 mg l^{-1}) by Ndeh et al. [57]. It was reported that germination percentages of test plants were significantly improved, also there was a slight reduction in root and shoot elongations. However, phytotoxic assays showed that the synthesized Au NPs showed a minimal level of toxicity. Furthermore, authors have highlighted the potential benefits of Au NPs as nanocarriers in sustainable agriculture.

3 Gold Nanocomposites for Improved Soil System

Environmental pollution has become a serious issue that is required to be addressed immediately. Different environmental matrices have been contaminated with a range of contaminants, and a large number of contaminated sites have emerged globally leading to environmental pollution. Hence, it is essential to clean up the environment from these miscellaneous pollutants in order to sustain the ecological health as well as the health of living organisms. Possessing a clean environment is utterly important for all living organisms as they depend remarkably on clean air, water as well as on soil for their survival. Unfortunately, several contaminants generated due to natural and human activities have continuously threatened these resources. Because of this, there is a need for environmental remediation approaches to protect this environment [35, 40, 58].

Recently, nanotechnology has been emerged as a promising approach to remediate different environmental pollutants. Simply the use of nanotechnological approaches on remediating environmental pollutants can be defined as nanoremediation. Nanoremediation become successful in providing novel and effective solutions for environmental remediation via pollution prevention, detection, monitoring, and remediation [58]. Nanoremediation involves the application of highly reactive NMs including metallic or bi-metallic NMs, metal oxide NMs, cellulose-based NMs, carbon-based NMs, etc., for transformation and detoxification of pollutants either by in situ or ex situ approaches [28, 34, 43, 58]. Moreover, nanoremediation aids in the development of techniques that will enable to implement specific and cost-effective remediation strategies. Also, nanotechnology allows the in situ application to perform better remediation sites which are inaccessible to operate via traditional operations [40].

Anthropogenic and technogenic activities are rapidly introducing diverse toxics to the environment where they finally end up in soil. Toxics including heavy metals, agrochemicals such as fertilizers, pesticides, and sewage sludge, and farm slurries excess of nutrients ultimately leading to soil pollution. Uneven accumulation of toxics in soil ultimately results in changes in physical, chemical, and biological properties of healthy soil leading to adverse effects on both flora and fauna. Healthy soil is utterly essential for agricultural production because it allows plants to absorb nutrients and other necessary elements for their growth. Therefore, soil remediation is incurred as one of the approaches to prevent and alleviate soil pollution [22].

Currently available remediation approaches are specifically based on physical and chemical methods, and these are laborious, time-consuming, and significantly expensive. Alternative strategies such as bioremediation which utilizes living organisms specially microorganisms to pollutant elimination have been reported. However, certain limitations of bioremediation have reduced the efficacy of these approaches. Since bioremediation requires living organisms to successfully minimize pollutant levels, it requires longer durations to analyze and optimize conditions favorable for the organism's growth in these strategies. In addition, presence of inhibitors or compounds which are not biodegradable (polymers, plastics) can diminish their efficiency [48]. Therefore, nanotechnology-based remediation approaches are rapidly emerging in the field of environmental remediation.

A composite membrane comprised of Au NPs and cellulose acetate membranes were tested over removal of radioactive iodine. A single filtration process using this membrane provided high removal efficiency within a very short period of time [55]. Alshehri (2018) biosynthetically developed gold nanoparticles using *Mucor indicus* CBS 226.29 ET for the nanoremediation of azo dyes. The high catalytic activity of Au NPs thus synthesized and enhanced the decolourization of azo dyes. Hence, it showed promising results proving the effectiveness of nanoremediation of azo dyes. Perera et al. [59] fabricated Au/TiO_2 nanocomposites via a green synthesis approach to photo-catalytically degrade methylene blue. It was observed that the developed composites' higher absorption capacity (8185 mg g^{-1}) efficiently degrades MB under UV radiation. Liu et al. [47] developed an aptamer-functionalized Au NPs-modified glass carbon electrode to detect mercury ions in contaminated environs. It was mentioned that the developed nanosensor had the detection limit of

0.005 nM. Wan et al. [68] fabricated an electrochemical sensor to simultaneously detect Pd and Cu using screen-printed gold electrodes having Au NPs. Au NPs-modified screen-printed gold electrodes demonstrated a broad detection range of lead from 20 to 200 ppb with a detection limit of 2.2 ppb and copper from 20 to 300 ppb with a detection limit of 1.6 ppb. Furthermore, Lu et al. [49] fabricated a screen-printed carbon electrode (SPCE) co-modified with Bi and Au NPs for the differential pulse anodic stripping voltammetric determination of Zn (II), Pb (II), and Cu (II). It was reported that the higher surface area and electrical conductivity of Au NPs had significantly enhanced the detection limits of the analytes. Accordingly the detection limits (at a S/N ratio of 3) are 50 ngL^{-1} for Zn(II), 20 ngL^{-1} for Pb(II), and 30 ngL^{-1} for Cu(II). Yang et al. [69] developed a non-enzymatic electrochemical sensor based on a $Au–ZrO_2–SiO_2$ modified electrode for the selective detection of paraoxon-ethyl which is a nitroaromatic organophosphate pesticides. The $Au-ZrO_2-SiO_2$ modified electrode can extract organophosphates within short durations (in 5 min) and can detect paraoxon-ethyl with a detection limit of 0.5 ng ml^{-1}.

4 Factors Affecting the Interaction of Gold Nanomaterials with Plant-Soil Systems

Nanotechnology is just like a double-edged sword where the same NM can show beneficial and harmful effects directly and indirectly on living organisms. In the same way, Au NPs have found to exhibit both beneficial and detrimental effects [13, 64]. Therefore, assessing the impacts of different nanocomposites is too important prior to its applications in any of the fields. Au nanocomposites have increased their utility and industrial applications. As elsewhere mentioned in the text unique physicochemical properties of Au nanocomposites enable to utilize in numerous fields. Therefore, understanding how Au NPs interact with plants is imperative to relate its effect. Not only that, interactions with soil system are also important to understand. Uptake of Au NPs is affected by several factors including the surface charge, surface functionalized ligand, size, shape, etc. It is noteworthy to mention that uptake and accumulation of Au NPs in biological systems depends on particle size, shape, chemical composition, surface modification, and stability [31]. For example, Li et al. [45] examined the effect of Au NPs (hydrodynamic size 8–12 nm) modified with cysteamine, cysteine, and thioglycolic acid on the uptake and translocation in *Oryza sativa* L. and *Solanum lycopersicum*. It was found that the negatively charged cysteine-modified Au NPs are more effectively absorbed by test plant roots and translocated to shoot regions in comparison to cysteamine and thioglycolic acid-modified Au NPs. Furthermore, the uptake of modified Au NPs involved the clathrin-dependent and -independent mechanisms. Clathrin-dependent endocytosis is a process that initiates with the concentration of high-affinity transmembrane receptors into specialized plasma membrane invaginations coated with clathrin [44]. Zhu et al. [71] also reported that positively charged Au NPs (6–10 nm) under hydroponic conditions are most effectively up taken by roots and negatively charged particles are most effectively translocated into the plant shoots. Feichtmeier et al. [31] studied the effect of citrate stabilized Au NPs on Barley plants and it was reported that the translocation Au NPs within Barley plants is apparently controlled by chemical nature of the NP itself. Malejko et al. [52] studied the effects of Au NPs from soil system and their distribution, translocation and chemical speciation in different plants. It was reported that plants thus tested are capable of altering the properties of Au NPs within their rhizospheres before they are up taken by plants. Also, these plants were capable of bio-transforming metallic ions into NPs within their tissues. Therefore, it is evident that plants can modify or chemically alter the properties of NMs prior to their uptake. Size is another critical factor that affects the uptake and translocation of NMs. Sabo-Attwood et al. [62] showed that the uptake, translocation of Au NPs occur in a size-dependent manner. Avellan et al. [24] demonstrated that the foliar application of Au NPs with diverted sizes (3, 10, or 50 nm) and coatings (citrate and Polyvinylpyrrolidone) on wheat plants shows Au NPs with smaller size and Polyvinylpyrrolidone modification accumulate more on leaf surfaces than citrate modified ones. Also, it was shown that proportion of citrate-Au NPs entered the test plants had effectively translocated throughout the plants. Irrespective of the NM type most of the Au NPs thus translocated had accumulated in younger shoots and roots. Furthermore, it was reported that Au NPs in size less than 50 nm had exuded into the rhizosphere soil. Also, it is important to note that the bioavailability of NPs (Au NPs) also depends on the species of root exudates [70]. Malejko et al. [52] found that plants have the ability to modify the characteristics of Au NPs in their rhizospheres. This implies that plants and gold nanocomposites interact dynamically, perhaps modifying the nanomaterials' characteristics prior to absorption.

5 Conclusion and Future Perspectives

Nanotechnology has revolutionized various fields, leading to the development of nanocomposites, a class of composite materials with superior properties. These materials, embedded with NPs, have several applications in medicine, energy, electronics, the environment, waste management, food, and agriculture. Au-based NMs, particularly Au NPs, have become versatile in various applications, including electronics, textiles, sensors, therapeutic agents, and biomedicine. Gold nanocomposites, combining gold with other materials, have shown promise in agriculture, particularly in improving plant-soil interactions. These nanocomposites are used in plant disease management, stress tolerance, and growth and seed germination, contributing to sustainable agriculture. Nanosensors developed using gold nanocomposites are applied for early detection of pathogens and offer improved sensitivity and efficiency compared to conventional immune assays. Gold nanocomposites have shown potential for enhancing plant stress tolerance, promoting growth and seed germination, and addressing soil pollution. Studies have shown that Au nanocomposites can alleviate stress conditions and modulate gene expression for stress responses. They have also been used as nanopriming agents, promoting seed emergence and seedling vigor. Gold nanocomposites have also been used in soil remediation, thus promoting environmental cleanup. Understanding the factors influencing the interaction of gold nanomaterials with plant-soil systems is crucial for their potential use. However, their widespread application raises concerns about their environmental fate and ecological impact. Further research into their specific applications, interactions with different plant species, and potential risks will help harness the full potential of nanotechnology for sustainable practices in agriculture and environmental management. The advancement of gold nanocomposites requires a delicate balance between innovation and responsible usage to maximize their benefits while minimizing potential harm to ecosystems and human health.

Acknowledgements The authors would like to thank the National Research Council of Sri Lanka (Investigator Driven Grant-NRC IDG 22-114) for the financial support.

References

1. ANZ A (2018) Using a novel mucor indicus CBS 226.29 ET for biosynthesis of gold nanoparticles and applying them in nanoremediation of azo dyes. SOJ Microbiol Infect Dis 6, 1–7. https://doi.org/10.15226/SOJMID/6/2/00192
2. Bachheti RK, Bachheti A, Husen A (2024) Metal and metal-oxide based nanomaterials (*Synthesis, Agricultural, Biomedical and Environmental Interventions*). Springer Nature Singapore Pte Ltd., 152 Beach Road, #21-01/04 Gateway East, Singapore 189721, Singapore
3. Chawla S, Singh S, Husen A (2023) Smart nanomaterials targeting pathological hypoxia. Springer Nature Singapore Pte Ltd., 152 Beach Road, #21-01/04 Gateway East, Singapore 189721, Singapore ISBN: 978-981-99-1717-4. https://doi.org/10.1007/978-981-99-1718-1
4. Husen A (2019) Natural product-based fabrication of zinc oxide nanoparticles and their application. In: Husen A, Iqbal M (Eds.) N*anomaterials* and plant potential Springer International Publishing AG, Gewerbestrasse 11, 6330 Cham, pp 193–291. https://doi.org/10.1007/978-3-030-05569-1_7
5. Husen A (2020) Introduction and techniques in nanomaterials formulation. In: Husen, A, Jawaid M (Eds.) Nanomaterials for agriculture and forestry applications. Elsevier Inc. 50 Hampshire St., 5th Floor, Cambridge, MA 02139, USA, pp 1–14. https://doi.org/10.1016/B978-0-12-817 852-2.00001-9
6. Husen A (2022) Engineered nanomaterials for sustainable agricultural production, soil improvement and stress management. Elsevier Inc., 50 Hampshire St., 5th Floor, Cambridge, MA 02139, USA. https://doi.org/10.1016/C2021-0-00054-7
7. Husen A (2023a) Secondary metabolites based green synthesis of nanomaterials and their applications. Springer Nature Singapore Pte Ltd., 152 Beach Road, #21-01/04 Gateway East, Singapore 189721, Singapore. https://doi.org/10.1007/978-981-99-0927-8
8. Husen A (2023b) Nanomaterials from agricultural and horticultural products. Springer Nature Singapore Pte Ltd., 152 Beach Road, #21-01/04 Gateway East, Singapore 189721, Singapore. https://link.springer.com/book/9789819934348
9. Husen A (2023c) Nanomaterials and nanocomposites exposures to plants (Response, Interaction, Phytotoxicity and Defense Mechanisms). Springer Nature Singapore Pte Ltd., 152 Beach Road, #21-01/04 Gateway East, Singapore 189721, Singapore. https://doi.org/10.1007/978-981-99-2419-6

10. Husen A, Bachheti RK, Bachheti A (2023) Current trends in green nano-emulsions (Food, Agriculture and Biomedical Sectors). Springer Nature Singapore Pte Ltd., 152 Beach Road, #21-01/04 Gateway East, Singapore 189721, Singapore
11. Husen A, Iqbal M (2019) Nanomaterials and plant potential. Springer International Publishing AG, Gewerbestrasse 11, 6330 Cham, Switzerland. https://doi.org/10.1007/978-3-030-05569-1
12. Husen A, Jawaid M (2020) Nanomaterials for agriculture and forestry applications. Elsevier Inc., 50 Hampshire St., 5th Floor, Cambridge, MA 02139, USA. https://doi.org/10.1016/C2018-0-02349-X
13. Husen A, Siddiqi KS (2014) Phytosynthesis of nanoparticles: concept, controversy and application. Nanoscale Res Lett 9(229):1–24. https://doi.org/10.1186/1556-276X-9-229
14. Husen A, Siddiqi KS (2023) Advances in smart nanomaterials and their applications. Elsevier Inc., 50 Hampshire St., 5th Floor, Cambridge, MA 02139, USA. https://doi.org/10.1016/C2021-0-02202-1
15. Jin-Chul K, Madhusudhan A, Husen A (2021) Smart nanomaterials in biomedical applications. Springer Nature Switzerland AG, Gewerbestrasse 11, 6330 Cham, Switzerland. https://doi.org/10.1007/978-3-030-84262-8
16. Madhusudhan A, Purohit SD, Prasad R, Husen A (2024) Functional smart nanomaterials and their theranostics approaches. Springer Nature Singapore Pte Ltd., 152 Beach Road, #21-01/04 Gateway East, Singapore 189721, Singapore
17. Mandal AK, Ghorai S, Husen A (2023) Functionalized smart nanomaterials for point-of-care testing. Springer Nature Singapore Pte Ltd., 152 Beach Road, #21-01/04 Gateway East, Singapore 189721, Singapore
18. Taghiyari HR, Morrell JF, Husen A (2023) Emerging nanomaterials (*Opportunities and Challenges in Forestry Sectors*). Springer Nature Switzerland AG, Gewerbestrasse 11, 6330 Cham, Switzerland. https://doi.org/10.1007/978-3-031-17378-3
19. Wassie AT, Ahmed IN, Bachheti A, Husen A, Bachheti RK (2024a) Plant extract-assisted green synthesis and characterization of copper nanocomposite. In: Bachheti RK, Bachheti A, Husen A (eds) Metal and metal-oxide based nanomaterials. Smart Nanomaterials Technology. Springer, Singapore, pp 43–66. https://doi.org/10.1007/978-981-99-7673-7_3
20. Wassie AT, Ahmed, IN, Bachheti A, Husen A, Bachheti RK (2024c) Role of zinc oxide nanomaterials for photocatalytic degradation of environmental pollutants. In: Bachheti RK, Bachheti A, Husen A (eds) Metal and metal-oxide based nanomaterials. smart nanomaterials technology. Springer, Singapore, pp 287–311. https://doi.org/10.1007/978-981-99-7673-7_14
21. Wassie AT, Pandey DP, Bachheti A, Husen A, Deepti Singh Vashishth, Bachheti RK (2024b) Use of green synthesized platinum nanoparticles for biomedical applications. In: Bachheti RK, Bachheti A, Husen A (eds) Metal and metal-oxide based nanomaterials. Smart Nanomaterials Technology. Springer, Singapore, pp 271–285. https://doi.org/10.1007/978-981-99-7673-7_13
22. Aqeel M, Jamil Mohd, Yusoff I (2014) Soil contamination, risk assessment and remediation. Environmental Risk Assessment of Soil Contamination. https://doi.org/10.5772/57287
23. Asefpour Vakilian K (2019) Gold nanoparticles-based biosensor can detect drought stress in tomato by ultrasensitive and specific determination of miRNAs. Plant Physiol Biochem 145:195–204. https://doi.org/10.1016/j.plaphy.2019.10.042
24. Avellan A, Yun J, Zhang Y, Spielman-Sun E, Unrine JM, Thieme J, Li J, Lombi E, Bland G, Lowry GV (2019) Nanoparticle size and coating chemistry control foliar uptake pathways, translocation, and leaf-to-rhizosphere transport in wheat. ACS Nano 13:5291–5305. https://doi.org/10.1021/ACSNANO.8B09781/SUPPL_FILE/NN8B09781_SI_001.PDF
25. Bhambure S, Rao AS (2019) A review on bio composites in industrial applications. SSRN Electron J. https://doi.org/10.2139/SSRN.3346958
26. Callister WD, Rethwisch DG (2007) Material science and engineering: an introduction, 10th Edition (10th ed.). Wiley
27. Chen D, Zhu H, Yang S, Li N, Xu Q, Li H, He J, Lu J (2016) Micro-nanocomposites in environmental management. Adv Mater 28:10443–10458. https://doi.org/10.1002/ADMA.201601486

28. Cheng JM, Liu YZ, Wang HW (2014) Effects of surface-modified nano-scale carbon black on Cu and Zn fractionations in contaminated soil. Int J Phytoremed 16:86–94. https://doi.org/10.1080/15226514.2012.759530
29. Dai S, Wang B, Song Y, Xie Z, Li C, Li S, Huang Y, Jiang M (2021) Astaxanthin and its gold nanoparticles mitigate cadmium toxicity in rice by inhibiting cadmium translocation and uptake. Sci Total Environ 786. https://doi.org/10.1016/j.scitotenv.2021.147496
30. Ebrahimi M, Norouzi P, Safarnejad MR, Tabaei O, Haji-Hashemi H (2019) Fabrication of a label-free electrochemical immunosensor for direct detection of Candidatus Phytoplasma Aurantifolia. J Electroanal Chem 851. https://doi.org/10.1016/j.jelechem.2019.113451
31. Feichtmeier NS, Walther P, Leopold K (2015) Uptake, effects, and regeneration of barley plants exposed to gold nanoparticles. Environ Sci Pollut Res 22:8549–8558. https://doi.org/10.1007/S11356-014-4015-0/METRICS
32. Feldman D (2016) Polymer nanocomposites in medicine. J Macromol Sci A 53:55–62. https://doi.org/10.1080/10601325.2016.1110459
33. Ferrari E, Barbero F, Busquets-Fité M, Franz-Wachtel M, Köhler H-R, Puntes V, Kemmerling B (2021) Growth-promoting gold nanoparticles decrease stress responses in Arabidopsis seedlings. Nanomaterials 11:3161. https://doi.org/10.3390/nano11123161
34. Fiorati A, Grassi G, Graziano A, Liberatori G, Pastori N, Melone L, Bonciani L, Pontorno L, Punta C, Corsi I (2020) Eco-design of nanostructured cellulose sponges for sea-water decontamination from heavy metal ions. J Clean Prod 246:119009. https://doi.org/10.1016/J.JCLEPRO.2019.119009
35. Grieger K, Hjorth R, Rice J, Kumar N, IUCN JB-B entry for, 2015, undefined, 2015. Nanoremediation: tiny particles cleaning up big environmental problems. core.ac.ukKD Grieger, R Hjorth, J Rice, N Kumar, J BangBlog entry for IUCN, 2015 core.ac.uk.
36. Gupta G, Kumar A, Tyagi R, Kumar S, Professor A, Pradesh U (2007) Application and future of composite materials: a review. Int J Innovative Res Sci Eng Technol (An ISO 3297). https://doi.org/10.15680/IJIRSET.2016.0505041
37. Hassan H, Sharma P, Hasan MR, Singh S, Thakur D, Narang J (2022) Gold nanomaterials—the golden approach from synthesis to applications. Mater Sci Energy Technol 5:375–390. https://doi.org/10.1016/J.MSET.2022.09.004
38. Husen A, Siddiqi KS (2014) Phytosynthesis of nanoparticles: concept, controversy and application. Nanoscale Res Lett 9:1–24. https://doi.org/10.1186/1556-276X-9-229
39. Idumah CI, Zurina M, Ogbu J, Ndem JU, Igba EC (2020) A review on innovations in polymeric nanocomposite packaging materials and electrical sensors for food and agriculture. Compos Interfaces 27:1–72. https://doi.org/10.1080/09276440.2019.1600972
40. Ingle AP, Seabra AB, Duran N, Rai M (2014) Nanoremediation: a new and emerging technology for the removal of toxic contaminant from environment. Microb Biodegrad Bioremed 233–250. https://doi.org/10.1016/B978-0-12-800021-2.00009-1
41. Ji L, Meduri P, Agubra V, Xiao X, Alcoutlabi M (2016) Graphene-based nanocomposites for energy storage. Adv Energy Mater 6:1502159. https://doi.org/10.1002/AENM.201502159
42. Jian Y-S, Lee C-H, Jan F-J, Wang G-J (2018) Detection of odontoglossum ringspot virus infected phalaenopsis using a nano-structured biosensor. J Electrochem Soc 165:H449–H454. https://doi.org/10.1149/2.0351809JES/XML
43. Kuang Y, Du J, Zhou R, Chen Z, Megharaj M, Naidu R (2015) Calcium alginate encapsulated Ni/Fe nanoparticles beads for simultaneous removal of Cu (II) and monochlorobenzene. J Colloid Interface Sci 447:85–91. https://doi.org/10.1016/J.JCIS.2015.01.080
44. Lamaze C, Johannes L (2006) Intracellular trafficking of bacterial and plant protein toxins. Comprehen Sourcebook Bacterial Protein Toxins 135–153. https://doi.org/10.1016/B978-012088445-2/50013-5
45. Li H, Ye X, Guo X, Geng Z, Wang G (2016) Effects of surface ligands on the uptake and transport of gold nanoparticles in rice and tomato. J Hazard Mater 314:188–196. https://doi.org/10.1016/J.JHAZMAT.2016.04.043
46. Liu H, Jian R, Chen H, Tian X, Sun C, Zhu J, Yang Z, Sun J, Wang C (2019) Application of biodegradable and biocompatible nanocomposites in electronics: current status and future directions. Nanomaterials 9:950 9, 950. https://doi.org/10.3390/NANO9070950

47. Liu Y, Deng Y, Li T, Chen Z, Chen H, Li S, Liu H (2018) Aptamer-based electrochemical biosensor for mercury ions detection using aunps-modified glass carbon electrode. J Biomed Nanotechnol 14:2156–2161. https://doi.org/10.1166/JBN.2018.2655
48. Lü JC, Li ZT, Hussain K, Yang GK (2011) Bioremediation: the new directions of oil spill cleanup. Middle East J Sci Res 7:738–740
49. Lu Z, Zhang J, Dai W, Lin X, Ye J, Ye J (2017) A screen-printed carbon electrode modified with a bismuth film and gold nanoparticles for simultaneous stripping voltammetric determination of Zn(II), Pb(II) and Cu(II). Microchim Acta 184:4731–4740. https://doi.org/10.1007/S00604-017-2521-8/METRICS
50. Luna-Moreno D, Sánchez-álvarez A, Islas-Flores I, Canto-Canche B, Carrillo-Pech M, Villarreal-Chiu JF, Rodríguez-Delgado M (2019) Early detection of the fungal banana black sigatoka pathogen *Pseudocercospora fijiensis* by an SPR immunosensor method. Sensors 19:465. https://doi.org/10.3390/S19030465
51. Mahakham W, Theerakulpisut P, Maensiri S, Phumying S, Sarmah AK (2016) Environmentally benign synthesis of phytochemicals-capped gold nanoparticles as nanopriming agent for promoting maize seed germination. Sci Total Environ 573:1089–1102. https://doi.org/10.1016/J.SCITOTENV.2016.08.120
52. Malejko J, Godlewska-Żyłkiewicz B, Vanek T, Landa P, Nath J, Dror I, Berkowitz B (2021) Uptake, translocation, weathering and speciation of gold nanoparticles in potato, radish, carrot and lettuce crops. J Hazard Mater 418:126219. https://doi.org/10.1016/J.JHAZMAT.2021.126219
53. Malhotra BD, Ali MdA (2018) Nanomaterials in biosensors: fundamentals and applications. Nanomater Biosensors 1. https://doi.org/10.1016/B978-0-323-44923-6.00001-7
54. Mishra K, Devi N, Siwal SS, Zhang Q, Alsanie WF, Scarpa F, Thakur VK (2022) Ionic liquid-based polymer nanocomposites for sensors, energy, biomedicine, and environmental applications: roadmap to the future. Adv Sci 9:2202187. https://doi.org/10.1002/ADVS.202202187
55. Mushtaq S, Yun SJ, Yang JE, Jeong SW, Shim HE, Choi MH, Park SH, Choi YJ, Jeon J (2017) Efficient and selective removal of radioactive iodine anions using engineered nanocomposite membranes. Environ Sci Nano 4:2157–2163. https://doi.org/10.1039/C7EN00759K
56. Mutiara T, Fahrurrozi M, Sulistyo H, Hidayat M (2023) Green synthesis methods and characterization of bacterial cellulose/silver nanoparticle composites. Green Process Synthesis 12. https://doi.org/10.1515/gps-2023-0067
57. Ndeh NT, Maensiri S, Maensiri D (2017) The effect of green synthesized gold nanoparticles on rice germination and roots*. Adv Nat Sci Nanosci Nanotechnol 8:035008. https://doi.org/10.1088/2043-6254/AA724A
58. Patil SS, Shedbalkar UU, Truskewycz A, Chopade BA, Ball AS (2016) Nanoparticles for environmental clean-up: a review of potential risks and emerging solutions. Environ Technol Innov 5:10–21. https://doi.org/10.1016/J.ETI.2015.11.001
59. Perera M, Wijenayaka LA, Siriwardana K, Dahanayake D, Nalin De Silva KM (2020) Gold nanoparticle decorated titania for sustainable environmental remediation: green synthesis, enhanced surface adsorption and synergistic photocatalysis. RSC Adv 10:29594–29602. https://doi.org/10.1039/D0RA05607C
60. Rahmati S, Doherty W, Amani Babadi A, Akmal Che Mansor A, Julkapli NM, Hessel V, Ostrikov K (2021) Gold–carbon nanocomposites for environmental contaminant sensing. Micromachines (Basel). https://doi.org/10.3390/mi12060719
61. Razo SC, Panferova NA, Panferov VG, Safenkova IV, Drenova NV, Varitsev YA, Zherdev AV, Pakina EN, Dzantiev BB (2019) Enlargement of gold nanoparticles for sensitive immunochromatographic diagnostics of potato brown rot. Sensors 19:153. https://doi.org/10.3390/S19010153
62. Sabo-Attwood T, Unrine JM, Stone JW, Murphy CJ, Ghoshroy S, Blom D, Bertsch PM, Newman LA (2012) Uptake, distribution and toxicity of gold nanoparticles in tobacco (*Nicotiana xanthi*) seedlings. Nanotoxicology 6:353–360. https://doi.org/10.3109/17435390.2011.579631

63. Sarraf M, Vishwakarma K, Kumar V, Arif N, Das S, Johnson R, Janeeshma E, Puthur JT, Aliniaeifard S, Chauhan DK, Fujita M, Hasanuzzaman M (2022) Metal/metalloid-based nanomaterials for plant abiotic stress tolerance: an overview of the mechanisms. Plants. https://doi.org/10.3390/plants11030316
64. Siddiqi KS, Husen A (2016) Engineered gold nanoparticles and plant adaptation potential. Nanoscale Res Lett. https://doi.org/10.1186/s11671-016-1607-2
65. Song Y, Zheng C, Basnet R, Li S, Chen J, Jiang M (2022) Astaxanthin synthesized gold nanoparticles enhance salt stress tolerance in rice by enhancing tetrapyrrole biosynthesis and scavenging reactive oxygen species in vitro. Plant Stress 6:100122. https://doi.org/10.1016/j.stress.2022.100122
66. Teixeira MC, Santini A, Souto EB (2017) Delivery of antimicrobials by chitosan-composed therapeutic nanostructures. Nanostructures Antimicrob Ther: Nanostructures Ther Med Ser 203–222. https://doi.org/10.1016/B978-0-323-46152-8.00008-1
67. Wahid I, Rani P, Kumari S, Ahmad R, Hussain SJ, Alamri S, Tripathy N, Khan MIR (2022) Biosynthesized gold nanoparticles maintained nitrogen metabolism, nitric oxide synthesis, ions balance, and stabilizes the defense systems to improve salt stress tolerance in wheat. Chemosphere 287:132142. https://doi.org/10.1016/J.CHEMOSPHERE.2021.132142
68. Wan H, Sun Q, Li H, Sun F, Hu N, Wang P (2015) Screen-printed gold electrode with gold nanoparticles modification for simultaneous electrochemical determination of lead and copper. Sens Actuators B Chem 209:336–342. https://doi.org/10.1016/J.SNB.2014.11.127
69. Yang Y, Tu H, Zhang A, Du D, Lin Y (2012) Preparation and characterization of Au–ZrO2–SiO2 nanocomposite spheres and their application in enrichment and detection of organophosphorus agents. J Mater Chem 22:4977–4981. https://doi.org/10.1039/C2JM15129D
70. Ye X, Li H, Wang Q, Chai R, Ma C, Gao H, Mao J (2018) Influence of aspartic acid and lysine on the uptake of gold nanoparticles in rice. Ecotoxicol Environ Saf 148:418–425. https://doi.org/10.1016/J.ECOENV.2017.10.056
71. Zhu ZJ, Wang H, Yan B, Zheng H, Jiang Y, Miranda OR, Rotello VM, Xing B, Vachet RW (2012) Effect of surface charge on the uptake and distribution of gold nanoparticles in four plant species. Environ Sci Technol 46:12391–12398. https://doi.org/10.1021/ES301977W/SUPPL_FILE/ES301977W_SI_001.PDF

Effect of Gold Nanoparticles on Seed Germination, Plant Growth, and Plant Protection

Mayur Mukut Murlidhar Sharma, Divya Kapoor, Atul Loyal, Rahul Kumar, Pankaj Sharma, and Azamal Husen

Abstract Nanotechnology has emerged as a revolutionary force in the realm of agriculture, presenting inventive solutions to elevate plant performance and alleviate a variety of stressors. Gold nanoparticles (AuNPs), distinguished by their distinctive physico-chemical properties, have gained considerable attention due to their potential role in modulating plant responses. They have the capacity to influence growth, optimize production, and offer protection against challenges posed by biotic stressors. The impact of AuNPs on plants spans various processes, including seed germination, photosynthetic apparatus and antioxidant system activity, and the expression of genes crucial for plant function under both optimal and adverse conditions. It is noteworthy that these influences and growth enhancement effects exhibit variability across different plant species. Research indicates that factors such as dosage, size, shape, surface charge, and capping agent significantly influence the uptake, translocation, distribution, and growth enhancement effects of AuNPs in plants. This chapter delves into the intricate interactions between gold nanoparticles and plants, elucidating the underlying mechanisms that contribute to the diverse outcomes observed.

M. M. M. Sharma
Department of Agriculture and Industry, Kangwon National University, Chuncheon, Korea

D. Kapoor · P. Sharma (✉)
Department of Microbiology, CCS Haryana Agricultural University, Hisar, Haryana 125004, India
e-mail: psk14495@gmail.com; sharmap@uic.edu

A. Loyal
Department of Genetics and Plant Breeding, CCS Haryana Agricultural University, Hisar, Haryana 125004, India

R. Kumar
Department of Zoology, College of Agriculture, RRS Bawal, Haryana 123501, India

P. Sharma
Department of Ophthalmology & Visual Sciences, University of Illinois Chicago, Chicago 60612, USA

A. Husen
Wolaita Sodo University, PO Box 138, Wolaita, Ethiopia

Department of Biotechnology, Graphic Era (Deemed to Be University), Dehradun, Uttarakhand 248002, India

A. Husen (ed.), *Plant Response to Gold Nanoparticles*, Smart Nanomaterials Technology, https://doi.org/10.1007/978-981-97-7174-5_8

The discussion extends to the impact of AuNPs on plant growth and development, emphasizing their potential to enhance nutrient uptake and facilitate targeted delivery, thereby augmenting plant productivity and yield. Furthermore, the influence of gold nanoparticles on the plant's defense mechanisms against pathogens and pests is thoroughly investigated. AuNPs exhibit promising antimicrobial and antifungal properties, presenting an environmentally friendly alternative to traditional pesticides. The multifaceted nature of these interactions highlights the potential of gold nanoparticles to revolutionize agricultural practices and contribute to sustainable and eco-friendly crop management.

Keywords Seed germination · Plant protection · Plant growth promotion · Plant growth · Plant vigor

1 Introduction

Agricultural systems are facing a lot of challenges across the globe. The quest to generate cleaner food production systems along with the maintenance of plant yields and agricultural sustainability is seeming to be the demand of the modern era. The traditional and current systems seem to be partially ineffective in fulfilling such demands [4, 73, 74, 91, 92]. Addressing the challenges of increasing food demands, the risk of resistant pathogens, depleting soil carbon reserves, and reliance on agrochemicals requires the development of a robust agricultural model. This model should aim to minimize wastage and maximize bioavailability. Strategies such as sustainable farming practices, precision agriculture, and the use of technology to optimize resource use can contribute to achieving these goals. Additionally, promoting agroecological approaches, crop diversity, and soil conservation measures can help create a more resilient and sustainable food production system. It's crucial to integrate scientific advancements and innovative solutions to ensure the long-term viability of agriculture while minimizing its environmental impact [76, 79, 83–87].

Nano-technological interventions strategically place the stimulus to transform various realms of diagnostics, encompassing health, medication, food, environment, and the agricultural sector [10, 17, 32–41, 45, 58, 62, 95, 101–103]. This transition effectively converts speculative attributes into tangible, practical outcomes [13, 48, 80, 82, 82, 86, 89]. A nanoparticle is characterized as a minute particle with a size ranging from 1 to 100 nm. The expanding array of applications attributed to various nanoparticles (NPs) is facilitated by their distinctive features, including heightened reactivity due to a favorable surface area-to-volume ratio, elevated surface potential, adjustable physical and chemical properties, and molecular manipulation, as compared to their bulk counterparts. It is noteworthy that the nano-form of any metal exhibits unique attributes that are not present in its bulk form [55, 56, 76, 79, 84, 92]. Documentation of AuNPs (Gold Nanoparticles) dates to 1996. Despite gold being inert and universally acknowledged as biocompatible, until recently, it was primarily known as a metal. The emergence of nanotechnology and the exploration of the physico-chemical properties of gold have positioned it as a superior material in advancing fields [47]. Michael Faraday initially observed tiny assemblies, with minimum of one dimension shorter than 100 nm, in 1857 when he created gold nanoparticles. Notably, he witnessed that nanostructured gold emanated a distinctive red color. He commented on the phenomenon, noting that as the preparations became more dilute, the solution became clearer, revealing dispersed gold. The particles, when illuminated by sunlight or a candle through a lens, created a visible cone when dropped into a liquid. Although the individual particles couldn't be discerned clearly, their radiance appeared golden and was proportional to the amount of solid gold present. Faraday's elucidation marked the inception of what is now known as the Tyndall effect. The unique feature of gold nanoparticles, perhaps owing to their miniature size, is the absorption of light in the blue-green spectrum (~450 nm) and the reflection of red light (~700 nm), resulting in a vibrant red color known as ruby gold. Faraday's finding marked the dawn of modern nanotechnology, attracting numerous scientists due to its promising research prospects. The concept of "nanoparticles" signifies an intermediate state between individual atoms and bulk materials, permitting substances to exhibit specific photonic, electronic, catalytic, and therapeutic properties. While the word encompasses particles with diameters smaller than approximately 1 μm (1–100 nm), it is most employed in biology laboratories to describe particles in this size range [21, 93]. Nanoparticle conformations encompass a diverse array, incorporating those with essential organic molecules like liposomes, fatty acids, viruses, DNA, and micelles. Additionally, there are configurations involving inorganic molecules like iron oxide, gold, and carbon nanotubes. Hybrid configurations, combining two or more of these elements, further contribute to the versatility and functionality of nanoparticle structures [30]. In laboratory settings, nanoparticles find utility in various applications, including the labeling of biomolecules or cells, transfection processes, and as repetitive imaging elements during in vitro experiments. Moving to living animal models, nanoparticles exhibit remarkable efficiency as targets for medicinal imaging, utilizing magnetic nanoparticles and nanobubbles. They also play a crucial role in medicine dissemination and cell stalking. Studies have demonstrated the effectiveness of using AuNPs to tag diverse cell types, including stem cells and immune cells, without compromising their therapeutic efficacy. Noble metals, particularly gold (Au), stand out as standout categories of metal nanoparticles. AuNPs have garnered considerable consideration because of their unique capacity to interact with light through surface plasmon resonance (SPR). Recent advancements in nanoscience suggest that gold nanoparticles (Au NPs) hold potential as building blocks for future plasmonic and optical devices. The applications of gold nanoparticles (NPs) have expanded exponentially across various fields, including chemistry, material sciences, physics, medicine, and life sciences, owing to their complex attributes in therapeutics, detection, imaging, and surface modification. In the realm of cancer therapy, metallic nanoparticles like gold play a pivotal role in theragnostic applications. Gold nanoparticles (AuNPs) find applications in extremely impressionable biomolecular screening, targeted eradication of cancer cells through photothermal therapy, protein labeling, and the delivery of cellular therapeutics. The transportation of drug molecules mediated by AuNPs primarily depends on passive diffusion (nonspecific passive transport) or active transport organizations, contingent on the availability of energy. When administered orally, these nanoparticles undergo absorption through the gastrointestinal tract and are subsequently circulated throughout the body via the circulatory system. An alternative route involves direct transportation to intravenous target organs. Following their introduction into circulation, AuNPs may be quickly eradicated by defensive appliances, such as discarding by the reticuloendothelial system (RES). The duration of circulation can also vary based on the size, shape, and surface features of the nanoparticles, allowing for extended presence in the systemic circulation [6]. Moreover, it facilitates easy customization of size, shape, and attachment of diverse functional groups. These nanoparticles are chemically biocompatible and possess tunable optical properties, providing a versatile platform for a range of applications (Meena et al. 2020). The production of these nanoparticles is achievable through numerous methods involving plants, fungi, and bacteria, employing diverse physical, chemical, and biological techniques. This constitutes a dynamic and notably significant academic field [47]. Addressing the encounter of determining suitable particle size, shape, and surface chemistry for probable applications necessitates further research into the interactions between nanoparticles and cells or tissues [31]. Gold nanoparticles (Au NPs) have garnered considerable attention from researchers across various disciplines, including physics, catalysis, and development of anticancer drugs, antioxidants, larvicides, antimicrobials, nanofluids, agriculture, and drug delivery [12]. The extensive dispersion of nanomaterials and their growing incorporation into human activities and the surrounding environment have sparked valid concerns within the ecological community. Regrettably, the penalties of potential dealings amid nanoparticles and living organisms, particularly their effects on plants, remain ambiguous. Given their stationary nature, plants are especially susceptible to adverse influences, underscoring the imperative need for inquiries into the character and outcomes of interactions between plants and nanoparticles. The influence of nanoparticles on plants is contingent on various factors, encompassing the type, size, and concentration of nanoparticles, the specific plant species under consideration, and the conditions prevailing in the experimental setup [99]. In this chapter, our focus will be on examining the impact of gold nanoparticles on higher plants (Fig. 1). Special attention will be given to exploring the potential pathways through which gold nanoparticles enter the plant organism, elucidating their primary impact, and discussing the scenarios of their use in various aspects of plant growth, production, and protection.

2 Seed Germination

Seed dormancy can be mitigated through various treatments, including soaking seeds in salt solutions (halo-priming), water (hydro-conditioning), osmotic agents (osmo-priming), solutions with plant hormones (hormonal priming), beneficial microbe solutions (bio-priming), exposure to a magnetic field (magneto-priming), a solution with a solid carrier (matri-conditioning), and solutions containing nanoparticles

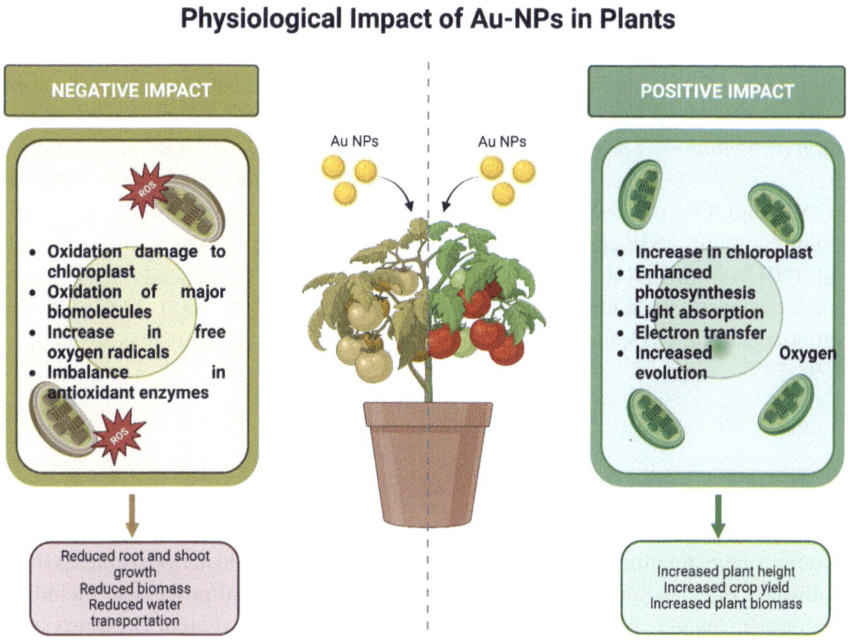

Fig. 1 A portrayal depicting physiological impact of gold nanoparticles in plant systems

(NPs) (nano-priming). Nano-priming stands as an advanced seed priming technology designed to enhance seed germination, seedling growth, and overall crop yield. It achieves this by imparting resistance to a variety of stresses experienced by plants. Nano-priming proves significantly more effective than other seed priming methods. The key attributes of nanoparticles (NPs) in seed priming involve fostering electron exchange and improving surface reaction capabilities associated with diverse components of plant cells and tissues. Nano-priming initiates the creation of nanopores in the shoot, facilitating water absorption. NPs can readily traverse the plasma membrane through a direct diffusion mechanism. It triggers reactive oxygen species (ROS)/ antioxidant mechanisms in seeds and generates hydroxyl radicals. These radicals play a role in loosening cell walls and act as inducers for the swift hydrolysis of starch. Additionally, nano-priming prompts the expression of aquaporin genes responsible for water intake and mediates the dispersion of H_2O_2 or ROS across biological membranes. Nano-priming further stimulates starch degradation by activating amylase, ultimately promoting seed germination [67]. Gold nanoparticles are known to enhance seed germinations and are largely being reported as important agents for nano-priming the seeds. For instance, Mahakham et al. [59] evaluated the efficacy of priming phytochemicals-capped gold nanoparticles on seed germination in maize plants. Priming using 5 ppm GNPs exhibited the most favorable outcomes in enhancing the emergence percentage, with a rate of 83%, surpassing both the unprimed control (43%) and hydroprimed groups (56%). Priming seeds with both 5 and 10 ppm GNPs resulted in a threefold increase in seedling vigor index compared to the control. Additionally, priming with 10 ppm GNPs was identified as particularly effective in enhancing the physiological and biochemical properties of maize seedlings. Similarly, Kumar et al. [50] observed a significant increase in seed germination of *Arabidopsis thaliana* in the media supplemented with gold nanoparticles. After 7 days, the germination rate was 90% on medium supplemented with 10 μg/ml of GNPs and 85% on medium with 80 μg/ml of GNPs. In contrast, the control MS medium without GNPs showed the lowest germination rate at 75%. Similarly, Ndeh et al. (2017) evaluated the efficacy of gold nanoparticles in rice seeds. Rice grains underwent treatment at various concentrations (0, 10, 100, 500, 1000, 2000 mg/l) for a duration of one week that resulted in a high germination percentage equivalent to 95–98.38%.

3 Plant Growth

Numerous investigations have highlighted the capability of gold nanoparticles (GNP) to infiltrate plant organisms and impact various physiological processes, including development, water altercation, photosynthesis, and the movement of antioxidant enzymes (Table 1). Considering these findings, we propose that GNP can be recommended for application in plant physiology as growth and development stimulants in a dose-dependent manner. For instance, Kumar et al. [50] conducted intriguing experiments involving 2-week-old *A. thaliana* seedlings, immersing their seeds in nanogold solutions (24 nm; 10 and 80 μg/mL). The outcomes revealed enhanced progression, an increased quantity of lateral roots, elevated water level in the foliage, enhanced action of antioxidant enzymes, and a higher seed count under the influence of gold nanoparticles (GNP). Significantly, the study also observed alterations in the expression of various microRNAs (miR398, miR408, miR164, miR167, and miR169) linked to the directive of *A. thaliana* development. For example, the well-documented involvement of miR398 in regulating the expression of genes encoding superoxide dismutase, along with genes controlling seed germination and plant growth, was noted. Deviations in the expression of miR167 and miR169 were identified to potentially impact seedling size, early flowering, and seed maturation. Moreover, the observed modifications could be attributed to the influence of GNP on auxin signal transduction pathways, particularly given that one of the targets of miR167 is factors responding to auxin, specifically ARF8 [50]. An elevation in the concentration of nanogold solutions has the potential to induce toxic effects. In the case of brown mustard experiments, a 10-day exposure to GNP solutions at concentrations of 50 and 100 μg/mL did not result in the accumulation of H2O2 in the leaves. However, a noticeable accumulation of H2O2 was observed when treated with nanogold at a concentration of 400 μg/mL [29]. Even lower concentrations of GNP can exert adverse effects on the plant organism, especially with extended exposure. For instance, experiments involving brown mustard demonstrated that treatment with GNP, even at concentrations as modest as 50 and 100 μg/mL, led to an increase in the content of MDA and H2O2 in the leaves. In simpler terms, prolonged exposure for up to 30 days resulted in heightened oxidative stress [7]. Crucial considerations also encompass the distinctive characteristics of plant species. Investigations involving soybean have revealed that a nanogold concentration of up to 1000 μg/mL not only stimulated growth but also elevated the content of photosynthetic pigments [18]. The repercussions of gold nanoparticles (AuNPs), synthesized via the utilization of *Terminalia arjuna* fruit extract, on the sprouting of *Gloriosa superba* seeds and the subsequent proliferation of the plant's foliage have been scrutinized. The findings elucidated a beneficial impact of gold nanoparticles (AuNPs) on the initiation of seed germination, extension of nodes, and the proliferative development of the plants. Notably, a robust correlation was discerned between the concentration of AuNPs in the surroundings and both the germination of *Gloriosa superba* seeds and the vegetative expansion [27].

In the pursuit of scientific inquiry, the immediate ramifications of silica, palladium, gold, and copper nanoparticles on a soil microbial community and the germination of lettuce seeds were meticulously examined across two discrete concentrations of nanoparticles. The findings unveiled a statistically inconsequential impact of the nanoparticles in the soil on parameters such as the abundance of colony forming units, peak areas of methyl ester of fatty acids in the FAME profile, or the overall metabolic fingerprint of the soil community ($P > 0.05$). Nevertheless, it was discerned that all nanoparticles scrutinized in the investigation wielded a salient influence on the growth of lettuce seeds, as evinced by the shoot/root ratios of the germinated

Table 1 Review of literature highlighting the role of gold nanoparticles in plant growth and protection

Type/Size of AuNPs	Activity	Target species	References
12 nm	Growth promotion	Arabidopsis	[25]
27.5 ± 4.8 nm	Improved biomass	*Lavandula angustifolia*	[43]
14–35 nm	Enhanced seedling growth	*Pennisetum glaucum*	[70]
30 – 113	Improved yield parameters	*Allium cepa*	[2]
14.9 ± 0.8	Improved in vitro proliferation	*Nardostachys jatamansi*	[46]
24.2 ± 2.4 nm	Enhanced seed germination	*Brassica napus*	[43]
-	Stimulated seed germination	*Dacus Carota*	[15]
3–28 nm	Mortality of plant pathogen	Pine Wilt Nematode	[20]
Gold-conjugated chitosan NPs	Reduction of fungal colony	*Fusarium oxysporum*	[54]
20–80 nm	Inhibitory effect	*Fusarium solani*	[68]

plants ($P < 0.05$) [78]. The assessment of the toxicity of these nanoparticles involved standard toxicity tests, encompassing germination tests for cucumber and lettuce, bioluminescent testing using *Photobacterium phosphoreum*, and anaerobic toxicity tests. Germination trials were executed at nanoparticle concentrations of 62, 100, and 116 $\mu g\ mL^{-1}$ for Au, Ag, and Fe_3O_4, respectively. Bioluminescent assessments (utilizing *P. phosphoreum*) were implemented at doses of 28, 45, and 52 $\mu g\ mL^{-1}$ for Au, Ag, and Fe3O4, respectively. Anaerobic analyses were conducted at nanoparticle levels of 10, 16, and 18 $\mu g\ mL^{-1}$ for Au, Ag, and Fe3O4, respectively. Throughout all the evaluations, minimal or negligible adverse effects were discerned [14]. A scientific inquiry was conducted to scrutinize the reversibility and ramifications of the internalization of citrate-coated aurous nanoparticles (Au-NPs) in the exemplar organism barley (*Hordeum vulgare* L.). Barley seeds were cultivated in a nutrient solution imbued with Au-NPs for a span of 2 weeks, after which the seedlings were translocated to a medium bereft of Au-NPs and afforded the opportunity to undergo additional proliferation for an extended period of 3 weeks. The durability of Au-NPs in the cultivation media was evaluated throughout the 2-week exposure duration. The gold content in the leaves and roots of the plants was quantified after the initial 2 weeks of exposure and at intervals of 7, 14, and 21 days during the regeneration phase, employing total reflection X-ray fluorescence (TXRF) analysis after microwave-assisted digestion. Moreover, the localization of Au-NPs within plant material was determined through transmission electron microscopy (TEM) of ultrathin cross sections. The results indicated a proclivity for the accumulation of Au-NPs in the plant roots. Concentration-dependent impacts on the absorption of macronutrients and micronutrients, along with biomass production in exposed plants, particularly affecting root growth, were discerned. Despite a certain level of regeneration demonstrated by the exposed barley plants, their root growth exhibited a persistent decline [24]. Remarkably, a remarkably low concentration of 1 μg gold mL^{-1} of gold nanoparticles in the nutrient medium demonstrated a growth-stimulating effect. There is speculation that this minimal dose might activate the functioning of hormones [14]. Conversely, higher doses and larger nanoparticles were observed to exert a detrimental effect on plant growth and biomass production. This adverse impact could be attributed to the adsorption of gold nanoparticles onto the cell wall surfaces of the primary root system, causing a reduction in pore size and consequently inhibiting water transport capacity. Consequently, this led to decreased plant growth and related attributes. Similar observations have been documented by previous researchers [8].

The rhizospheric region functions as an abode for a myriad of soil microorganisms, notably plant growth-promoting rhizobacteria (PGPR), celebrated for their adeptness in amplifying plant growth [88]. In a comprehensive investigation, the repercussions of gold nanoparticles (GNPs) at diverse concentrations on the proliferation of PGPR were systematically examined. The GNPs were synthetically generated through the reduction of HAuCl4 and subjected to characterization employing methodologies such as UV–Vis spectroscopy, X-ray diffraction (XRD), and transmission electron microscopy (TEM). The impact of GNPs on PGPR was meticulously assessed utilizing the Clinical Laboratory Standards Institute (CLSI)-endorsed

Broth-Microdilution technique against four specifically chosen PGPR strains: *Pseudomonas fluorescens*, *Bacillus subtilis*, *Paenibacillus elgii*, and *Pseudomonas putida*. Intriguingly, no significant acceleration or diminution in growth was discerned in *P. putida* owing to the presence of GNPs. In contradistinction, a conspicuous augmentation was witnessed in the case of *P. fluorescens*, *P. elgii*, and *B. subtilis* [90]. A novel study assessed the effectiveness of gold nanoparticles (GNPs) as a nano-priming agent for aged maize seeds. Utilizing environmentally realistic doses (5–15 ppm), GNPs significantly enhanced seed germination and early seedling growth. The 5 ppm treatment achieved the highest emergence percentage (83%), surpassing both the unprimed control (43%) and hydroprimed groups (56%). Additionally, priming with both 5 and 10 ppm GNPs led to a substantial increase in seedling vigor index, reaching up to three times the control values. Interestingly, the 10 ppm treatment resulted in the most optimal physiological and biochemical characteristics in maize seedlings [59]. A separate study employed *Cassia auriculata* leaf extract at room temperature to biogenically synthesize gold nanoparticles. Their characterization involved UV–vis spectroscopy, X-ray diffraction, and transmission electron microscopy. This research aimed to assess the impact of these bio-nanogold particles on *Pennisetum glaucum*, a crucial crop for food and biofuel production. The study found favorable outcomes, including augmented seed germination rates and improved seedling growth [70]. Abd El-Aziz et al. [1], harnessed *Zingiber officinale* (ginger) to synthesize biogenic gold nanoparticles (AuNPs) and explore their impact on lentil (*Lens culinaris*) seedlings, a model for legumes. They tested five AuNP concentrations (0, 10, 25, 50, and 100 ppm) to assess their influence on both physiological parameters (seed germination, plant height, leaf number, and biomass) and biochemical parameters (total soluble sugars, phenols, chlorophyll, carotenoids, amino acids, and protein content). The authors observed a significant increase in all the growth parameters under test. A separate study unveiled a simple, efficient, and affordable method for green-synthesizing AuNPs using *T. arjuna* leaf extracts. Notably, these nanoparticles, ranging from 20 to 50 nm in size and exhibiting a spherical shape, displayed enhanced efficacy in promoting both mitotic cell division and pollen germination. Crucially, no harmful effects were observed on cells or pollen, suggesting promising applications of green-synthesized nanoparticles in stimulating cell division and pollen germination, especially for vulnerable plant species [28].

3.1 Plant Protection

AuNPs emerged as versatile champions for a thriving agricultural landscape. From directly combating plant pathogens with their antibacterial, antifungal, and nematicidal abilities to indirectly boosting plant resilience, they offer a powerful weapon against agricultural threats. They even extend their service to pesticide detection, acting as color-changing sensors that reveal the presence of harmful chemicals with a simple visual cue. This dual threat and detection capability makes AuNPs a promising avenue for sustainable and efficient agriculture. For example, Bai et al.

[11] undertook an investigation into gold nanoparticles as colorimetric probes for the screening of the insecticide pymetrozine. The study elucidated that compounds harboring nitrogen heterocycles and amine groups possess the capability to bind to the surface of metal nanoparticles, thereby instigating their accumulation [3, 26]. Geared with a versatile structure featuring an exocyclic amine and a nitrogen-rich ring, pymetrozine becomes a magnet for gold nanoparticles. This captivating interaction not only paints a mesmerizing visual spectacle but also empowers these tiny sensors to track down pymetrozine residues in plants and food, ensuring a healthier harvest on our plates. For instance, Bai et al. [11] achieved a remarkable feat, pinpointing pymetrozine at a minuscule concentration of $1 \times 10^{\wedge-6}$ M, solidifying its superior sensitivity compared to 11 other pesticides. Adding salt and lowering the pH further amplified the system's detection prowess, showcasing its exceptional potential for precise pymetrozine analysis [60]. Such kind of nanosensors can be important tools for monitoring the levels of residual pesticides in agro-ecosystems that can be a significant check on the bioaccumulation of these residues. The scarcity of water for irrigation is indeed another significant challenge for agro-ecosystems. The scarcity of irrigation water is forcing the farmers to irrigate their fields with sewage water that can be a significant source of harmful xenobiotic compounds. The continuous usage of such water poses significant concerns for both human health and the environment that can lead to their bioaccumulation in the plant systems. Addressing this issue requires a multifaceted approach to ensure sustainable water management in agriculture. Gold nanoparticles can play a crucial role in water purification. For example, Zhang et al. [104] armed gold nanoparticles with the keen sense of smell for imidacloprid, a common water contaminant. Not only can these clever sensors pinpoint its presence, but their gold-powered clumping act offers a promising strategy for its removal. This innovative approach adds another weapon to the water-cleaning arsenal, joining forces with fungus-made gold nanoparticles that also target both pesticides and pathogens [19]. Gold nanoparticles are more than just bling for nanobots; they're nature's tiny warriors against microbial threats. Their ability to combat bacteria and fungi makes them exciting prospects for the future of plant disease management, food safety, and even personalized medicine. Imagine crops thriving without harmful pathogens, food free from contamination, and targeted treatments tailored to individual needs. Jayaseelan et al. [44] harnessed the reducing and capping powers of *Abelmoschus esculentus* seed extract to create gold nanoparticles, revealing their potent antifungal activity against a spectrum of pathogens, including *Candida albicans*, *Aspergillus niger*, *Aspergillus flavus*, and *Puccinia graminis tritici*. Interestingly, these nanoparticles also exhibit antibacterial activity against a range of common bacterial pathogens, including *Klebsiella pneumoniae*, *Bacillus subtilis*, *Pseudomonas aeruginosa*, and *Escherichia coli* [5, 64]. Furthermore, biosynthesized gold nanoparticles derived from *Trichoderma atroviride* have shown promise in controlling plant fungal diseases like *Phomopsis canker* in tea plants [71]. This success in field trials paves the way for broader applications in both agricultural pest control and public health settings [96] (Sundararajan and Kumari 2017). Thakur et al. [96] delved into the world of root-knot nematodes (*Meloidogyne incognita*) plaguing tomato crops, investigating the nematicidal

potential of biosynthesized gold nanoparticles. These tiny warriors proved effective against the nematodes, leaving plant growth unscathed. While the insecticidal prowess of gold nanoparticles against pesky mosquitoes has been well-documented, their ability to combat crop-threatening nematodes presents a groundbreaking opportunity. Bacterial culprits like *Acidovorax avenae* subspecies *citrulli,* responsible for fruit blotch, can be apprehended through a rapid on-site DNA sensor. This ingenious tool harnesses the power of colloidal gold nanoparticles (AuNPs) attached to single-stranded DNA probes designed to zero in on a specific genetic sequence within the bacterial villain. The probe, meticulously crafted to target the internal transcribed spacer (ITS) region, can reveal the presence of even small amounts of the pathogen (just 4 nM!). For a more precise analysis, the sensor delves deeper, allowing for semi-quantitative detection based on optical density and target DNA concentration, reaching an impressive limit of detection of 0.48 nM [105]. To rapidly detect *Pseudomonas syringae* pathovars, a notorious plant pathogen, researchers developed a colorimetric assay using AuNP-labeled DNA probes. These probes, designed against the conserved N-terminal region of the hrcV gene, changed color from red to purple upon binding to the pathogen's DNA in a sample, allowing for visual identification of the culprit bacteria. This innovative approach offers a simple and sensitive tool for early detection and control of crop diseases [98]. Echoing the success with *Pseudomonas syringae*, researchers deployed a similar gold nanoparticle (AuNP) strategy to directly detect the notorious potato wilt culprit, *Ralstonia solanacearum*, in soil samples. This innovative assay bypassed the need for time-consuming DNA amplification by employing a colorimetric trick. A specific AuNP-bound probe, designed to latch onto its unique target within the R. solanacearum genome, held the key. When the probe found its match, the solution remained a vibrant red under acidic conditions, signifying the presence of the villainous bacterium. But in the absence of its target, the lonely probes caused the AuNPs to clump together, transforming the solution to a telltale purple. This simple yet ingenious color shift offered a rapid, sensitive, and specific tool for identifying *R. solanacearum*, paving the way for early disease diagnosis and potentially saving countless potato crops from the wilt's deadly grip [49]. Black Sigatoka, a global scourge on banana plantations caused by the hemi-biotrophic fungus *Pseudocercospora fijiensis*, demands rapid and accurate detection to mitigate losses and prevent its relentless spread. Luna-Moreno et al. [57] took a novel approach to tackling Black Sigatoka by developing a highly specific surface plasmon resonance (SPR) immuno-sensor. Imagine this sensor as a microscopic bouncer at a nightclub, only instead of checking IDs, it's trained to recognize a specific protein (HF1) found on the surface of *P. fijiensis* cells. This bouncer, built on a gold chip, uses light to detect the fungal pathogen with exceptional accuracy. This innovative method surpasses previous detection techniques, boasting a limit of detection (LOD) as low as one single copy of *L. maculans* DNA. This remarkable sensitivity can help quickly identify *L. maculans* outbreaks before they cause significant damage to crop [52]. This study investigated the efficacy of DNA-tagged gold nanoparticles in controlling *Spodoptera litura* Fab. (fall armyworm), a notorious polyphagous pest. Different concentrations (200, 300, 400, and 500 ppm) of the nanoparticles were diluted, and 10 μl of each solution were dispensed onto chickpea-based semi-synthetic diet placed in 5-ml glass vials. Twenty-minute surface-treated diets were then infested with second-instar *S. litura* larvae of consistent age and size. Each concentration of the nanoparticles was tested on 20 *Spodoptera litura* larvae for 30 s, alongside a control group with untreated larvae. Statistical analysis (ANOVA) of larval mortality data confirmed the effectiveness of the particles, with concentrations exceeding 500 ppm inducing 50% mortality. A positive correlation was observed between larval mortality and both particle concentration and time elapsed after treatment. The analysis revealed statistically significant differences in larval mortality across various nanoparticle concentrations. The LC50 values (lethal concentration required for 50% mortality) ranged from 216.91 to 938.95 ppm (95% CI) and increased with larval age [16]. *Bursaphelenchus xylophilus*, the insidious pine wood nematode, relies on a cohort of symbiotic bacteria to wreak havoc on pine trees. To counter this microbial alliance, researchers turned to a fascinating eco-friendly solution: gold nanoparticles (AuNPs) mediated by natural extracts from the *Parkia biglobosa* tree. Both leaf-derived (PL-AuNPs) and stem-derived (PS-AuNPs) AuNPs proved to be potent microbial warriors. Against *Pseudomonas syringae*, *Escherichia coli*, and *Staphylococcus aureus*, PS-AuNPs and PL-AuNPs formed impressive inhibition zones, ranging from 46 to 66 mm—considerably larger than the 40 to 51 mm achieved against *Bacillus anthracis*. Notably, these biosynthesized AuNPs outperformed their inert gold salt counterpart, which showed no antibacterial activity. This study reveals a promising avenue for combating *B. xylophilus* and protecting pine forests. By hindering the growth of its symbiotic bacteria, PS-AuNPs and PL-AuNPs could effectively shorten the lifespan of the nematode, mitigating its destructive impact. Moreover, this research reinforces the value of exploring ethnobotanical knowledge in the development of eco-friendly pesticides, paving the way for a sustainable future for our forests [20]. Phytosynthesized gold nanoparticles (GNPs) have emerged as a game changer in pest control. When combined with the Bt-deltamethrin duo, these tiny plant-powered warriors significantly boost their effectiveness against pesky *Ae. aegypti* and *An. stephensi* mosquitoes. This remarkable feat likely stems from GNPs' remarkable ability to disarm trypsins and proteases, acting as a shield for deltamethrin against microbial foes. By embracing this green approach, we can not only slash the number of chemical pesticides needed to curb larval populations, but also breathe easier knowing we're protecting our environment and potentially stalling the development of pesky pesticide resistance [65].

4 Yield and Production

The potential of gold nanoparticles (GNPs) to elevate plant yield is undeniable, although limited practical applications have been documented. Cost remains a significant hurdle, as the expense of spraying them at 10 mg L^{-1} often outweighs crop output. Yet, the *B. juncea* seedling study, where foliar GNP application stimulated both growth parameters and seed yield, paints a promising picture for a future where engineered GNPs could play a transformative role in enhancing global food security [7]. Studies on the impact of gold nanoparticles on plant growth revealed intriguing results. In *Brassica juncea*, a modest foliar spray with 25 ppm gold nanoparticles boosted oil, reducing sugar, and total sugar content, while also improving the plant's redox status. Interestingly, seed germination thrived at this concentration but faltered at higher nanoparticle doses. This "antagonistic effect" might hinder ethylene activity, potentially through binding or absorption onto the nanoparticles, leading to increased leaf production. Moreover, treating *B. juncea* with around 10 ppm gold nanoparticles yielded a noteworthy 19% increase in seed production. However, the tale diverges for *Hordeum vulgare*. Here, rising concentrations of gold nanoparticles (0–10 mg $L-^{1}$) painted a grim picture. Fresh biomass dwindled, leaves donned a sickly yellow hue, and roots darkened ominously. Clearly, the optimal dosage and impact vary greatly between species, highlighting the delicate dance between beneficial and detrimental effects of gold nanoparticles in the plant kingdom [24]. While concerns linger about the impact of gold nanoparticles on some plant species, their potential to boost biomass in controlled environments cannot be ignored. A study explored this potential by introducing different ratios of gold (Au) and silver (Ag) nanoparticles to submerged plant cell cultures. When combined with a plant growth hormone (NAA), a specific ratio of 3 parts Ag to 1 part Au proved to be the magic recipe. Compared to the control group, this combination yielded a stunning 40% increase in fresh biomass and nearly 200% in dry biomass. Interestingly, the presence of Au, even without Ag, seemed to consistently promote biomass accumulation throughout the entire culture cycle, suggesting a more independent role in this process. Moreover, specific combinations of AgAu nanoparticles with NAA further enhanced protein content and the activity of antioxidant enzymes, hinting at a possible mechanism beyond simple cell proliferation. These findings, while confined to lab settings, offer a tantalizing glimpse into the potential of harnessing gold nanoparticles for targeted biomass optimization in agriculture, paving the way for future research to bridge the gap between in vitro and in vivo applications [23]. The impact of gold nanoparticles (GNPs) on plant growth and yield has been a subject of intense research, with some intriguing results. In a study involving Arabidopsis thaliana, a small flowering plant widely used in scientific research, exposure to 24 nm size GNPs at a concentration of 10 μg/ml resulted in a threefold increase in total seed yield compared to the control. This remarkable boost suggests that GNPs could potentially play a role in enhancing agricultural productivity [50]. Gold nanoparticles (AuNPs) seem to hold surprising potential for boosting plant growth and yield, as evidenced by studies across various species. Take onions, for example. Researchers treated onion seeds with AuNPs and observed a remarkable 63.2% emergence rate, compared to just 37.4% for untreated seeds. That's a nearly 70% improvement in seedling emergence! (Image showcasing a vibrant patch of healthy onion seedlings from AuNP-treated seeds compared to a sparser group from untreated seeds would be great here.) But the benefits don't stop there. Onions treated with AuNPs also enjoyed a 23.9% increase in average yield, meaning they produced significantly more bulbs than their untreated counterparts [2]. The potential of gold nanoparticles (AuNPs to boost plant growth isn't just a glimmer,it's rapidly blossoming into a field of exciting possibilities. Take Calendula officinalis, for instance. When treated with a mere 10 ppm of AuNPs, these vibrant marigolds experienced a growth spurt across nearly every parameter imaginable (Srivastava et al., 2021. Furthermore, the utilization of AuNPs at a concentration of 10 mg/L markedly boosted the activities of antioxidant enzymes (ascorbate peroxidase, catalase, and guaiacol peroxidase in wheat. This enhancement translated into improved growth parameters, including increased fresh and dry biomass, expanded leaf area, and elevated levels of chlorophyll (a, b, as well as total chlorophyll content (Manaf et al., 2021. In an investigation focusing on rhizospheric *Pseudomonas monteilii*, the concentration-dependent impact of biogenic gold nanoparticles (AuNPs) on the production of indole-3-acetic acid (IAA) was examined. The AuNPs, synthesized by *Bacillus subtilis* SJ15, were characterized and observed to amplify IAA production by *P. monteilii* specifically at a concentration of 50 μg/mL. Furthermore, the presence of 50 μg/mL AuNPs led to an augmentation in the growth of *Vigna unguiculata* seedlings [69].

5 Pathways of GNPs Arrival in Plants

Despite ongoing debate about the exact routes and mechanisms of nanoparticle uptake and movement within plants, existing research offers fascinating insights. Our review delves into key hypotheses and findings on this topic, tracing the historical exploration of gold nanoparticle penetration in plants since the 1970s. Notably, a growing body of evidence, including studies utilizing radio-fluorescence and mass spectrometry, consistently affirms the ability of gold nanoparticles to enter plant tissues, navigate internal pathways, and accumulate within cells. For example, experiments have shown that even nanoparticles as small as 3 nm can traverse wheat seedlings from leaves to roots within a week, with some even escaping into the soil [29]. In experiments conducted with watermelon seedlings, mass spectrometry analysis detected the presence of gold nanoparticles (GNPs) in both leaves and roots as early as 48 h following the application of solutions containing GNPs of various sizes and shapes [72]. In tobacco roots, incubated with solutions containing 3.5 nm and 18 nm GNPs, X-ray microscopy detected their presence after 3 days [75]. Electron microscopy showed the presence of GNPs in the root cells of brown mustard and watermelon grown in nanoparticle-containing solutions [22]. As a result, gold nanoparticles (GNPs) can infiltrate plant tissues through both the root system and leaves. The mechanisms governing this phenomenon are not fully elucidated and are primarily discussed in a hypothetical manner. It is presumed that nanoparticles enter the root system through water absorption via lateral roots, root hairs, and rhizodermis cells [97]. The primary route for nanoparticle penetration into leaves is thought to be through stomata [51]. The insights provided in the referenced papers may contribute to identifying the pathway for the translocation of nanoparticles from xylem to phloem. It is theorized that, owing to their amphiphilic properties, nanoparticles might overcome the lipophilic cuticular layer or disrupt the wax layer of the cuticle, facilitating their reach to the leaf [9]. Beyond root and leaf surfaces, various above-ground structures like hydathodes, trichomes, and lenticels provide additional entry points for nanoparticles into plant organs. The cell wall, however, represents a significant barrier, with pores ranging from 5 to 20 nm in size permitting access only to nanoparticles smaller than 20 nm. Larger nanoparticles can potentially overcome this limitation via two mechanisms. Firstly, inducing the disruption of cell wall filaments to create temporary, size-accommodating channels, and directly breaching the plasma membrane. Subsequent intraplantar transport occurs through either the apoplastic pathway, utilizing the extracellular space between cell walls, or the symplastic pathway, involving movement through the interconnected cytoplasmic network via plasmodesmata with diameters of approximately 50 nm. This latter route necessitates the ability of nanoparticles to traverse the plasma membrane and plasmodesmata checkpoints [100]. While traversing cell walls is a significant hurdle, the crux of nanoparticle translocation within plants lies in breaching the final frontier: the plasma membrane (PM). Endocytosis reigns supreme as the likely conqueror, engulfing nanoparticles in membrane vesicles for cellular entry [53]. Moscatelli et al. [63] beautifully demonstrate this in tobacco pollen tubes, where electron microscopy captured GNPs internalized through swift endocytic dances. Intriguingly, amphiphilic nanoparticles might waltz directly through the PM's lipid bilayer, bypassing the endocytic tango altogether [66]. But their internal voyage remains a tightrope walk, dictated by size, surface chemistry, and charge. Positively charged GNPs find eager partners in plant roots, while their negatively charged comrades gracefully migrate towards stems and leaves [9]. This intricate waltz continues within the plant, where nanoparticles don a "protein-lipid-carbohydrate crown" [53], charming their way through subsequent biological barriers.

6 Conclusion

Nanotechnology, an evolving scientific discipline, has unveiled a multitude of potential applications, with gold nanoparticles emerging as particularly promising. The application of engineered gold nanoparticles in plant production and enhancement has exhibited considerable potential, manifesting both favorable and adverse effects on plant systems. Gold nanoparticles permeate plants through a mechanism dependent on their size, exerting a dual influence on growth and biomass. They possess the capability to either stimulate growth or hinder it by inducing physiological, biochemical, and molecular imbalances that result in oxidative stress. Exposure to gold nanoparticles has been observed to exhibit heightened inhibition of reactive oxygen species (ROS), indicating the efficient scavenging of free radicals by these nanoparticles. Additionally, the activity of antioxidant enzymes is elevated, facilitating stress modulation in plants. Beyond their impact on ROS and antioxidant systems, gold nanoparticle exposure has been found to alter microRNA and gene expression in plants. This suggests a complex regulatory role of gold nanoparticles at the molecular level, influencing key genetic processes within the plant. The application of gold nanoparticles in fruiting plants holds the potential to enhance both the quality and quantity of fruits and vegetables. However, economic feasibility considerations must precede their widespread use. The calculation of costs associated with the production and application of engineered gold nanoparticles is essential to assess the viability of their incorporation into agricultural practices. Despite the promising aspects, the escalating production, use, and disposal of engineered gold nanoparticles raises concerns about their release into the ecosystem. Consequently, there is an imperative need for systematic investigations to comprehensively evaluate the impact of gold nanoparticles on plant systems and their surrounding ecosystems. Such studies are essential for establishing a robust understanding of the potential risks and benefits associated with the use of gold nanoparticles in agriculture and ensuring their responsible application in the broader environment.

References

1. Abd El-Aziz AR, Al-Othman MR (2019) Gold nanoparticles biosynthesis using *Zingiber officinale* and their impact on the growth and chemical composition of lentil (*Lens culinaris* medic.). Pak J Bot 51(2):443–50
2. Acharya P, Jayaprakasha GK, Crosby KM, Jifon JL, Patil BS (2019) Green-synthesized nanoparticles enhanced seedling growth, yield, and quality of onion (*Allium cepa* L.). ACS Sustain Chem & Eng 7(17):14580–90
3. Ai K, Liu Y, Lu L (2009) Hydrogen-bonding recognition-induced color change of gold nanoparticles for visual detection of melamine in raw milk and infant formula. J Am Chem Soc 131(27):9496–9497
4. Anamika, Patra A, Shehzad S, Rani A, Sharma P, Mohammad KF, Mehta S (2023) Microbial elicitors for priming plant defense mechanisms. InSustainable Agric Rev 60. Microbial Processes in Agriculture, pp 175–196. Springer Nature Switzerland, Cham
5. Annamalai A, Christina VL, Sudha D, Kalpana M, Lakshmi PT (2013) Green synthesis, characterization and antimicrobial activity of Au NPs using *Euphorbia hirta* L. leaf extract. Colloids Surfaces B Biointerfaces 108:60–5
6. Ansari MJ (2019) Transport phenomenon of nanoparticles in animals and humans. Diffusion Foundations 9(23):173–186
7. Arora S, Sharma P, Kumar S, Nayan R, Khanna PK, Zaidi MGH (2012) Gold-nanoparticle induced enhancement in growth and seed yield of *Brassica juncea*. Plant Growth Regul 66:303
8. Asli S, Neumann PM (2009) Colloidal suspensions of clay or titanium dioxide nanoparticles can inhibit leaf growth and transpiration via physical effects on root water transport. Plant, Cell Environ 32(5):577–584
9. Avellan A, Yun J, Zhang Y, Spielman-Sun E, Unrine JM, Thieme J, Li J, Lombi E, Bland G, Lowry GV (2019) Nanoparticle size and coating chemistry control foliar uptake pathways, translocation, and leaf-to-rhizosphere transport in wheat. ACS Nano 13(5):5291–5305
10. Bachheti RK, Bachheti A, Husen A (2024) Metal and metal-oxide based nanomaterials (*Synthesis, Agricultural, Biomedical and Environmental Interventions*). Springer Nature Singapore Pte Ltd., 152 Beach Road, #21-01/04 Gateway East, Singapore 189721, Singapore
11. Bai LY, Zhang YP, Chen J, Zhou XM, Hu LF (2010) Rapid, sensitive and selective detection of pymetrozine using gold nanoparticles as colourimetric probes. Micro Nano Lett 5(5):304–308
12. Balalakshmi C, Gopinath K, Govindarajan M, Lokesh R, Arumugam A, Alharbi NS, Kadaikunnan S, Khaled JM, Benelli G (2017) Green synthesis of gold nanoparticles using a

cheap *Sphaeranthus indicus* extract: impact on plant cells and the aquatic crustacean *Artemia nauplii*. J Photochem Photobiol, B 1(173):598–605
13. Bandi R, Dadigala R, Alle M (2023) Emerging role of gold nanoparticles for healthier crop plants growth and enhanced yield. In: Engineered nanomaterials for sustainable agricultural production, soil improvement and stress management , pp 125–143. Academic Press
14. Barrena R, Casals E, Colón J, Font X, Sánchez A, Puntes V (2009) Evaluation of the ecotoxicity of model nanoparticles. Chemosphere 75(7):850–857
15. Bodale I, Teliban G, Ursu EL, Stoleru V, Cazacu A (2019) The influence of gold nanoparticles on germination of carrot seeds. Int Multidiscip Sci GeoConf SGEM 19(6.1):451–8
16. Chakravarthy AK, Bhattacharyya A, Shashank PR, Epidi TT, Doddabasappa B, Mandal SK (2012) DNA-tagged nano gold: a new tool for the control of the armyworm, Spodoptera litura Fab.(Lepidoptera: Noctuidae). Afr J Biotechnol 11(38):9295–301
17. Chawla S, Singh S, Husen A (2023) Smart Nanomaterials targeting pathological hypoxia. Springer Nature Singapore Pte Ltd., 152 Beach Road, #21-01/04 Gateway East, Singapore 189721, Singapore ISBN: 978-0981-99-1717-4. https://doi.org/10.1007/978-981-99-1718-1
18. Das S, Debnath N, Pradhan S, Goswami A (2017) Enhancement of photon absorption in the light-harvesting complex of isolated chloroplast in the presence of plasmonic gold nanosol—a nanobionic approach towards photosynthesis and plant primary growth augmentation. Gold Bull 50(3):247–257
19. Das SK, Das AR, Guha AK (2009) Gold nanoparticles: microbial synthesis and application in water hygiene management. Langmuir 25(14):8192–8199
20. Davids JS, Ackah M, Okoampah E, Fometu SS, Guohua W, Jianping Z (2021) Biocontrol of bacteria associated with pine wilt nematode, *Bursaphelenchus xylophilus* by using plant mediated gold nanoparticles. Int J Agric Biol 26(4):517–526
21. Di Pietro P, Strano G, Zuccarello L, Satriano C (2016) Gold and silver nanoparticles for applications in theranostics. Curr Top Med Chem 16(27):3069–3102
22. Dykman L, Khlebtsov N (2017) Gold nanoparticles in biomedical applications. CRC Press
23. Fazal H, Abbasi BH, Ahmad N, Ali M, Shujait Ali S, Khan A, Wei DQ (2019) Sustainable production of biomass and industrially important secondary metabolites in cell cultures of selfheal (*Prunella vulgaris* L.) elicited by silver and gold nanoparticles. Artificial Cells Nanomed Biotechnol 47(1):2553–61
24. Feichtmeier NS, Walther P, Leopold K (2015) Uptake, effects, and regeneration of barley plants exposed to gold nanoparticles. Environ Sci Pollut Res 22:8549–8558
25. Ferrari E, Barbero F, Busquets-Fité M, Franz-Wachtel M, Köhler HR, Puntes V, Kemmerling B (2021) Growth-promoting gold nanoparticles decrease stress responses in Arabidopsis seedlings. Nanomaterials 11(12):3161
26. Gittins DI, Caruso F (2001) Spontaneous phase transfer of nanoparticulate metals from organic to aqueous media. Angew Chem Int Ed 40(16):3001–3004
27. Gopinath K, Gowri S, Karthika V, Arumugam A (2014) Green synthesis of gold nanoparticles from fruit extract of *Terminalia arjuna*, for the enhanced seed germination activity of *Gloriosa superba*. J Nanostruct Chem 4:1–1
28. Gopinath K, Venkatesh KS, Ilangovan R, Sankaranarayanan K, Arumugam A (2013) Green synthesis of gold nanoparticles from leaf extract of *Terminalia arjuna*, for the enhanced mitotic cell division and pollen germination activity. Ind Crops Prod 1(50):737–742
29. Gunjan B, Zaidi MG, Sandeep A (2014) Impact of gold nanoparticles on physiological and biochemical characteristics of *Brassica juncea*. J Plant Biochem Physiol 2(3):1–6
30. Herizchi R, Abbasi E, Milani M, Akbarzadeh A (2016) Current methods for synthesis of gold nanoparticles. Artificial Cells Nanomed Biotechnol 44(2):596–602
31. Hoshyar N, Gray S, Han H, Bao G (2016) The effect of nanoparticle size on in vivo pharmacokinetics and cellular interaction. Nanomedicine 11(6):673–692
32. Husen A (2019) Natural product-based fabrication of zinc oxide nanoparticles and their application. In: Husen A, Iqbal M (Eds.) N*anomaterials* and plant potential. Springer International Publishing AG, Gewerbestrasse 11, 6330 Cham, pp 193–291. https://doi.org/10.1007/978-3-030-05569-1_7

33. Husen A (2020) Introduction and techniques in nanomaterials formulation. In: Husen A, Jaeaid M (Eds.) Nanomaterials for agriculture and forestry applications. Elsevier Inc. 50 Hampshire St., 5th Floor, Cambridge, MA 02139, USA, pp 1–14. https://doi.org/10.1016/B978-0-12-817852-2.00001-9
34. Husen A (2022) Engineered nanomaterials for sustainable agricultural production, soil improvement and stress management. Elsevier Inc., 50 Hampshire St., 5th Floor, Cambridge, MA 02139, USA. https://doi.org/10.1016/C2021-0-00054-7
35. Husen A (2023a) Secondary metabolites based green synthesis of nanomaterials and their applications. Springer Nature Singapore Pte Ltd., 152 Beach Road, #21-01/04 Gateway East, Singapore 189721, Singapore. https://doi.org/10.1007/978-981-99-0927-8
36. Husen A (2023b) Nanomaterials from agricultural and horticultural products. Springer Nature Singapore Pte Ltd., 152 Beach Road, #21-01/04 Gateway East, Singapore 189721, Singapore. https://link.springer.com/book/9789819934348
37. Husen A (2023c) Nanomaterials and Nanocomposites Exposures to Plants (Response, Interaction, Phytotoxicity and Defense Mechanisms). Springer Nature Singapore Pte Ltd., 152 Beach Road, #21–01/04 Gateway East, Singapore 189721, Singapore. https://doi.org/10.1007/978-981-99-2419-6
38. Husen A, Bachheti RK, Bachheti A (2023) current trends in green nano-emulsions (Food, Agriculture and Biomedical Sectors). Springer Nature Singapore Pte Ltd., 152 Beach Road, #21-01/04 Gateway East, Singapore 189721, Singapore
39. Husen A, Iqbal M (2019) Nanomaterials and Plant Potential. Springer International Publishing AG, Gewerbestrasse 11, 6330 Cham, Switzerland. https://doi.org/10.1007/978-3-030-05569-1
40. Husen A, Jawaid M (2020) Nanomaterials for agriculture and forestry applications. Elsevier Inc., 50 Hampshire St., 5th Floor, Cambridge, MA 02139, USA. https://doi.org/10.1016/C2018-0-02349-X
41. Husen A, Siddiqi KS (2014) Phytosynthesis of nanoparticles: concept, controversy and application. Nanoscale Res Lett 9(229):1–24. https://doi.org/10.1186/1556-276X-9-229
42. Husen A, Siddiqi KS (2023) Advances in smart nanomaterials and their applications. Elsevier Inc., 50 Hampshire St., 5th Floor, Cambridge, MA 02139, USA. https://doi.org/10.1016/C2021-0-02202-1
43. Jadczak P, Kulpa D, Bihun M, Przewodowski W (2019) Positive effect of AgNPs and AuNPs in in vitro cultures of *Lavandula angustifolia* Mill. Plant Cell Tissue Organ Culture (PCTOC) 139:191–7
44. Jayaseelan C, Ramkumar R, Rahuman AA, Perumal P (2013) Green synthesis of gold nanoparticles using seed aqueous extract of Abelmoschus esculentus and its antifungal activity. Ind Crops Prod 1(45):423–429
45. Jin-Chul K, Madhusudhan A, Husen A (2021) Smart nanomaterials in biomedical applications. Springer Nature Switzerland AG, Gewerbestrasse 11, 6330 Cham, Switzerland. https://doi.org/10.1007/978-3-030-84262-8
46. Joshi S, Dar AI, Acharya A, Joshi R (2022) Charged gold nanoparticles promote in vitro proliferation in nardostachys jatamansi by differentially regulating chlorophyll content, hormone concentration, and antioxidant activity. Antioxidants 11(10):1962
47. Kalimuthu K, Cha BS, Kim S, Park KS (2020) Eco-friendly synthesis and biomedical applications of gold nanoparticles: a review. Microchem J 1(152):104296
48. Kapoor D, Yadav S, Sharma MM, Sharma P (2023) Interaction between metal nanoparticles and pgpr on the plant growth and development. In: Nanomaterials and nanocomposites exposures to plants: response, interaction, phytotoxicity and defense mechanisms, pp 327–351. Singapore: Springer Nature Singapore
49. Khaledian S, Nikkhah M, Shams-bakhsh M, Hoseinzadeh S (2017) A sensitive biosensor based on gold nanoparticles to detect *Ralstonia solanacearum* in soil. J Gen Plant Pathol 83:231–239
50. Kumar V, Guleria P, Kumar V, Yadav SK (2013) Gold nanoparticle exposure induces growth and yield enhancement in *Arabidopsis thaliana*. Sci Total Environ 1(461):462–468

51. Larue C, Castillo-Michel H, Sobanska S, Cécillon L, Bureau S, Barthès V, Ouerdane L, Carrière M, Sarret G (2014) Foliar exposure of the crop *Lactuca sativa* to silver nanoparticles: evidence for internalization and changes in Ag speciation. J Hazard Mater 15(264):98–106
52. Lee JI, Jang SC, Chung J, Choi WK, Hong C, Ahn GR, Kim SH, Lee BY, Chung WJ (2021) Colorimetric allergenic fungal spore detection using peptide-modified gold nanoparticles. Sens Actuators, B Chem 15(327):128894
53. Li H, Ye X, Guo X, Geng Z, Wang G (2016Aug) Effects of surface ligands on the uptake and transport of gold nanoparticles in rice and tomato. J Hazard Mater 15(314):188–196
54. Lipşa FD, Ursu EL, Ursu C, Ulea E, Cazacu A (2020) Evaluation of the antifungal activity of gold–chitosan and carbon nanoparticles on *Fusarium oxysporum*. Agronomy 10(8):1143
55. Loyal A, Pahuja SK, Rani N, Pooja, Srivastava RK, Sharma P (2023a) Titanium oxide nanoparticles: plant response, interaction, phytotoxicity, and defence mechanisms. In: Nanomaterials and nanocomposites exposures to plants: response, interaction, phytotoxicity and defense mechanisms, pp. 263–284. Singapore: Springer Nature Singapore
56. Loyal A, Pahuja SK, Sharma P, Malik A, Srivastava RK, Mehta S (2023b) Potential environmental and human health implications of nanomaterials used in sustainable agriculture and soil improvement. In: Engineered nanomaterials for sustainable agricultural production, soil improvement and stress management, pp 387–412. Academic Press
57. Luna-Moreno D, Sánchez-Álvarez A, Islas-Flores I, Canto-Canche B, Carrillo-Pech M, Villarreal-Chiu JF, Rodríguez-Delgado M (2019) Early detection of the fungal banana black sigatoka pathogen *Pseudocercospora fijiensis* by an SPR immunosensor method. Sensors 19(3):465
58. Madhusudhan A, Purohit SD, Prasad R, Husen A (2024) Functional smart nanomaterials and their theranostics approaches. Springer Nature Singapore Pte Ltd., 152 Beach Road, #21-01/04 Gateway East, Singapore 189721, Singapore
59. Mahakham W, Theerakulpisut P, Maensiri S, Phumying S, Sarmah AK (2016) Environmentally benign synthesis of phytochemicals-capped gold nanoparticles as nanopriming agent for promoting maize seed germination. Sci Total Environ 15(573):1089–1102
60. Malarkodi C, Rajeshkumar S, Annadurai G (2017) Detection of environmentally hazardous pesticide in fruit and vegetable samples using gold nanoparticles. Food Control 1(80):11–18
61. Manaf A, Wang X, Tariq F, Jhanzab HM, Bibi Y, Sher A, Razzaq A, Fiaz S, Tanveer SK, Qayyum A (2021) Antioxidant enzyme activities correlated with growth parameters of wheat sprayed with silver and gold nanoparticle suspensions. Agronomy 11(8):1494
62. Mandal AK, Ghorai S, Husen A (2023) Functionalized smart nanomaterials for point-of-care testing. Springer Nature Singapore Pte Ltd., 152 Beach Road, #21–01/04 Gateway East, Singapore 189721, Singapore
63. Moscatelli A, Ciampolini F, Rodighiero S, Onelli E, Cresti M, Santo N, Idilli A (2007) Distinct endocytic pathways identified in tobacco pollen tubes using charged nanogold. J Cell Sci 120(21):3804–3819
64. Muthuvel A, Adavallan K, Balamurugan K, Krishnakumar N (2014) Biosynthesis of gold nanoparticles using Solanum nigrum leaf extract and screening their free radical scavenging and antibacterial properties. Biomed Prev Nutr 4(2):325–332
65. Narkhede CP, Suryawanshi RK, Patil CD, Borase HP, Patil SV (2016) Use of protease inhibitory gold nanoparticles as a compatibility enhancer for Bt and deltamethrin: A novel approach for pest control. Biocatal Agric Biotechnol 1(8):8–12
66. Navarro E, Baun A, Behra R, Hartmann NB, Filser J, Miao AJ, Quigg A, Santschi PH, Sigg L (2008) Environmental behavior and ecotoxicity of engineered nanoparticles to algae, plants, and fungi. Ecotoxicology 17:372–386
67. Nile SH, Thiruvengadam M, Wang Y, Samynathan R, Shariati MA, Rebezov M, Nile A, Sun M, Venkidasamy B, Xiao J, Kai G (2022) Nano-priming as emerging seed priming technology for sustainable agriculture—recent developments and future perspectives. Journal of nanobiotechnology. 20(1):1–31
68. Osonga FJ, Eshun GB, Sadik OA (2022) Ligand effect on controlling the synthesis of branched gold nanomaterials against fusarium wilt diseases. RSC Adv 12(49):31855–31868

69. Panichikkal J, Thomas R, John JC, Radhakrishnan EK (2019) Biogenic gold nanoparticle supplementation to plant beneficial *Pseudomonas monteilii* was found to enhance its plant probiotic effect. Curr Microbiol 1(76):503–509
70. Parveen A, Mazhari BB, Rao S (2016) Impact of bio-nanogold on seed germination and seedling growth in *Pennisetum glaucum*. Enzyme Microb Technol 1(95):107–111
71. Ponmurugan P (2017) Biosynthesis of silver and gold nanoparticles using *Trichoderma atroviride* for the biological control of Phomopsis canker disease in tea plants. IET Nanobiotechnol 11(3):261–267
72. Raliya R, Franke C, Chavalmane S, Nair R, Reed N, Biswas P (2016) Quantitative understanding of nanoparticle uptake in watermelon plants. Front Plant Sci 26(7):1288
73. Rani K, Rani A, Sharma P, Dahiya A, Punia H, Kumar S, Sheoran S, Banerjee A (2022) Legumes for agroecosystem services and sustainability. In: Advances in legumes for sustainable intensification, pp 363–380. Academic Press
74. Rani K, Sharma P, Kumar S, Wati L, Kumar R, Gurjar DS, Kumar D, Kumar R (2019) Legumes for sustainable soil and crop management. Sustain Manage Soil Environ 193–215.
75. Sabo-Attwood T, Unrine JM, Stone JW, Murphy CJ, Ghoshroy S, Blom D, Bertsch PM, Newman LA (2012) Uptake, distribution and toxicity of gold nanoparticles in tobacco (*Nicotiana xanthi*) seedlings. Nanotoxicology 6(4):353–360
76. Sangwan S, Sharma P, Wati L, Mehta S. Effect of chitosan nanoparticles on growth and physiology of crop plants. In: Engineered nanomaterials for sustainable agricultural production, soil improvement and stress management, pp 99–123. Academic Press
77. Shacklette HT, Lakin HW, Hubert AE, Curtin GC (1949) Absorption of gold by plants. US Department of the Interior, Geological Survey; Washington, DC
78. Shah V, Belozerova I (2009) Influence of metal nanoparticles on the soil microbial community and germination of lettuce seeds. Water Air Soil Pollut 197:143–148
79. Sharma MM, Kapoor D, Rohilla R, Sharma P (2023) Nanomaterials and their toxicity to beneficial soil microbiota and fungi associated plants rhizosphere. In: Nanomaterials and nanocomposites exposures to plants: response, interaction, phytotoxicity and defense mechanisms, pp 353–380. Springer Nature Singapore, Singapore
80. Sharma MM, Sharma P, Kapoor D, Beniwal P, Mehta S (2021) Phytomicrobiome community: an agrarian perspective towards resilient agriculture. Plant Performance Under Environmental Stress: Hormones, Biostimulants and Sustainable Plant Growth Management 493–534
81. Sharma P, Meyyazhagan A, Easwaran M, Sharma MM, Mehta S, Pandey V, Liu WC, Kamyab H, Balasubramanian B, Baskaran R, Klemeš JJ (2022) Hydrogen sulfide: a new warrior in assisting seed germination during adverse environmental conditions. Plant Growth Regul 98(3):401–420
82. Sharma P, Pandey V, Sharma MM, Patra A, Singh B, Mehta S, Husen A (2021) A review on biosensors and nanosensors application in agroecosystems. Nanoscale Res Lett 16:1–24
83. Sharma P, Patra A, Singh B, Mehta S. Microbial rejuvenation of soils for sustainable agriculture. In: Sustainable agriculture reviews 60: microbial processes in agriculture, pp 293–323). Springer Nature Switzerland, Cham
84. Sharma P, Sangwan S, Kaur H, Patra A, Anamika, Mehta S (2023) Diversity and evolution of nitrogen fixing bacteria. Sustainable Agriculture Reviews 60: Microbial Processes in Agriculture 95–120
85. Sharma P, Sangwan S, Kaur H, Patra A, Mehta S (2023) Microbial remediation of agricultural residues. InSustainable Agriculture Reviews 60: Microbial Processes in Agriculture, pp 325–358. Springer Nature Switzerland, Cham
86. Sharma P, Sangwan S, Kumari A, Singh S, Kaur H (2022) Impact of climate change on soil microorganisms regulating nutrient transformation. In: Plant stress mitigators: action and application, pp 145–172. Singapore: Springer Nature Singapore.
87. Sharma P, Sangwan S, Mehta S (2023) Emerging role of phosphate nanoparticles in agriculture practices. In: Engineered nanomaterials for sustainable agricultural production, soil improvement and stress management, pp 71–97. Academic Press

88. Sharma P, Sharma MM, Anamika, Kapoor D, Rani K, Singh D, Barkodia M (2020) Role of microbes for attaining enhanced food crop production. Microbial Biotechnol Basic Res Appl 55–78
89. Sharma P, Sharma MM, Malik A, Vashisth M, Singh D, Kumar R, Singh B, Patra A, Mehta S, Pandey V (2021) Rhizosphere, rhizosphere biology, and rhizospheric engineering. In: Plant growth-promoting microbes for sustainable biotic and abiotic stress management, pp 577–624. Springer International Publishing, Cham
90. Shukla SK, Kumar R, Mishra RK, Pandey A, Pathak A, Zaidi MG, Srivastava SK, Dikshit A (2015) Prediction and validation of gold nanoparticles (GNPs) on plant growth promoting rhizobacteria (PGPR): a step toward development of nano-biofertilizers. Nanotechnol Rev 4(5):439–448
91. Singh A, Sharma P, Kumari A, Kumar R, Pathak DV (2019) Management of root-knot nematode in different crops using microorganisms. Plant Biotic Interactions: State of the Art 85–99
92. Singh S, Sangwan S, Sharma P, Devi P, Moond M (2021) Nanotechnology for sustainable agriculture: an emerging perspective. J Nanosci Nanotechnol 21(6):3453–3465
93. Soloviev M (ed) (2012) Nanoparticles in biology and medicine: methods and protocols. Humana Press, New York
94. Srivastava RK, Chand S, Bhuj BD, Kumar A, Sharma R, Belwal S (2021) Assessment of growth, flowering and seed yield in Calendula (*Calendula officinalis* L.) as influenced by gold-nanoparticle. Assessment 45(1)
95. Taghiyari HR, Morrell JF, Husen A (2023) Emerging Nanomaterials (*Opportunities and Challenges in Forestry Sectors*). Springer Nature Switzerland AG, Gewerbestrasse 11, 6330 Cham, Switzerland. https://doi.org/10.1007/978-3-031-17378-3
96. Thakur RK, Dhirta B, Shirkot P (2018) Studies on effect of gold nanoparticles on Meloidogyne incognita and tomato plants growth and development. BioRxiv 26:428144
97. Tripathi DK, Singh S, Singh S, Pandey R, Singh VP, Sharma NC, Prasad SM, Dubey NK, Chauhan DK (2017) An overview on manufactured nanoparticles in plants: uptake, translocation, accumulation and phytotoxicity. Plant Physiol Biochem 1(110):2–12
98. Vaseghi A, Safaie N, Bakhshinejad B, Mohsenifar A, Sadeghizadeh M (2013) Detection of *Pseudomonas syringae* pathovars by thiol-linked DNA–Gold nanoparticle probes. Sens Actuators, B Chem 1(181):644–651
99. Venzhik YV, Moshkov IE, Dykman LA (2021) Gold nanoparticles in plant physiology: principal effects and prospects of application. Russ J Plant Physiol 68(3):401–412
100. Wang P, Lombi E, Zhao FJ, Kopittke PM (2016) Nanotechnology: a new opportunity in plant sciences. Trends Plant Sci 21(8):699–712
101. Wassie AT, Ahmed IN, Bachheti A, Husen A, Bachheti RK (2024a) Plant extract-assisted green synthesis and characterization of copper nanocomposite. In: Bachheti RK, Bachheti A, Husen A (eds) Metal and metal-oxide based nanomaterials. Smart Nanomaterials Technology. Springer, Singapore, pp 43–66. https://doi.org/10.1007/978-981-99-7673-7_3
102. Wassie AT, Ahmed, IN, Bachheti A, Husen A, Bachheti RK (2024c) Role of zinc oxide nanomaterials for photocatalytic degradation of environmental pollutants. In: Bachheti RK, Bachheti A, Husen A (eds) Metal and metal-oxide based nanomaterials. smart nanomaterials technology. Springer, Singapore. pp 287–311. https://doi.org/10.1007/978-981-99-7673-7_14
103. Wassie AT, Pandey DP, Bachheti A, Husen A, Deepti Singh Vashishth, Bachheti RK (2024b) Use of green synthesized platinum nanoparticles for biomedical applications. In: Bachheti RK, Bachheti A, Husen A (eds) Metal and metal-oxide based nanomaterials. Smart nanomaterials technology. Springer, Singapore, pp 271–285. https://doi.org/10.1007/978-981-99-7673-7_13
104. Zhang X, Sun Z, Cui Z, Li H (2014) Ionic liquid functionalized gold nanoparticles: Synthesis, rapid colorimetric detection of imidacloprid. Sens Actuators, B Chem 1(191):313–319
105. Zhao W, Lu J, Ma W, Xu C, Kuang H, Zhu S (2011) Rapid on-site detection of *Acidovorax avenae* subsp. citrulli by gold-labeled DNA strip sensor. Biosensors Bioelectronics 26(10):4241–4

Mitigation of Metal Toxicity in Plants Using Gold Nanoparticles

Mohammad Areeb Siddiqui, Archana Vimal, Swati Sharma, Poonam Sharma, and Reena Vishvakarma

Abstract In recent times, the emergence of trace elements in the environment at potentially harmful amounts has become a global concern. High potentially hazardous levels of these elements in the environment are building up due to anthropogenic activities like construction, mining, and population growth. Plants that are exposed to settings contaminated with metals suffer significant effects on their vegetative and reproductive growth, which ultimately affects crop productivity and performance. Finding ways to lessen the stress that hazardous substances create in plants that are important to agriculture is therefore imperative. In this regard, it is well known that metal nanoparticles, such as those of zinc and iron, can reduce metal toxicity and encourage plant development under a variety of stressful circumstances. Using gold nanoparticles (AuNPs) to mitigate the deadly effects of metals and promote crop development is a relatively new strategy. Astaxanthin synthesized gold nanoparticles (Ast-AuNPs) and AuNP produced from melatonin (Mel-AuNPs) are two examples of nanocomposites that contain AuNPs, however they have been only researched in rice to reduce metal toxicity, primarily of cadmium. With an emphasis on potential mechanisms to lessen metal toxicity in plants, this chapter aims to help researchers understand the potential of AuNPs to reduce metal stress and enhance plant growth and foster future investigation of full potential of these nanoparticles for the alleviation of metal stress in crops and other agricultural fields.

Keywords Metal toxicity · Nanoparticle · Gold nanoparticle · Plant growth · Reactive oxygen species

M. A. Siddiqui · A. Vimal · P. Sharma · R. Vishvakarma (✉)
Department of Bioengineering, Integral University, Lucknow, UP 226026, India
e-mail: reenav@iul.ac.in

S. Sharma
Department of Biosciences, Integral University, Lucknow, UP 226026, India

A. Husen (ed.), *Plant Response to Gold Nanoparticles*, Smart Nanomaterials Technology, https://doi.org/10.1007/978-981-97-7174-5_9

1 Introduction

Heavy metals in the environment are a major worldwide concern due to anthropogenic and natural events that lead to their entry into water bodies and soil. Considering that majority of them play no substantial role in the functions of either plant or human body, they are equally detrimental to the survival of animals and plants alike when present in excess. Once they enter the body, they may lead to several problems due to accumulation for extended periods due to their toxic and non-biodegradable nature. For instance, in plants, heavy metals harm the roots, leaves, and cellular components by obstructing vital biochemical processes like photosynthesis and mineral absorption [55]. Heavy metal concentrations that are too high have a negative impact on plant metabolism, which in turn degrades the quality and quantity of food produced. When heavy metals enter human tissues through a variety of absorption mechanisms, including oral consumption, cutaneous contact, direct ingestion, diet through the soil-food cycle, and inhalation, their health may be adversely affected [64].

Some metals in low concentrations are needed for the plant metabolism, termed as essential heavy metals [71] including zinc (Zn), copper (Cu), molybdenum (Mo), cobalt (Co), iron (Fe), nickel (Ni), and manganese (Mn) [71]. However, when these essential metals are present in high concentrations, these tend to exhibit toxicity in plants, potentially impeding their overall development [43]. Alternatively, certain metals like arsenic (As), mercury (Hg), lead (Pb), cadmium (Cd), and arsenic (As) are categorized as non-essential heavy metals. Even in small quantities, these non-essential metals can have severe detrimental effects on plants [5].

In order to minimize metal toxicity, a variety of management strategies are being employed to lower the quantity of metal that is available to the plants. Certain conventional methods are expensive or limited to isolating contaminated sites. The application of biology-based technologies to address heavy metal contamination—such as transgenic technology or the use of naturally existing hypermetal accumulator plants—has received a lot of attention in recent times [45, 64]. In the contemporary scientific landscape, nanoparticles are attracting widespread attention owing to their remarkable adsorption and mechanical features along with enormous specific surface area, electrical property, and chemical stability, usage of nanoparticles for metal cleanup in plants have attracted a lot of study interest [47]. Gold nanoparticles (AuNPs) are particularly noteworthy in agricultural research, with experiments revealing both positive and negative impacts on plants [25, 26, 62]. At lower concentrations, AuNPs have been found to have positive effects on plants, enhancing chlorophyll production, plant size, stem diameter, yield, and leaf number. Conversely, higher concentrations of AuNPs can exert opposite effects on plants [4]. Interestingly, recent research suggests that AuNPs have the potential to mitigate metal toxicity in plants, thereby diminishing the harmful effects of metals. However, there is still a need for more experiments in this direction. This chapter, therefore, will be mainly focusing on delineating the twin roles of AuNPs, firstly, its positive aspects in plant growth and metabolism and secondly, combating the metal toxicity in plants.

2 Synthesis of AuNPs

Nanoparticle synthesis can be accomplished via two methods: the "top-down" and "bottom-up" approaches. The latter approach involves the spontaneous self-assembly of smaller atomic or molecular components, driven by natural physical principles or external forces, leading to the creation of larger and well-organized systems. In contrast, the top-down approach begins with a larger piece and progressively employs increasingly precise tools to generate correspondingly smaller structures (Fig. 1).

There are several methods we can choose for AuNPs' synthesis, which include electrochemical, thermal, sono chemical, and chemical methods, of which chemical stands at the top of the list in utility [38, 41, 46, 74]. In chemical method, the chemical reduction of gold precursors is achieved in two ways: (1) reduction through agents and (2) stabilization using agents [76]. In Turkevich method, reduction of $HAuCl_4$ was done by citrate leading to formation of AuNPs of 20 nm diameter size [22, 69]. In this process citrate acted as both reducing and stabilizing agent. Turkevich method was later modified into Turkevich-Frens to give AuNPs in the size range of 15–150 nm (FRENS 1973). In the Brust-Schiffrin method, tetraoctylammonium bromide (TOAB) served as a phase-transfer agent which transferred $AuCl_4$ to toluene phase and is then reduced by $NaBH_4$ which led to the formation of AuNPs which exhibited thermal and air stability [10].

Another method called electrochemical method involves the use of two simple electrode cells, in which redox reaction takes place. This method is believed to be superior in comparison to other discussed methods in this section, because of its affordability, minimal machinery requirement, low processing temperature, and superior quality [23, 53, 54]. In the seeding growth method, AuNPs of size range of 5–40 nm in diameter can be synthesized. In this process, the source of OH ions is trisodium citrate ion and the reducing agent is sodium borohydrate [66].

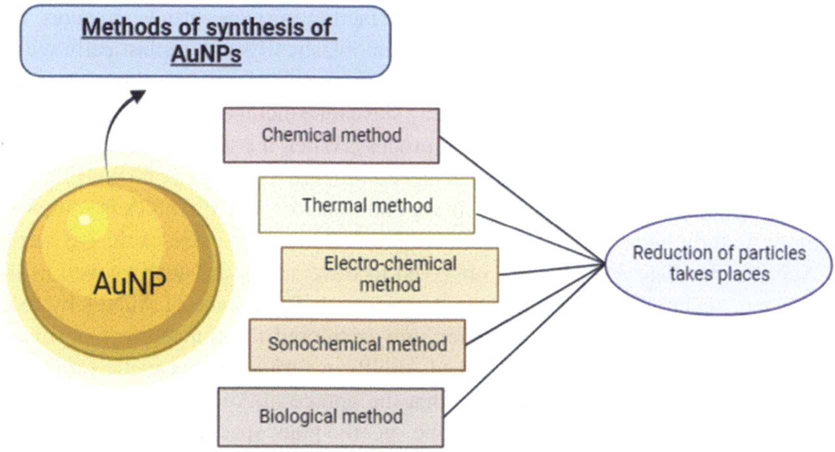

Fig. 1 Methods of synthesis of AuNPs

It is evident from previous studies that nanoparticles obtained via chemical methods are not free from harmful consequences in biomedical applications which limits their use due to which biological synthesis of NPs has been seen as an eco-friendly and cheap alternative. In biological method, nanoparticles are synthesized by plants and their extracts and enzymes and microorganisms [42, 48, 58, 65]. Benefits of using sono chemical method include rapid rate reaction and synthesis of very small nanoparticles. Ionic liquids such as thiol-functionalized ionic liquids act as stabilizing agent for AuNP synthesis [50].

2.1 Plant's Response Towards AuNPs

Presence of AuNPs in plants in miniscule quantities is reported to show enhanced vegetative growth and better fruit yield [51]. Any variation in the size and concentration of the nanoparticles substantially affects the physiology and biochemistry of the plant. The effect also varies with variation in the plant species. Therefore, to understand the plant response towards AuNPs, this section has been incorporated to focus on the concept of uptake of AuNPs by plants and how AuNPs affect the plants in general.

2.2 Uptake and Translocation of AuNPs

The plant cell wall serves as a protective barrier with pores ranging from 2 to 20 nm, posing a challenge for nanoparticles to enter plant cells. This limited pore size makes it difficult for larger particles to traverse. Nevertheless, nanoparticles can find their way into plant cells through various mechanisms. Carrier proteins, aquaporins, endocytosis, and ion channels play essential roles in facilitating nanoparticle transport. The movement of nanoparticles can occur either apoplastically or symplastically within the cell. Additionally, nanoparticles can move from cell-to-cell via plasmodesmata, which are microscopic channels connecting sieve tube members and companion cells via an endoplasmic reticulum-mediated process. When it comes to reaching leaves, the transportation of nanoparticles is achieved through xylem and phloem [56, 75].

Our current understanding of the uptake and translocation of AuNPs is somewhat limited. A study indicates that the root epidermis can act as a size-exclusive barrier to AuNP exposures. AuNPs were observed to traverse the cell wall as well as the membrane, accumulating inside the cell cytoplasm. The uptake of AuNPs by roots appears to be size-dependent. Another study explored the transport of positively charged, neutral, and negatively charged AuNPs in various plants such as ryegrass, pumpkin, radish, and rice. It revealed that the uptake of AuNPs is affected by the surface charge on the nanoparticle and the specific plant species involved [77].

Considering the electrostatic interactions, the negatively charged epidermis cells influence uptake. Positively charged AuNPs exhibit high uptake due to attractive electrostatic interactions, while negatively charged AuNPs show low uptake due to repulsive electrostatic interactions. Neutrally charged AuNPs demonstrate intermediate uptake. Surprisingly, it was found that the translocation of negatively charged AuNPs to the shoot is more efficient compared to positively charged AuNPs [7].

Furthermore, the uptake of AuNPs is influenced by the plant species. For instance, roots of radishes exhibit the highest concentration of AuNPs irrespective of nanoparticle charge, plausible explanation being the dominance of long root hairs, but on the contrary, radishes also display a low translocation factor. Given the similar parameters, rice has the lowest concentration of AuNPs in roots with high translocation factor. Pumpkin shows low root accumulation, and ryegrass behaves similarly to rice. This intricate interplay between nanoparticle characteristics and plant species sheds light on the complexity of nanoparticle uptake and translocation in different plants [79].

2.3 Effect of AuNPs on Plants

In the realm of scientific exploration, the effects of AuNPs on plant growth have been found to be diverse, encompassing both positive and negative outcomes. When introduced in lower concentrations, AuNPs demonstrate a capacity to positively influence plant development. Their presence has been associated with heightened vegetative growth and increased yields of seeds and fruits; a phenomenon attributed to the activation of plant repair mechanisms (Fig. 2). However, it is essential to acknowledge the significance of engineered AuNPs on plant cells can vary, contingent upon factors such as particle shape, size, and concentration [62].

A noteworthy experiment involving *Brassica juncea* shed light on the positive aspects of AuNP treatment. In this study, a concentration of 25 ppm of AuNPs was observed to reduce the time required for seed germination, a phenomenon potentially linked to an increased absorptivity of the seed capsule. Furthermore, at a concentration of 10 ppm, an intriguing 25% increase in the leaves number was noted, though the average leaf area did not exhibit a corresponding increase. It was suggested that AuNPs might influence the ethylene hormone, potentially slowing down its effects. Impressively, plant size and stem diameter showed an increase in seedlings treated with AuNPs, a response correlated with elevated levels of gibberellic acid. The 10 ppm AuNP treatment also led to a higher number of secondary branches per plant, possibly owing to the presence of auxin. Total chlorophyll content saw a significant 29% increase, and there was a remarkable 19% boost in seed yield per plant. These findings collectively indicate that, at an optimal concentration, AuNPs can enhance plant growth by reducing oxidative stress through improved electron exchange/transport reactions, offering promising insights into their potential application [4].

Another exploration, this time involving barley plants, delved into the impact of citrate stabilized AuNPs. The findings indicated a nuanced response, with exposure resulting in varying elemental content. Specifically, there was an increase in

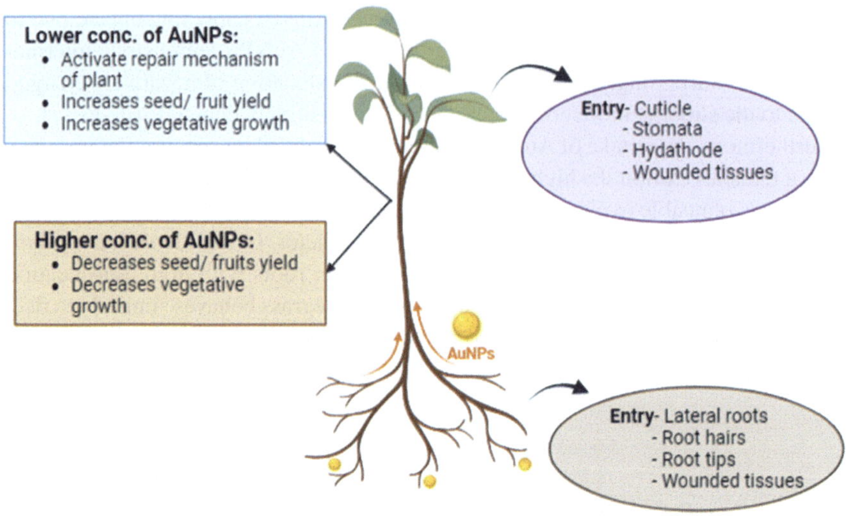

Fig. 2 Entry of AuNPs in plant and their effect on plant metabolism

iron (Fe), potassium (K), calcium (Ca), and zinc (Zn) content, while the content of manganese (Mn) exhibited a decrease. Importantly, this exposure did not lead to any adverse effects on barley, even in the short term. This underscores the complexity of AuNP interactions with plants and the need for a nuanced understanding, taking into consideration factors such as stabilizing agents and exposure duration [18].

In summary, the intricate relationship between AuNPs and plant growth involves a delicate interplay of concentration, particle characteristics, and plant species. Unraveling these complexities not only contributes to our understanding of nanomaterial-plant interactions but also holds promise for harnessing the potential benefits of AuNPs in agricultural practices.

2.4 Effects of Metals on Plants

Plants need an ideal concentration of metals in order to promote growth. However, toxicity in plants can result from concentrations greater than necessary (Fig. 3). There are essential and non-essential metals. Hg, Pb, As, and Cd are examples of non-essential metals whose uptake even at low concentrations can be hazardous to plants. Fe, Cu, Mo, Zn, Co, Fe, Ni, and Mn are examples of necessary micronutrients whose excess concentrations can cause toxicity in plants (Table 1) [5].

Excess Cu concertation in plants induces stress and injury. It can also cause leaf chlorosis, reactive oxygen species (ROS) generation leading to oxidative stress [31]. Root growth reduction in rhodes grass could be seen due to Cu. Similarly, in black bindweed mortality, seed production and biomass reduction could be seen due to

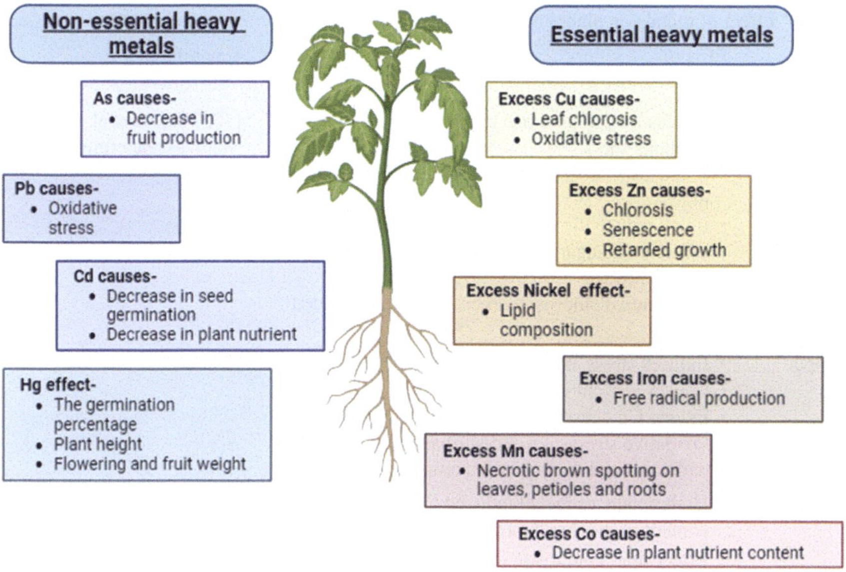

Fig. 3 Detrimental effect of metals on plant growth

Cu toxicity [32, 60]. Zn toxicity in plants could result in senescence, retardation in growth, and chlorosis in younger plants. Excess Zn could lead to deficiency of Cu and Mn in plant shoots [16] and reduction in carotenoid, chlorophyll, germination, sugar, amino acids, and growth in cluster beans. In rye grass, reduction in growth, nutrient content, and photosynthetic energy conversion could be seen [9, 11, 34], whereas reduction of chlorophyll content, photosynthesis, and growth could be seen in pea plants [15].

Reduced shoot growth and biomass depletion could be seen in plants such as barley, oilseed rape, and tomato due to Co toxicity. Excess of Co has a negative effect on the translocation of phosphorus (P), sulfur (S), Mn, Zn, Cu from root to shoot and leaves in cauliflower. Additionally, increased concentration of Co in the leaves of cauliflower can also restrict the concentration of Fe, chlorophyll, catalase activity, and protein. Decrease in plant nutrient content could be seen in tomatoes. Reduction in plant sugars, proteins, and amino acids was visible in radish and mung bean [27, 28]. In *Oryza sativa* shoots, Ni^{+2} ions have detrimental effect on H-ATPase activity and lipid constitution of plasma membrane [57]. Reduction in chlorophyll content and stomatal conductance has been spotted in pigeon pea. Also, Ni decreases enzyme activity which can affect the functioning of Calvin cycle and metabolic pathways involved in CO_2 fixation [61]. In tobacco, canola, soyabean and *Hydrilla verticillate*, Fe toxicity decreases plant photosynthesis and yield while increasing oxidative stress [3]. Excess Fe^{+2} causes free radical production which can damage membranes, DNA, and proteins. Mn and Fe are both essential for plants, absorption of one leads to the deficiency of the other. Commonly necrotic brown spotting on

Table 1 Effect of metals on plant growth and metabolism

Heavy metals	Effects	Key references
Zinc	• Excess Zn can give rise to Mn and Cu deficiencies in plants shoot • It can also cause appearance of purplish red color in leaves • It can reduce the germination, chlorophyll, carotenoid, sugar, amino acids, and growth of cluster beans • Also causes chlorosis, senescence, and retarded growth	[15, 71] Lee et al. (1996), Manivasagaperumal et al. (2011)
Copper	• Induces stress and causes stress to plants • Causes leaf chlorosis • Excess Cu causes oxidative stress and ROS • Oxidative stress causes disturbance of metabolic pathways and damage to macromolecule	[35–37, 67]
Nickel	• High uptake of nickel induced a decline in water content of dicot and monocot plant species • Effect lipid composition	[20, 49, 57]
Iron	• Causes free radical production that impairs cellular structure	[3]
Manganese	• Mn decreases the chlorophyll a and carotenoid content • Causes necrotic brown spotting on leaves, petioles, and roots	[6, 72]
Cobalt	• It has shown the adverse effect on shoot growth and biomass • High level of Co affects the translocation of P, S, Mn, Zn and Cu from roots to tops in cauliflower	[3]
Cadmium	• It reduces the absorption of nitrate and its transport from roots to shoots • It also interferes with the uptake, transport and use of several elements (Ca, Mg, P and K) and water by plants • Causes decrease in seed germination and plant nutrient	[2, 14, 21]
Arsenic	• Causes decrease in fruit production • Arsenic causes stunted growth, chlorosis and wilting	[1, 12, 40]
Lead	• It inhibits seed germination of *Spartiana alterniflora*, *Pinus helipensis* • Early seedling growth was also inhibited by lead in certain legumes • Causes oxidative stress	[44, 52, 68]
Mercury	• High level of mercury can induce visible injuries • Effect the germination percentage, plant height, flowering, and fruit weight	[59, 78]

stems, leaves, and petioles can be seen in Mn toxicity. 'Crinkle leaf' in youngest leaves, petioles are also common in Mn toxicity. Additionally, Mn in pea plant has reported to reduce the content of chlorophyll a and b and decrease the plant growth [17].

As decreases fruit production, reduces the leaf fresh weight in tomato, while its excess in canola leads to wilting and chlorosis [12]. Additionally, in rice, As reduces the dry matter, seed germination, and leaf area [1, 40]. Lead in soyabean, rice, maize, barley, tomato, and certain legumes can inhibit early seedling growth [24]. In maize germination, plant biomass, growth, and protein content could be reduced by lead toxicity. High Pb absorption in plants causes ROS generation which ultimately sums up to oxidative stress. Mercury in plants can affect the plant height, fruit weight and yield, and germination percentage in tomato, it also results in chlorosis in tomatoes [59]. High absorption of Cd in plants decreases seed germination, decrease in plant nutrients, decreased root and shoot length in wheat, while in garlic and maize it decreases the growth of shoot [30, 70]. It is also seen that Cd slows down the ATPase activity in plasma membrane of wheat and sunflower roots [73]. Nitrogen fixation in soyabean plants is also affected due to Cd. It further affects the physiology of plants by hindering the uptake of elements including Mg, K, P, and Ca as well as their transport and utility in plants.

2.4.1 Potential of AuNPs in Mitigation of Metal Toxicity in Plants

Cd, a heavy metal found in rice, can be harmful to plants when absorbed by cereal crops from the soil. This poses a significant health risk as it accumulates via food chain into the human body. Astaxanthin, a natural and antioxidative compound belonging to the carotenoid family, is commonly found in various microorganisms and seafood. A previous study explored the potential of using gold nanoparticles (Ast-AuNPs) synthesized with astaxanthin to reduce Cd toxicity in hydroponically grown rice plants. The research revealed that Ast-AuNPs were more effective in reducing Cd levels compared to using astaxanthin alone. Specifically, when the growth medium was supplemented with Ast-AuNPs (100 μg/mL), there was a significant reduction of 26.2% in Cd concentration in the roots and an impressive 85.9% decrease in the leaves of rice plants. Further observations demonstrated that Ast-AuNPs supplementation had positive effects on the rice plants. It not only restored chlorophyll biosynthesis but also mitigated the oxidative stress induced by Cd by significantly reducing the levels of hydrogen peroxide (H_2O_2), superoxide anion (O^{2-}), and malondialdehyde (MDA). Additionally, Ast-AuNPs played a pivotal role in enhancing the antioxidant enzymes' activity, specifically, peroxidase (POD), superoxide dismutase (SOD), and catalase (CAT). Digging deeper into the molecular level, the study found that the supplementation of Ast-AuNPs into the growth medium inhibited the Cd-induced expression of metal transporter genes including OsIRT1, OsIRT2, OsHMA2, OsHMA3, OsNramp1, and OsNramp5 in rice root which is the major site of Cd uptake. Moreover, Ast-AuNPs were identified as regulators of the free amino acid metabolism and were associated with an increase in the concentration of glutathione and ascorbic acid that function as non-enzymatic antioxidants. In summary, this research demonstrated the potential of Ast-AuNPs in mitigating Cd toxicity in rice seedlings. The mechanisms involved inhibiting the uptake of Cd, scavenging ROS, and combating oxidative stress by enhancing the functions of antioxidants. These findings not only contribute to our understanding of how plants respond to heavy metal pollution but also suggest a novel application for functional gold nanoparticles in addressing environmental challenges, specifically in alleviating heavy metal pollution in plants [13].

In another experimental investigation designed to explore the impact of AuNP synthesized from melatonin (Mel-AuNPs) on cadmium (Cd) toxicity in rice plants, a hydroponic system was employed. The study involved the administration of various treatments, including the introduction of Cd at a concentration of 100 uM, the application of SC-AuNPs at 200 uM, and the supplementation of melatonin at 50, 100, and 200 uM concentrations. Additionally, Mel-AuNPs were administered at the same concentration levels. Upon the introduction of 100 uM Cd, pronounced toxicity was observed in the rice seedlings with reduction in biomass, which droped to 68.8% and 55.1% of the levels observed in the non-treated control (CK). Furthermore, Cd exposed rice seedlings displayed evident symptoms of chlorosis in their young leaves, accompanied by a substantial decline of 51.6% in total chlorophyll content after Cd treatment (10 days) compared to the CK. Interestingly, when 200 uM SC-AuNPs were applied to the growth medium, they failed to reduce the effects of Cd toxicity in the rice plants. This indicates that the conventional application rate of AuNPs was ineffective in mitigating the toxic consequences of Cd exposure in these plants. Conversely, a different outcome emerged when addressing the growth inhibition triggered by Cd through the supplementation of the growth medium with varying concentrations of melatonin. Specifically, the introduction of 200 uM melatonin led to a noteworthy reduction in chlorosis symptoms in the rice leaves. Moreover, the total chlorophyll content experienced a notable restoration, reaching 85.1% of the levels observed in the CK. This level of chlorophyll recovery surpassed that in seedlings treated with Cd alone by a remarkable factor of 1.75. Similarly, rice seedling dry biomass saw a substantial increase, reaching 89.0% of the CK, outperforming seedlings subjected to Cd-only treatment by a factor of 1.61 [29]. The application of melatonin synthesized nanoparticles, Mel-AuNPs emerged as a significantly superior approach for combating the growth inhibition induced by Cd in rice seedlings when compared to other supplements. When treated with 200 uM Mel-AuNPs, the seedling's dry biomass (100.5%) and height (99.7%) were nearly fully restored in comparison to the CK. These outcomes markedly surpassed those seen in rice seedlings treated with melatonin alone, exceeding them by 1.18-fold and 1.23-fold, respectively. Additionally, the total chlorophyll content achieved an impressive recovery, reaching 99.4% of the CK, representing a substantial 1.27-fold increase compared to the treatment involving melatonin alone. Mel-AuNP supplements, therefore, significantly enhanced the rice plants' ability to resist the harmful effects of cadmium (Cd) toxicity. This improvement occurred through several means: reducing the buildup of Cd, restoring the production of chlorophyll, lowering the activity of genes associated with leaf aging, increasing the activity of oxidative stress combating enzymes, and reducing the levels of harmful ROS.

Furthermore, Mel-AuNPs helped alleviate the negative impacts of Cd toxicity by impeding the translocation of Cd to the leaves from the roots and reducing the Cd uptake by the roots. This was achieved by diminishing the activity of genes responsible for Cd transportation and absorption. It's worth noting that when melatonin was used alone without Mel-AuNPs, these beneficial effects were not observed. Consequently, the external application of Mel-AuNPs proves to be an effective method for mitigating the toxic effect of Cd in rice plants which is achieved by increasing melatonin absorption and inhibiting Cd uptake [29].

In an interesting experiment on *Brassica juncea*, it was found that in vivo synthesis of Ag^0 and Au^0 nanoparticles in plants reduces the concentration of Ag^+ and Au^{3+} ions in plants. A significant number of nanoparticles tends to be made within or near chloroplasts. Chloroplasts play a crucial role in producing and storing sugars, particularly glucose and fructose, which are reducing sugars. These sugars are thought to be involved in the process of converting silver and gold salts into metal nanoparticles. Metal nanoparticles are also detected in plant roots, like *B. juncea*, possibly due to sugar storage. Plants rich in reducing sugars in their roots, such as carrots (as opposed to beets, which mainly contain sucrose), might have the ability to generate more nanoparticles in their roots through a reduction process [8].

Other than AuNPs, rich literature is available on prior research to determine the heavy metal toxicity combating the effect of other metallic nanoparticles. For instance, a study was conducted to identify whether magnetic nanoparticles (Fe_3O_4) remove heavy metals and alleviate the negative impacts of heavy metal exposure on the growth and oxidative stress in wheat seedlings. The results clearly demonstrated that supplementation of nano-Fe_3O_4 significantly reduced the concentrations of all tested heavy metals in the solutions after five days, compared to their corresponding solutions without nano-Fe_3O_4 [39]. Previous research had already indicated that the concentration of cadmium could be reduced by increasing the concentration of iron oxide nanoparticles in the soil. In this study, the following order was observed for the efficient removal of the tested heavy metals: Pb > Cu > Cd > Zn [39]. According to another recent finding, the effectiveness of magnetic (Fe_3O_4) nanoparticles in adsorbing heavy metal cations could be attributed to various electrostatic attractions that occur between metal cations and adsorption sites bearing negative charge. Adsorption is considered a conventional but efficient method for environmental detoxification, as it helps eliminating toxic heavy metal ions and bacterial pathogens [63]. Using nano-Fe_3O_4 particles as adsorbents in, primarily, environmental treatments is a practical approach for contaminant removal by applying external magnetic fields. Toxicity reduction by accumulating the heavy metals in wheat seedlings can be attributed to the adsorption potential of metals by nano-Fe_3O_4 as well as to the enrichment of antioxidants in the plants such as POD and SOD [33].

Such comprehensive study on the detoxifying mechanism of AuNP on plants is still lacking. Some recent studies as previously mentioned have reported the potential of AuNPs in combating the heavy metal toxicity in plants but more research needs to be done as studies in this domain are still limited.

3 Conclusion

Metal toxicity in plants is a major concern in the current scenario. The excessive concentration of metal ions in plants alters their physiology and anatomy which ultimately leads to substantial risk to reduction in food production and crop yield. With technological advances, the use of several nanoparticles including AuNPs has been evident in detoxification of plants by removing excess metals. The interaction of nanoparticles with metal ions depends upon their physicochemical features like charge and size, where the anatomical features of plant cells foster the inward movement of nanoparticles within the cell. Though AuNPs have shown efficacy in combating metal toxicity, several recent studies have highlighted the potential toxic effect of AuNPs on plants. Also, the application of AuNPs in agriculture is limited due to high cost of nanoparticle synthesis, which can be corrected by switching to green synthesis methods. Furthermore, the effect of AuNP in combination with molecules such as H_2O_2 and auxin needs to be determined on toxicity reduction. By addressing these issues with comprehensive study in the future, we will be able to better understand how plants react to AuNP and how AuNPs help plants reduce metal toxicity, which is presently limited to a small number of plants, through sustainable methods.

References

1. Abedin MJ, Cotter-Howells J, Meharg AA (2002) Arsenic uptake and accumulation in rice (*Oryza sativa* L.) irrigated with contaminated water. Plant Soil 240:311–319
2. Ahmad I, Akhtar MJ, Zahir ZA, Jamil A (2012) Effect of cadmium on seed germination and seedling growth of four wheat (*Triticum aestivum* L.) cultivars. Pak J Bot 44:1569–1574
3. Arora A, Sairam RK, Srivastava GC (2002) Oxidative stress and antioxidative system in plants. Curr Sci 1227–1238
4. Arora S, Sharma P, Kumar S et al (2012) Gold-nanoparticle induced enhancement in growth and seed yield of Brassica juncea. Plant Growth Regul 66:303–310. https://doi.org/10.1007/s10725-011-9649-z
5. Asati A, Pichhode M, Nikhil K (2016) Effect of heavy metals on plants. Int J Appl Innov Eng Manage 5:56–66. https://doi.org/10.13140/RG.2.2.27583.87204
6. Asrar Z, Khavari-Nejad RA, Heidari H (2005) Excess manganese effects on pigments of Mentha spicata at flowering stage. Arch Agron Soil Sci 51:101–107
7. Balusamy SR, Joshi AS, Perumalsamy H, et al (2023) Advancing sustainable agriculture: a critical review of smart and eco-friendly nanomaterial applications. J Nanobiotechnol 21
8. Beattie IR, Haverkamp RG (2011) Silver and gold nanoparticles in plants: sites for the reduction to metal. Metallomics 3:628–632. https://doi.org/10.1039/c1mt00044f
9. Bonnet M, Camares O, Veisseire P (2000) Effects of zinc and influence of *Acremonium lolii* on growth parameters, chlorophyll a fluorescence and antioxidant enzyme activities of ryegrass (*Lolium perenne* L. cv Apollo). J Exp Bot 51:945–953. https://doi.org/10.1093/jxb/51.346.945
10. Brust M, Walker M, Bethell D, et al (1994) Synthesis of thiol-derivatised gold nanoparticles in a two-phase Liquid–Liquid system. J Chem Soc Chem Commun 801–802. https://doi.org/10.1039/C39940000801
11. Choi J, Pak C, Lee CW (1996) Micro nutrient toxicity in French marigold. J Plant Nutr 19:901–916. https://doi.org/10.1080/01904169609365169

12. Cox MS, Bell PF, Kovar JL (1996) Differential tolerance of canola to arsenic when grown hydroponically or in soil. J Plant Nutr 19:1599–1610
13. Dai S, Wang B, Song Y, et al. (2021) Astaxanthin and its gold nanoparticles mitigate cadmium toxicity in rice by inhibiting cadmium translocation and uptake. Sci Total Environ 786:147496. https://doi.org/10.1016/j.scitotenv.2021.147496
14. Das P, Samantaray S, Rout GR (1997) Studies on cadmium toxicity in plants: a review. Environ Pollut 98:29–36
15. Doncheva S, Stoynova Z, Velikova V (2001) Influence of succinate on zinc toxicity of pea plants. J Plant Nutr 24:789–804. https://doi.org/10.1081/PLN-100103774
16. Ebbs SD, Kochian L V (1997) Toxicity of zinc and copper to brassica species: implications for phytoremediation. J Environ Qual 26:776–781. https://doi.org/10.2134/jeq1997.00472425002600030026x
17. Elamin OM, Wilcox GE (1986) Effect of magnesium and manganese nutrition on muskmelon growth and manganese toxicity. J Am Soc Hortic Sci 111:582–587
18. Feichtmeier NS, Walther P, Leopold K (2015) Uptake, effects, and regeneration of barley plants exposed to gold nanoparticles. Environ Sci Pollut Res 22:8549–8558. https://doi.org/10.1007/s11356-014-4015-0
19. FRENS G (1973) Controlled nucleation for the regulation of the particle size in monodisperse gold suspensions. Nat Phys Sci 241:20–22. https://doi.org/10.1038/physci241020a0
20. Gajewska E, Skłodowska M, Słaba M, Mazur J (2006) Effect of nickel on antioxidative enzyme activities, proline and chlorophyll contents in wheat shoots. Biol Plant 50:653–659
21. Hernandez LE, Carpena-Ruiz R, Garate A (1996) Alterations in the mineral nutrition of pea seedlings exposed to cadmium. J Plant Nutr 19:1581–1598
22. Hu M, Chen J, Li Z-Y et al (2006) Gold nanostructures: engineering their plasmonic properties for biomedical applications. Chem Soc Rev 35:1084–1094. https://doi.org/10.1039/B517615H
23. Huang C-J, Chiu P-H, Wang Y-H et al (2006) Electrochemically controlling the size of gold nanoparticles. J Electrochem Soc 153:D193. https://doi.org/10.1149/1.2358103
24. Hussain A, Abbas N, Arshad F, et al. (2013) Effects of diverse doses of Lead (Pb) on different growth attributes of *Zea-Mays* L.
25. Husen A (2020) Interactions of metal and metal-oxide nanomaterials with agricultural crops: An overview. In: Husen A, Jawaid M (Eds.) Nanomaterials for agriculture and forestry applications. Elsevier Inc. 50 Hampshire St., 5th Floor, Cambridge, MA 02139, USA, pp 167–197
26. Husen A (2023) Nanomaterials and nanocomposites exposures to plants (Response, Interaction, Phytotoxicity and Defense Mechanisms). Springer Nature Singapore Pte Ltd., 152 Beach Road, #21-01/04 Gateway East, Singapore 189721, Singapore. https://doi.org/10.1007/978-981-99-2419-6
27. Jayakumar K, Jaleel CA, Vijayarengan P (2007) Changes in growth, biochemical constituents, and antioxidant potentials in radish (*Raphanus sativus* L.) under cobalt stress. Turk J Biol 31:127–136
28. Jayakumar K, Rajesh M, Baskaran L, Vijayarengan P (2013) Changes in nutritional metabolism of tomato (*Lycopersicon esculantum* Mill.) plants exposed to increasing concentration of cobalt chloride. Int J Food Nutr Saf 4:62–69
29. Jiang M, Dai S, Wang B et al (2021) Gold nanoparticles synthesized using melatonin suppress cadmium uptake and alleviate its toxicity in rice. Environ Sci Nano 8:1042–1056. https://doi.org/10.1039/d0en01172j
30. Jiang W, Liu D, Hou W (2001) Hyperaccumulation of cadmium by roots, bulbs and shoots of garlic (*Allium sativum* L.). Bioresour Technol 76:9–13
31. Katare J, Pichhode M, Kumar N (2015) Growth of Terminalia Bellirica [(Gaertn.) Roxb.] on the Malanjkhand copper mine overburden dump spoil material. Int J Res -GRANTHAALAYAH 3:14–24. https://doi.org/10.29121/granthaalayah.v3.i8.2015.2957
32. KjÆr C, Elmegaard N (1996) Effects of copper sulfate on black bindweed (*Polygonum convolvulus* L.). Ecotoxicol Environ Saf 33:110–117. https://doi.org/10.1006/eesa.1996.0014
33. Konate A, He X, Zhang Z et al (2017) Magnetic (Fe_3O_4) nanoparticles reduce heavy metals uptake and mitigate their toxicity in wheat seedling. Sustainability (Switzerland) 9:1–16. https://doi.org/10.3390/su9050790

34. Kumar N (2015) Effect of copper mining dust on the soil and vegetation in India: a critical review. Int J Modern Sci Eng Technol 2:73–76
35. Kumar N (2015) Growth of *Terminalia bellirica* [(gaertn.) roxb.] on the malanjkhand copper mine overburden dump spoil material. Int J Res Granthaalayah 3:14–24
36. Kumar N (2015) Effect of copper dust on photosynthesis pigments concentration in plants species. Int J Eng Res Manag 2:63–66
37. Lewis S, Donkin ME, Depledge MH (2001) Hsp70 expression in *Enteromorpha intestinalis* (Chlorophyta) exposed to environmental stressors. Aquat Toxicol 51:277–291
38. Mandal S (2014) Synthesis of radioactive gold nanoparticle in surfactant medium. J Radioanal Nucl Chem 299:1209–1212. https://doi.org/10.1007/s10967-013-2602-z
39. Manzoor N, Ahmed T, Noman M, et al (2021) Iron oxide nanoparticles ameliorated the cadmium and salinity stresses in wheat plants, facilitating photosynthetic pigments and restricting cadmium uptake. Sci Total Environ 769:145221. https://doi.org/10.1016/j.scitotenv.2021.145221
40. Marin AR, Pezeshki SR, Masschelen PH, Choi HS (1993) Effect of dimethylarsenic acid (DMAA) on growth, tissue arsenic, and photosynthesis of rice plants. J Plant Nutr 16:865–880
41. Mody V, Siwale R, Singh A, Mody H (2010) Introduction to metallic nanoparticles. J Pharm Bioallied Sci 2:282. https://doi.org/10.4103/0975-7406.72127
42. Mohanpuria P, Rana NK, Yadav SK (2008) Biosynthesis of nanoparticles: technological concepts and future applications. J Nanopart Res 10:507–517. https://doi.org/10.1007/s11051-007-9275-x
43. Monni S, Salemaa M, Millar N (2000) The tolerance of Empetrum nigrum to copper and nickel. Environ Pollut 109:221–229. https://doi.org/10.1016/S0269-7491(99)00264-X
44. Mrozek E Jr, Funicelli NA (1982) Effect of zinc and lead on germination of *Spartina alterniflora* Loisel seeds at various salinities. Environ Exp Bot 22:23–32
45. Mustafa G, Komatsu S (2016) Toxicity of heavy metals and metal-containing nanoparticles on plants. Biochimica et Biophysica Acta (BBA). Proteins Proteomics 1864:932–944. https://doi.org/10.1016/j.bbapap.2016.02.020
46. Nakanishi M, Takatani H, Kobayashi Y, et al. (2005) Characterization of binary gold/platinum nanoparticles prepared by sonochemistry technique. Appl Surf Sci 241:209–212. https://doi.org/10.1016/j.apsusc.2004.09.046
47. Newfang DA, Johnson GT, Harbison RD (2015) Nanoparticles. Hamilton and Hardy's Industrial Toxicology: Sixth Edition 5:1025–1028. https://doi.org/10.1002/9781118834015.ch97
48. Noruzi M, Zare D, Khoshnevisan K, Davoodi D (2011) Rapid green synthesis of gold nanoparticles using Rosa hybrida petal extract at room temperature. Spectrochim Acta A Mol Biomol Spectrosc 79:1461–1465. https://doi.org/10.1016/j.saa.2011.05.001
49. Pandey N, Sharma CP (2002) Effect of heavy metals Co2+, Ni2+ and Cd2+ on growth and metabolism of cabbage. Plant Sci 163:753–758
50. Park J-E, Atobe M, Fuchigami T (2006) Synthesis of multiple shapes of gold nanoparticles with controlled sizes in aqueous solution using ultrasound. Ultrason Sonochem 13:237–241. https://doi.org/10.1016/j.ultsonch.2005.04.003
51. Pradhan L, Yenurkar D, Mukherjee S (2023) Toxicity assessment of gold ions and gold nanoparticles on plant growth. In: Nanomaterials and nanocomposites exposures to plants: response, interaction, phytotoxicity and defense mechanisms. Springer, pp 175–189
52. Reddy AM, Kumar SG, Jyothsnakumari G et al (2005) Lead induced changes in antioxidant metabolism of horsegram (*Macrotyloma uniflorum* (Lam.) Verdc.) and bengalgram (*Cicer arietinum* L.). Chemosphere 60:97–104
53. Reetz MT, Helbig W (1994) Size-selective synthesis of nanostructured transition metal clusters. J Am Chem Soc 116:7401–7402. https://doi.org/10.1021/ja00095a051
54. Reetz MT, Helbig W, Quaiser SA et al (1979) Visualization of surfactants on nanostructured palladium clusters by a combination of STM and high-resolution TEM. Science 267:367–369. https://doi.org/10.1126/science.267.5196.367

55. Rehman AU, Nazir S, Irshad R, et al. (2021) Toxicity of heavy metals in plants and animals and their uptake by magnetic iron oxide nanoparticles. J Mol Liq 321:114455. https://doi.org/10.1016/j.molliq.2020.114455
56. Roberts AG, Oparka KJ (2003) Plasmodesmata and the control of symplastic transport. Plant Cell Environ 26:103–124. https://doi.org/10.1046/j.1365-3040.2003.00950.x
57. ROS ROC, COOK DT, Martinez-Cortina C, PICAZO I (1992) Nickel and cadmium-related changes in growth, plasma membrane lipid composition, ATPase hydrolytic activity and proton-pumping of rice (*Oryza sativa* L. cv. Bahia) shoots. J Exp Bot 43:1475–1481
58. Shankar SS, Rai A, Ankamwar B et al (2004) Biological synthesis of triangular gold nanoprisms. Nat Mater 3:482–488. https://doi.org/10.1038/nmat1152
59. Shekar CHC, Sammaiah D, Shasthree T, Reddy KJ (2011) Effect of mercury on tomato growth and yield attributes. Int J Pharma Bio Sci 2
60. Sheldon AR, Menzies NW (2005) The effect of copper toxicity on the growth and root morphology of rhodes grass (*Chloris gayana* Knuth.) in resin buffered solution culture. Plant Soil 278:341–349. https://doi.org/10.1007/s11104-005-8815-3
61. Sheoran IS, Singal HR, Singh R (1990) Effect of cadmium and nickel on photosynthesis and the enzymes of the photosynthetic carbon reduction cycle in pigeonpea (*Cajanus cajan* L.). Photosynth Res 23:345–351. https://doi.org/10.1007/BF00034865
62. Siddiqi KS, Husen A (2016) Engineered gold nanoparticles and plant adaptation potential. Nanoscale Res Lett 11. https://doi.org/10.1186/s11671-016-1607-2
63. Singh A, Chaudhary S, Dehiya BS (2021) Fast removal of heavy metals from water and soil samples using magnetic Fe_3O_4 nanoparticles. Environ Sci Pollut Res 28:3942–3952. https://doi.org/10.1007/s11356-020-10737-9
64. Singh A, Prasad SM (2015) Remediation of heavy metal contaminated ecosystem: an overview on technology advancement. Int J Environ Sci Technol 12:353–366. https://doi.org/10.1007/s13762-014-0542-y
65. Singh M, Kalaivani R, Manikandan S et al (2013) Facile green synthesis of variable metallic gold nanoparticle using *Padina gymnospora*, a brown marine macroalga. Appl Nanosci 3:145–151. https://doi.org/10.1007/s13204-012-0115-7
66. Siti RM, Khairunisak AR, Aziz AA, Noordin R (2013) Green synthesis of 10 nm gold nanoparticles via seeded-growth method and its conjugation properties on lateral flow immunoassay. Adv Mat Res 686:8–12. https://doi.org/10.4028/www.scientific.net/AMR.686.8
67. Stadtman ER, Oliver CN (1991) Metal-catalyzed oxidation of proteins. Physiological consequences. J Biol Chem 266:2005–2008
68. Sudhakar C, Syamalabai L, Veeranjaneyulu K (1992) Lead tolerance of certain legume species grown on lead ore tailings. Agric Ecosyst Environ 41:253–261
69. Turkevich J, Stevenson PC, Hillier J (1951) A study of the nucleation and growth processes in the synthesis of colloidal gold. Discuss Faraday Soc 11:55–75. https://doi.org/10.1039/DF9511100055
70. Wang M, Zou J, Duan X et al (2007) Cadmium accumulation and its effects on metal uptake in maize (*Zea mays* L.). Bioresour Technol 98:82–88
71. Wintz H, Fox T, Vulpe C (2002) Responses of plants to iron, zinc and copper deficiencies. Biochem Soc Trans 30:766–768. https://doi.org/10.1042/BST0300766
72. Wu S (1994) Effect of manganese excess on the soybean plant cultivated under various growth conditions. J Plant Nutr 17:991–1003
73. Yourtchi MS, Bayat HR (2013) Effect of cadmium toxicity on growth, cadmium accumulation and macronutrient content of durum wheat (Dena CV.). Int J Agric Crop Sci (IJACS) 6:1099–1103
74. Yu C-S, Lee C-L, Wang CRC (1997) Gold nanorods: electrochemical synthesis and optical properties. J Phys Chem B 101:6661–6664. https://doi.org/10.1021/jp971656q
75. Zhai G, Walters KS, Peate DW et al (2014) Transport of gold nanoparticles through plasmodesmata and precipitation of gold ions in woody poplar. Environ Sci Technol Lett 1:146–151. https://doi.org/10.1021/ez400202b

76. Zhao P, Li N, Astruc D (2013) State of the art in gold nanoparticle synthesis. Coord Chem Rev 257:638–665. https://doi.org/10.1016/j.ccr.2012.09.002
77. Zhou P, Adeel M, Shakoor N et al (2021) Application of nanoparticles alleviates heavy metals stress and promotes plant growth: an overview. Nanomaterials 11:1–18
78. Zhou ZS, Huang SQ, Guo K et al (2007) Metabolic adaptations to mercury-induced oxidative stress in roots of Medicago sativa L. J Inorg Biochem 101:1–9
79. Zhu ZJ, Wang H, Yan B et al (2012) Effect of surface charge on the uptake and distribution of gold nanoparticles in four plant species. Environ Sci Technol 46:12391–12398. https://doi.org/10.1021/es301977w

Mitigation of Drought Stress in Plants Using Gold Nanoparticles

Mayur Mukut Murlidhar Sharma, Divya Kapoor, Atul Loyal, Rahul Kumar, Pankaj Sharma, and Azamal Husen

Abstract Amidst unprecedented shifts in global climate dynamics, the exacerbating impact of drought-induced stress on plant ecosystems presents formidable challenges to both agriculture and food security. This chapter delves into the captivating realm of nanotechnology, with a specific focus on the efficacy of gold nanoparticles in ameliorating the challenges induced by drought stress in plants. The narrative unfolds by scrutinizing the current landscape of drought-induced challenges, underscoring the pressing need for innovative solutions. Subsequently, it navigates through the distinctive properties of gold nanoparticles that position them as promising contenders for alleviating drought stress. The chapter conducts a thorough review of recent research endeavors and breakthroughs, elucidating the underlying mechanisms through which gold nanoparticles interact with plant physiology to augment stress tolerance. This comprehensive exploration aspires to contribute to the evolving discourse on harnessing nanotechnology for resilient and sustainable plant ecosystems amidst the mounting challenges posed by climate change.

M. M. M. Sharma
Department of Agriculture and Industry, Kangwon National University, Chuncheon, Korea

D. Kapoor · P. Sharma (✉)
Department of Microbiology, CCS Haryana Agricultural University, Hisar, Haryana 125004, India
e-mail: psk14495@gmail.com; sharmap@uic.edu

A. Loyal
Department of Genetics and Plant Breeding, CCS Haryana Agricultural University, Hisar, Haryana 125004, India

R. Kumar
Department of Zoology, College of Agriculture, RRS Bawal, Haryana 123501, India

P. Sharma
Department of Ophthalmology & Visual Sciences, University of Illinois Chicago, Chicago 60612, USA

A. Husen
Wolaita Sodo University, PO Box 138, Wolaita, Ethiopia

Department of Biotechnology, Graphic Era (Deemed to Be University), Dehradun, Uttarakhand 248002, India

A. Husen (ed.), *Plant Response to Gold Nanoparticles*, Smart Nanomaterials Technology, https://doi.org/10.1007/978-981-97-7174-5_10

Keywords AuNPs · Abiotic stress · Drought stress management · Plant protection

1 Introduction

In recent decades, a myriad of challenges, including enduring demographic strains, incessantly shifting climatic conditions, and heightened competition for resources, have unfolded. These multifaceted issues collectively present a formidable threat, underscoring an imperative requirement to secure global food sustainability [63, 65, 72]. Contemporary agricultural systems confront numerous challenges arising from the ever-changing environmental conditions, diminished efficiency, nutrient inadequacy, and a decline in organic matter content [41, 75, 77] Fig. 1. Furthermore, the repercussions of the green revolution continue to cast a shadow over agroecosystems, contributing to ongoing distress. Simultaneously, the global population has experienced rapid growth, projected to reach 8.1 billion by 2025, and further escalate to 9.6 billion in 2050 and 10.9 billion by the close of the twenty-first century [49, 50]. This exponential surge in population has significantly altered the landscape of food consumption, depleting the world's available food resources. Consequently, the world grapples with a pervasive state of hunger, with the number of undernourished individuals rising from approximately 804 million in 2016 to 821 million in 2017. Numerous factors contribute to the current situation, with one notable aspect being the demanding conditions imposed on plants. In their native surroundings, plants grapple with a variety of biotic and abiotic stresses, either individually or in tandem [64, 69, 73]. Simultaneously, the escalating impact of human interventions and the continual surge in anthropogenic activities within natural plant environments have heightened the susceptibility of plant systems to a diverse range of abiotic stresses. These include heat stress, heavy metal stress, salinity stress, drought stress, flood stress, metalloid stress, oxidative stress, and others, collectively influencing the optimal productivity of plants [59]. Exposing plants to environmental stresses results in various cellular-level alterations, including lipid membrane peroxidation, protein denaturation, loss of enzyme activity, DNA damage, and so on [6, 25, 26]. The perception and transduction of abiotic stresses require the involvement of signal molecules to facilitate adaptive responses through diverse physiological processes and transcription factors. Different signal molecules have been identified to mitigate plant stresses by augmenting the expression levels of various genes. Meteorological drought, characterized by a prolonged shortage of precipitation, in conjunction with heightened evapotranspiration demand, gives rise to agricultural drought. This form of drought is defined by the inadequate moisture levels necessary for the normal growth and development of plants, hindering their ability to complete their life cycle.

The adverse effects of drought on plant growth and development are profound, causing substantial decreases in both crop growth rate and biomass accumulation. Key consequences of drought in crop plants encompass a reduction in the rate of cell division and expansion, diminished leaf size, constrained stem elongation and root proliferation, as well as disrupted stomatal oscillations. These disruptions collectively

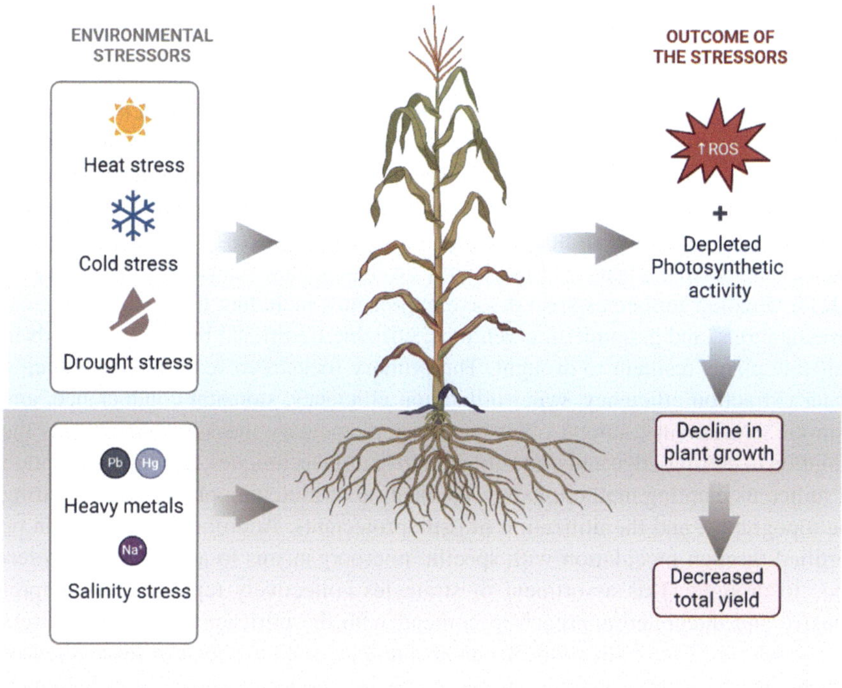

Fig.1 Different environmental factors leading to decline in plant productivity

impact plant water and nutrient relations, leading to a decline in crop productivity and a decrease in water use efficiency [60, 66–68, 70]. Crop growth models predict an intensification of this challenge in the future. Drought not only disrupts normal plant growth, interferes with water relations, and diminishes water use efficiency but also triggers a sophisticated array of physiological and biochemical responses at both cellular and organismal levels, constituting a complex phenomenon. The reduction in photosynthetic rate primarily arises from factors such as stomatal closure, membrane damage, and disruptions in the activity of various enzymes, particularly those crucial for ATP synthesis. Plants employ a diverse repertoire of mechanisms to endure drought, encompassing the reduction of water loss through heightened diffusive resistance, increased water uptake facilitated by prolific and deep root systems, and the development of smaller, succulent leaves to mitigate transpirational loss. Low-molecular-weight osmolytes, including glycinebetaine, proline, other amino acids, organic acids, and polyols, assume pivotal roles in sustaining cellular functions during drought. Plant growth substances such as salicylic acid, auxins, gibberellins, cytokinins, and abscisic acid modulate plant responses to drought [15]. Additionally, polyamines, citrulline, and various enzymes function as antioxidants, mitigating the adverse effects of water deficit. Effectively managing plant drought stress involves implementing strategies such as mass screening and breeding, marker-assisted selection, and the exogenous application of hormones and osmoprotectants to seeds or growing plants. Furthermore, engineering for drought resistance emerges as a pragmatic approach in addressing the challenges posed by water scarcity. Enacting optimal management practices concerning variables such as sowing timing, plant density, plant genotype, and soil and nutrient handling proves instrumental in mitigating grain yield losses in field crops experiencing drought stress. Yet, the most prominently highlighted strategy for alleviating drought stress involves the deployment of transgenic plants endowed with drought-tolerant attributes (Ilycas et al. 2021). Ongoing initiatives span diverse approaches including breeding, molecular investigations, and genomic interventions, utilizing traditional breeding methods to cultivate plants resilient to drought. The primary focus revolves around enhancing water extraction efficiency, water utilization efficiency, stomatal conductance, and osmotic adjustments, among other traits. Supplementary measures encompass the adoption of cutting-edge and more efficient irrigation techniques, the implementation of judicious planting methodologies, application of mulching practices, contouring the topography, and the utilization of osmoprotectants. Additionally, plants can be fortified through inoculation with specific microorganisms to augment their tolerance to drought. This assortment of strategies collectively represents a comprehensive and integrated approach to contend with the intricacies of drought stress in agriculture. The application of nanotechnological approaches in the pursuit of enhancing global food production seems to offer solutions to prevailing challenges [92]. The recent introduction of nanomaterials in agriculture has demonstrated significant promise for the sustainable evolution of agroecosystems in the long run[27–33]. The alleviation of drought stress using nanoparticles has been studied widely during the last decade. Research has underscored the advantageous role of carbon-based nanomaterials (NMs), including multiwalled carbon nanotubes (MWNTs), single-walled carbon nanotubes (SWNTs), graphene, fullerene, and metal-based nanoparticles (NPs) such as Ag, Au, Cu, Fe_2O_3, TiO_2, and ZnO, in supporting plants under adverse conditions like drought [19]. These nanoparticles assist plants in coping with drought stress by enhancing various growth indicators and promoting biomass. They contribute to improved water and nutrient uptake and utilization, facilitating water retention through alterations in cell walls and the regulation of stomatal closure. Gold nanoparticles (AuNPs) are an emerging category of the NPs with their role in agricultural attributes including plant stress tolerance [57]. This chapter highlights the importance of AuNPs for extenuating drought stress in crop plants.

2 Reasons for Drought Stress

In numerous scenarios, climate change manifests deleterious ramifications on both natural and agricultural ecosystems. The escalation of air temperatures, in particular, has the potential to induce the thawing of glaciers, thereby posing a significant hazard of inundation to agricultural lands, especially those characterized by minimal or absent slopes [8]. Moreover, the recession of glaciers is engendering the reduction of water reservoirs, curbing the accessibility of water to crops a trajectory that is amplifying progressively. Indeed, in diverse rain-fed agricultural locales worldwide, the annual aggregation of precipitation has attenuated due to the impact of global warming [89]. The depletion of water attributed to global warming extends beyond the soil, encompassing the plant level as well. Substantial amounts of internal water within plants are being relinquished to the atmosphere, propelled by elevated temperatures induced by global warming. This phenomenon serves to compound the preexisting challenges of water deficit in diverse agricultural systems worldwide [80]. In the event that the projected elevation in air temperature, approximately 2 °C beyond current levels, materializes by the conclusion of this century, roughly one-fifth of the world's population is anticipated to grapple with severe water deficit challenges [58]. Greater stress is anticipated in regions reliant solely on rainfall for crop production compared to areas benefiting from irrigation through canals, rivers, and water channels. Consequently, in rain-fed areas, episodes of drought exhibit a robust correlation with the distribution of rainfall throughout the year, leading to heightened probabilities of water stress in certain years over specific periods [46]. The predominant anthropogenic activities of industrialization, deforestation, and urbanization exert significant influence on rainfall patterns, thereby affecting water accessibility to plants through their contributions to climate change [16]. During the summer months, the combination of heightened atmospheric thirst for plants, escalated rates of evaporation and transpiration, and diminished rainfall availability characteristic of this season exacerbates the unfavorable impacts of drought stress on the growth and development of plants. However, the dispersion and intensity of rainfall, spanning individual years and overarching periods, profoundly shape the management of water resources for plants and the prevalence of drought stress in the majority of instances [42]. Furthermore, Anticipations suggest that in rain-fed regions, the volume of summer precipitations is projected to decrease by 70% by the onset of the twenty-second century, should current conditions persist. Notably, a substantial two-thirds of the world's population currently confront food insecurity, a predicament primarily arising from the pronounced fluctuations in seasonal rainfall, attributed to shifts in the monsoon patterns [4].

3 Effect on Plants from Drought Stress

Observable manifestations of water deficit during the vegetative stage encompass diminished plant height, wilting of leaves, and modifications in both the quantity and surface area of leaves. The substantial reduction in plant height, profoundly influenced by drought, is intricately connected to processes involving diminished cell expansion and the onset of leaf senescence. The decline in plant height primarily stems from reduced cell expansion, increased shedding of leaves, and impaired mitosis under the conditions of drought [93]. Apart from changes in plant height, there are notable variations in the morphology of different plant organs [18, 34, 35].

Leaves, serving as direct indicators of the level of water scarcity, play a pivotal role in plant assimilation and transpiration. In response to drought, plant leaves typically undergo adaptations, characterized by reduced leaf areas, increased leaf thickness, and higher leaf tissue density [90]. The alteration in leaf area, a pivotal determinant of plant photosynthetic processes and ultimate yield, emerges as one of the conspicuously discernible traits of plant leaves experiencing the impact of drought stress. Previous studies have elucidated that the primary determinants influencing the fluctuation in plant leaf area encompass leaf turgor pressure, canopy temperature dynamics, and the availability of photoassimilates [81]. Notwithstanding, varied plant species manifest disparate reactions to drought stress, as evidenced by the marginal elongation observed in sugarcane leaves under such environmental constraints [52]. Beyond foliage, plant root systems, instrumental in direct water absorption, wield significant influence in the context of drought stress. For instance, drought conditions can hinder the growth of cotton seedlings, spur the elongation and attenuation of fine roots, abbreviate the longevity of fine roots with diverse diameters, prompt the elongation of root hairs, and expedite their subsequent demise [91]. Additionally, there is a discernible alteration in the root-to-stem ratio of plants. Notably, soybean plants evince a substantial diminution in both shoot and root biomass when subjected to the rigors of drought stress. Concomitantly, maize plants, under conditions of water scarcity, manifest pronounced reductions in height, leaf dimensions, and stem girth [44]. In addition to external morphological changes, the internal structure of plants undergoes notable alterations. Particularly, a distinct development of cuticle on the outer surface of leaf epidermis becomes evident. This cuticle, comprised of a lipid membrane, functions to reduce water loss to the atmosphere and acts as a barrier against plant water evaporation. A robust cuticle, distinguished by its thickness, not only enhances energy reflection but also mitigates transpiration, thereby strengthening the plant's resistance to drought. Under conditions of drought stress, plants tend to manifest a substantial increase in cuticular lipids content [47]. Stomatal development emerges as a pivotal metric in the context of water stress. The drought process appears to induce an increase in both stomatal length and width, along with elevated stomatal density and a broader extent of stomatal opening [36]. Photosynthesis, a pivotal biological process, undergoes nuanced modulation in response to drought stress. The synthesized products stemming from leaf photosynthesis constitute the elemental substrates vital for plant growth. The net photosynthetic rate, acting as a direct gauge of material productivity per unit leaf area, theoretically positions itself as a reliable metric for assessing the biological production level of plants. As soil relative water content experiences a decline, there is a parallel reduction in both the photosynthetic rate and transpiration rate. Past research has delineated that the decline in photosynthetic rate during drought stress is intricately influenced by two distinct factors: stomatal and non-stomatal limitations. Stomatal limitation predominantly governs the reduction in photosynthetic rate under mild drought conditions. In contrast, under severe drought circumstances, non-stomatal factors emerge as the primary drivers behind the diminishing photosynthetic rate. The inadequacy of water directly precipitates a constriction of photosynthesis by diminishing CO_2 availability, thereby resulting in diffusion constraints at both stomatal and mesophyll levels [94].

4 Molecular Mechanism Behind Drought Stress

Plants confronted with drought challenges orchestrate a myriad of adaptive mechanisms at the molecular level to modulate drought stress. In these conditions, a sophisticated interplay of up and down-regulation events unfolds at the transcriptional level, orchestrated in response to stress. The accrual of stress proteins assumes a pivotal role in bolstering tolerance to drought. Signaling pathways implicated in abiotic stress, prominently drought, demonstrate the involvement of a diverse array of dehydration-responsive element-binding genes [43]. Plants undergoing drought stress undergo detrimental outcomes such as cell harm facilitated by reactive oxygen species (ROS), elevated cellular temperature, and heightened viscosity of cellular contents. Additionally, there are modifications in the interaction, aggregation, and denaturation of proteins under these challenging circumstances [14]. Research has revealed that under water stress conditions, plants regulate the expression of diverse genes responsive to drought, and this regulation transpires through a intricate transcriptome network [76]. A myriad of physiological, cellular, and molecular mechanisms contribute to the enhancement of drought tolerance in plants. These encompass the nuanced regulation, be it up-regulation or down-regulation, of gene expression responsible for osmolyte accumulation. Additionally, there is an augmentation in both non-enzymatic and enzymatic antioxidant systems, a decline in the transpiration rate, and the facilitation of shoot growth and tillering [54]. During instances of water stress, an augmented concentration of abscisic acid (ABA) assumes a pivotal role in orchestrating the closure of stomata. Furthermore, it governs the expression of an array of stress-responsive genes. However, it is imperative to underscore that the regulation of genes receptive to drought is also intricately governed by a system independent of ABA [1]. A myriad of protein-coding pivotal genes undergo expression modulation in response to drought, assuming crucial roles in both metabolic and regulatory pathways. Notably, genes involved in detoxification, osmolyte biosynthesis, water channel regulation, ion transportation, heat shock proteins, and the proteolysis of cellular substrates are intricately associated with the latter stages of embryogenesis [40]. Moreover, the genes tasked with regulatory functions predominantly encompass transcription factors such as AREB, NAC, AP2/ERF, MYC, MYB, and bZIP. In addition, the presence of mitogen-activated protein kinases (MAPK) orchestrating diverse cellular signaling pathways, protein kinases intricately connected to ribosomal processes, receptors, and transcriptionally regulated systems further contribute to the multifaceted regulatory network. Furthermore, proteins that govern the synchronization of signal transduction pathways play indispensable roles in these intricate molecular processes [62]. The signaling pathways associated with diverse abiotic pressures, drought stress included, share fundamental sequences. Primarily, this involves the detection of stress stimuli by detectors or receptors, situated either in the plasma membrane or freely dispersed in the cytosol. Following this, the signaling is transduced through secondary couriers. Upon the initiation of stress perception, it triggers the activation of secondary couriers such as calcium ions, cyclic nucleotides (cAMP, cGMP), nitric oxide, sugars, ROS, and others. These secondary couriers subsequently activate the signaling pathway by adjusting the expression or repression of various genes responsive to different stressors [53]. Protein kinases and phosphatases orchestrate the phosphorylation and de-phosphorylation, respectively, of the proteins intricately involved in the signaling pathway, encompassing receptors, secondary messengers, and transcription factors. Remarkably, within this regulatory network responsive to drought stress, MAPKs and CDPKs emerge as pivotal protein kinases that wield significant influence over the intricate orchestration of the signaling cascade [51]. Specific transcription factors (TFs) engage in direct communication with cis-regulatory elements situated in the promoter region of a cluster of downstream genes, thereby coordinating their expression. Furthermore, multiple TFs are subject to regulation through enhancers positioned in the upstream region of their respective genes [10].

5 Nanoparticles in Alleviating Drought Stress

The utilization of nanoparticles (NPs) serves as a protective shield for cellular membranes, maintains water relations, and amplifies nutrient and water absorption, resulting in a substantial augmentation in plant growth during drought stress. NPs act as defenders for the photosynthetic apparatus, enhancing photosynthetic efficiency and fostering the accumulation of osmolytes, hormones, and phenolics. Additionally, they fortify antioxidant activities and modulate gene expression, ultimately furnishing plants with heightened resistance against the adverse effects of drought stress [57]. Carbonaceous nanomaterials such as graphene, fullerene, fullerol, and carbon nanotubes, alongside metallic nanoparticles like ZnO, TiO_2, Fe, and Cu NPs, are widely utilized to mitigate the impact of drought stress. Their effectiveness lies in enhancing both water and nutrient absorption through stress tolerance mechanisms, coupled with the up-regulation of genes associated with cellular growth [74]. Nanoparticles (NPs) have been noted for their capacity to elevate germination parameters, expedite growth rates, amplify biomass and yield, regulate stomatal conductance, and modulate transpiration rates. Simultaneously, they enhance photosynthetic parameters, curbing membrane ion leakage, and facilitating the assimilation of carbon dioxide in leaves [7]. In another study, formulations incorporating three micronutrient nanoparticles ZnO, B_2O_3, and CuO coupled with their respective salts ($ZnSO_4{\cdot}7H_2O$, H_3BO_3, and $CuSO_4{\cdot}5H_2O$) were administered to soybean crops either through soil or foliar application three weeks post seed germination. Drought stress, implemented at 50% field moisture capacity, persisted for 19 weeks. Parameters related to growth, yield, and nutrient uptake dynamics were meticulously measured throughout the experiment. Results delineate that the imposed drought led to a consequential 27% reduction in soybean shoot growth and an even more substantial 54% decline in grain yield. Notably, the composite formulations, on average, alleviated the detrimental impacts of drought by bolstering shoot growth by 33% and enhancing grain yield by 36% [13]. In another study on wheat, Silica nanoparticles (SiO_2 NPs) were introduced into the soil and administered through foliar application at distinct stages of plant growth. The outcomes demonstrated a marked reduction in the majority of assessed traits under drought stress compared to regular irrigation. Particularly noteworthy was the fact that, amid drought stress, soil application of NPs resulted in a significant augmentation of 12.54% in leaf greenness (SPAD) and a remarkable 84.04% increase in Relative Water Content (RWC) relative to the control (NPs = 0 ppm). Furthermore, wheat yield witnessed a substantial surge of 25.35% and 17.81%, respectively, with foliar and soil applications of NPs in the face of drought stress. In the same irrigation conditions, soil application of NPs significantly elevated both plant height and biomass compared to the foliar application of NPs [5]. In another study, Drought-affected *Setaria italica* plants underwent treatment with iron oxide nanoparticles synthesized through green methods, both in their pristine and capped configurations. These nanoparticles, designated as FeNp, not only served as nanonutrients for the plants but also conferred resilience against the challenges of drought stress. A conspicuous overall augmentation in the growth of *Setaria italica* plants was discerned. Moreover, an increasing concentration of FeNp (Fe_3O_4) exhibited a concomitant elevation in chlorophyll and soluble sugar content in the seedlings, suggesting that the assimilated iron played a role in the synthesis of photoassimilates. The evaluation of iron, H2O2, and proline content in the plants further substantiated the effective assimilation of FeNp by the plants, enabling them to successfully overcome the adversities of drought stress [79]. Plants subjected to priming with copper nanoparticles demonstrated an elevated capacity for drought resilience, as evidenced by increased leaf water content and augmented plant biomass in comparison to plants treated solely with water under drought conditions. Furthermore, our findings indicated that the application of copper nanoparticles on plants led to an amplification of anthocyanin, chlorophyll, and carotenoid contents compared to water-treated plants facing drought stress. Additionally, histochemical analyses involving nitro blue tetrazolium and 3,3′-diaminobenzidine unveiled a reduction in the accumulation of reactive oxygen species in primed plants, attributed to the enhancement of activities in scavenging enzymes responsible for reactive oxygen species under drought conditions [83].

6 Role of AuNPs in Mitigating Drought Stress

The role of AuNPs has been studied for different types of abiotic and biotic stress. The inertness and less reactivity of Gold NPs make them one of the best NMs suitable for plant health. Barring the cost-effectiveness, Gold NPs' benefits to plant health are second to none. They are one of the least toxic NPs in vogue. Their low detrimental effect on plants has been the reason that despite their high cost, there have been ample studies based on Gold NPs for their role in the management of different kinds of stress (Fig. 2). Different stress mitigation activities as exhibited by AuNPs are listed in Table 1. For instance, in a recent study, delved into the repercussions of biofabricated gold nanoparticles (AuNPs) on wheat (*Triticum aestivum*) plants facing salt-induced stress. The green-synthesized AuNPs demonstrated their advantageous effects in regulating the K+/Na+ ratio, chlorophyll concentration, defense mechanisms, nitrogen assimilation, stomatal dynamics, and growth traits amidst salt stress conditions. Moreover, the undue buildup of markers indicating oxidative stress, encompassing reactive oxygen/nitrogen species, was effectively managed through the application of AuNPs treatment during salt stress [88]. An investigation into counteracting cadmium (Cd) toxicity in hydroponically cultured rice plants has demonstrated that melatonin, when presented in the guise of gold nanoparticles (Mel-AuNPs), surpasses the efficacy of standalone melatonin. In a medium enriched with 200 μM Mel-AuNPs, a substantial augmentation in melatonin content was evident in both roots (18.0%) and leaves (20.3%), concurrently with a marked reduction in Cd levels by 33.0% in roots and 46.2% in leaves. Furthermore, the supplementation of Mel-AuNPs not only reinstated chlorophyll biosynthesis but also ameliorated Cd-induced oxidative stresses, resulting in a significant upswing in antioxidant enzyme activities. Further scrutiny uncovered that the provision of Mel-AuNPs restrained the Cd-induced expression of genes associated with metal transporters (OsHMA2, OsHMA3, OsIRT1, OsIRT2, OsNramp1, OsNramp5, and OsLCT1) in the roots of rice plants [39]. In an alternative investigation, it was illustrated that gold nanoparticles exhibit adaptogenic qualities, fortifying the ability of plants to endure freezing conditions. The approach involved subjecting winter wheat seeds (*Triticum aestivum* L., var. Moskovskaya 39, Poaceae) to a one-day priming process in solutions containing gold nanoparticles (15-nm diameter) at concentrations varying from 5 to 50 μg/ml. This treatment resulted in a conspicuous improvement of freezing tolerance in 7-day-old wheat seedlings. The study established a correlation between the heightened freezing tolerance in wheat and changes in critical indicators pivotal for its development, including growth intensity, the activity of the photosynthetic apparatus, oxidative processes, and the accumulation of soluble sugars in the seedlings [86]. A cutting-edge optical biosensor, employing gold nanoparticles, has been introduced for the precise determination of miRNA-1886 levels in the roots of tomato plants. Investigations elucidated that manipulating irrigation levels, ranging from 100 to 60% of field capacity, triggered an elevation in the concentration of miRNA-1886, spanning from around 100 to 6800 fM (femtomolar).

This engendered a linear adjustment in the biosensor response (R2 = 0.97). Furthermore, the results highlighted a distinctive aspect: in contrast to the conventional morpho-physiological and biochemical traits of the plant, the concentration of miRNA-1886 remained notably unaltered ($P < 0.01$) in response to various other stressors (Vakilian et al., 2019). In times of arid conditions, plants showcase heightened performance and physiological enhancements through their interaction with gold particles. Notably, both growth rate and yield experience a boost when the synthesis of gold nanoparticles (Au-NPs) is facilitated using green tea extract (Wahab et al., 2023). In the midst of drought-induced stress, a phase where plants are particularly susceptible to microbial challenges, gold nanoparticles have showcased their effectiveness in addressing such issues. Their antimicrobial prowess is apparent in their capability to impede infection rates caused by pathogens. The fundamental

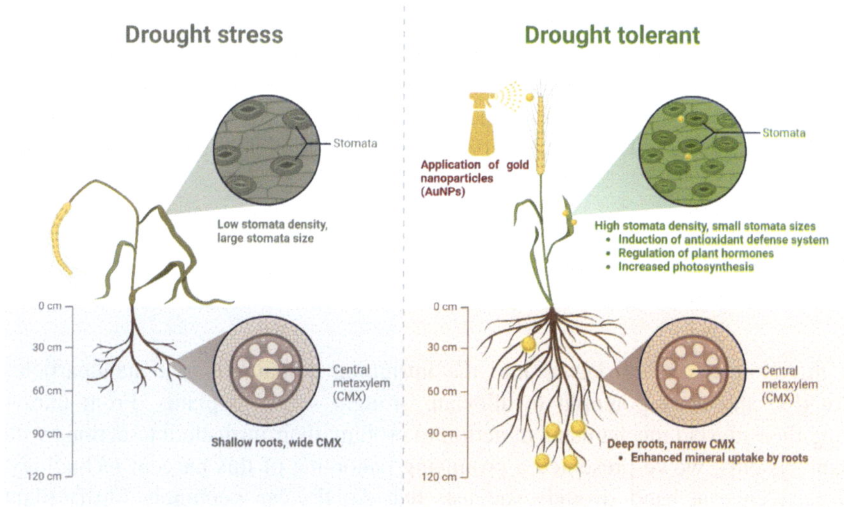

Fig. 2 A portrayal depicting mitigation of drought stress in plant systems on application of AuNPs

Table 1 Different stress mitigation activities of AuNPs in plants

Type of AuNP	Activity	Crop plant	References
Ast-AuNPs	Salinity stress amelioration	Rice	[78]
Au–Ag nanoparticles	Salinity stress amelioration	*Mentha piperita*	[2]
AuNPs	Oxidative stress modulation	*Arabidopsis*	[17]
AuNPs	Osmotic stress	*Nicotiana tabacum*	[38]
Ast-AuNPs	Heavy metal (cadmium) stress	Rice	[9]
AuNPs	Cold stress	Wheat	[86]
AuNPs	Antiphytopathogenoic fungi	Rice	[11]
AuNPs	Nematicidal (Pine wilt nematode)	Not mentioned	[12]
AuNPs	Cold stress	Wheat seedlings	[86]
AuNPs	Cold stress	*Lamprocapnos spectabilis*	[48]

attributes of Au-NPs revolve around their integral role in bolstering disease resistance and augmenting the immune system in plants [20]. Gold nanoparticles (Au-NPs), harvested from bioactive compounds within plants, have surfaced as a potent instrument for reinforcing vegetative robustness in arid climates. Their ability to amplify nutrient absorption, exhibit antioxidative attributes, heighten photosynthetic rates, and fortify resistance against diseases positions them as invaluable tools in agricultural contexts. Additionally, there is an exigency for fresh and comprehensively expounded research detailing their operational mechanisms, precisely regulated application methodologies, and environmentally conscientious characteristics to attain universal approval, particularly in the challenging milieu of drought conditions [45].

7 Conclusion

In this chapter, we've delved into the intriguing domain of gold nanoparticles (AuNPs) and their potential to alleviate drought stress in plants. From unraveling their myriad mechanisms of action to probing their intricate interactions with plant systems, we've presented a promising panorama of this nascent technology. The evidence at hand strongly suggests that AuNPs can genuinely fortify plant resilience against drought through various avenues. By boosting water uptake, modulating antioxidant activity, and influencing gene expression, these minuscule marvels provide a multifaceted approach to drought mitigation. While the specific pathways and optimal application methods continue to be subjects of investigation, the positive outcomes observed across a spectrum of plant species are heartening. Nevertheless, it is imperative to acknowledge the existing limitations. Concerns regarding potential environmental ramifications and the long-term destiny of nanoparticles necessitate further research and conscientious development. Additionally, surmounting challenges related to optimizing delivery methods and ensuring cost-effectiveness is pivotal before widespread adoption can become a reality. Despite these challenges, the trajectory of AuNP-mediated drought mitigation looks promising. Ongoing research endeavors hold the potential to refine application strategies, address environmental apprehensions, and unlock the complete potential of this technology. Moving forward, the golden touch of AuNPs might indeed hold the key to ensuring sustainable agricultural practices in a future marked by water scarcity.

References

1. Aguado A, Capote N, Romero F, Dodd IC, Colmenero-Flores JM (2014) Physiological and gene expression responses of sunflower (*Helianthus annuus* L.) plants differ according to irrigation placement. Plant Sci 227:37–44
2. Aliakbarpour F, Vaziri A, Mohseni M (2020) Green synthesis of Au-Ag nanoparticles using Mentha piperita and effects of Au-Ag alloy nanoparticles on the growth of Mentha piperita under salinity stress. J Plant Process Funct 9:48
3. Anamika, Patra A, Shehzad S, Rani A, Sharma P, Mohammad KF, Mehta S (2023) Microbial elicitors for priming plant defense mechanisms. In: Sustainable Agriculture Reviews 60: Microbial Processes in Agriculture, pp 175–196. Springer Nature Switzerland, Cham

4. Aryal JP, Sapkota TB, Khurana R, Khatri-Chhetri A, Rahut DB, Jat ML (2020) Climate change and agriculture in South Asia: adaptation options in smallholder production systems. Environ Dev Sustain 22(6):5045–5075
5. Behboudi F, Tahmasebi Sarvestani Z, Kassaee MZ, Modares Sanavi SA, Sorooshzadeh A (2018) Improving growth and yield of wheat under drought stress via application of SiO2 nanoparticles. J Agric Sci Technol 20(7):1479–1492
6. Bachheti A, Deepti, Bachheti RK, Husen A (2021) Medicinal plants and their pharmaceutical properties under adverse environmental conditions. In: Husen A (Ed.) Harsh environment and plant resilience. Springer International Publishing AG, Gewerbestrasse 11, 6330 Cham, pp 457–502. https://doi.org/10.1007/978-3-030-65912-7_19
7. Chandrashekar HK, Singh G, Kaniyassery A, Thorat SA, Nayak R, Murali TS, Muthusamy A (2023) Nanoparticle-mediated amelioration of drought stress in plants: a systematic review. 3 Biotech 13(10):336
8. Cook BI, Smerdon JE, Seager R, Coats S (2014) Global warming and 21st century drying. Clim Dyn 43:2607–2627
9. Dai S, Wang B, Song Y, Xie Z, Li C, Li S, Huang Y, Jiang M (2021) Astaxanthin and its gold nanoparticles mitigate cadmium toxicity in rice by inhibiting cadmium translocation and uptake. Sci Total Environ 10(786):147496
10. Danquah A, De Zélicourt A, Colcombet J, Hirt H (2014) The role of ABA and MAPK signaling pathways in plant abiotic stress responses. Biotechnol Adv 32(1):40–52
11. Das A, Dutta P (2021) Antifungal activity of biogenically synthesized silver and gold nanoparticles against sheath blight of rice. J Nanosci Nanotechnol 21(6):3547–3555
12. Davids JS, Ackah M, Okoampah E, Fometu SS, Guohua W, Jianping Z (2021) Biocontrol of bacteria associated with pine wilt nematode, *Bursaphelenchus xylophilus* by using plant mediated gold nanoparticles. Int J Agric Biol 26(4):517–526
13. Dimkpa CO, Bindraban PS, Fugice J, Agyin-Birikorang S, Singh U, Hellums D (2017) Composite micronutrient nanoparticles and salts decrease drought stress in soybean. Agron Sustain Dev 37:1–3
14. Farooq M, Basra SM, Wahid A, Cheema ZA, Cheema MA, Khaliq A (2008) Physiological role of exogenously applied glycinebetaine to improve drought tolerance in fine grain aromatic rice (*Oryza sativa* L.). J Agron Crop Sci 194(5):325–33
15. Farooq M, Hussain M, Wahid A, Siddique KH (2012) Drought stress in plants: an overview. Plant responses to drougt stress: from morphological to molecular features 2012:1–33
16. Fatima A, Farid M, Safdar K, Fayyaz A, Ali SM, Adnan S, Nawaz M, Munir H, Raza N, Zubair M (2020) Loss of agro-biodiversity and productivity due to climate change in continent Asia: a review. In: Plant ecophysiology and adaptation under climate change: mechanisms and perspectives I: general consequences and plant responses, pp 51–71
17. Ferrari E, Barbero F, Busquets-Fité M, Franz-Wachtel M, Köhler HR, Puntes V, Kemmerling B (2021) Growth-promoting gold nanoparticles decrease stress responses in Arabidopsis seedlings. Nanomaterials 11(12):3161
18. Getnet Z, Husen A, Fetene M, Yemata G (2015) Growth, water status, physiological, biochemical and yield response of stay green sorghum *Sorghum bicolor* (L.) Moench varieties-a field trial under drought-prone area in Amhara regional state, Ethiopia. J Agronomy 14:188–202
19. Heikal YM, El-Esawi MA, El-Ballat EM, Abdel-Aziz HM (2023) Applications of nanoparticles for mitigating salinity and drought stress in plants: an overview on the physiological, biochemical and molecular genetic aspects. N Z J Crop Hortic Sci 51(3):297–327
20. Hlapisi N, Motaung TE, Linganiso LZ, Oluwafemi OS, Songca SP (2019) Encapsulation of gold nanorods with porphyrins for the potential treatment of cancer and bacterial diseases: a critical review. Bioinorg Chem Appl 30:2019
21. Husen A (2021a) The harsh environment and resilient plants: an overview. In: Husen A (Ed.)Harsh environment and plant resilience. Springer International Publishing AG, Gewerbestrasse 11, 6330 Cham, pp 1–23. https://doi.org/10.1007/978-3-030-65912-7_1
22. Husen A (2021b) Morpho-anatomical, physiological, biochemical and molecular responses of plants to air pollution. In: Husen A (Ed.) Harsh environment and plant resilience. Springer

International Publishing AG, Gewerbestrasse 11, 6330 Cham, pp 203–234. https://doi.org/10.1007/978-3-030-65912-7_9
23. Husen A (2021c) Cross talk between autophagy and hormones for abiotic stress tolerance in plants. In: Husen A (eds) Plant performance under environmental stress. Springer International Publishing AG, Gewerbestrasse 11, 6330, Cham, pp 1–15. https://doi.org/10.1007/978-3-030-78521-5_1
24. Husen A (2021d) Harsh Environment and Plant Resilience (*Molecular and Functional Aspects*). Springer Nature Switzerland AG, Gewerbestrasse 11, 6330 Cham, Switzerland. https://doi.org/10.1007/978-3-030-65912-7
25. Husen A (2021e) Plant performance under environmental stress (*Hormones, Biostimulants and Sustainable Plant Growth Management*). Springer Nature Switzerland AG, Gewerbestrasse 11, 6330 Cham, Switzerland. https://doi.org/10.1007/978-3-030-78521-5
26. Husen A, Iqbal M (2023) Medicinal Plants - Their Response to Abiotic Stress. Springer Nature Singapore Pte Ltd., 152 Beach Road, #21-01/04 Gateway East, Singapore 189721, Singapore. https://doi.org/10.1007/978-981-19-5611-9
27. Husen A (2022) Engineered Nanomaterials for Sustainable Agricultural Production, Soil Improvement and Stress Management. Elsevier Inc., 50 Hampshire St., 5th Floor, Cambridge, MA 02139, USA. https://doi.org/10.1016/C2021-0-00054-7
28. Husen A (2023) Nanomaterials and Nanocomposites Exposures to Plants (Response, Interaction, Phytotoxicity and Defense Mechanisms). Springer Nature Singapore Pte Ltd., 152 Beach Road, #21-01/04 Gateway East, Singapore 189721, Singapore. https://doi.org/10.1007/978-981-99-2419-6
29. Husen A, Iqbal M (2019) Nanomaterials and Plant Potential. Springer International Publishing AG, Gewerbestrasse 11, 6330 Cham, Switzerland. https://doi.org/10.1007/978-3-030-05569-1
30. Husen A, Jawaid M (2020) Nanomaterials for Agriculture and Forestry Applications. Elsevier Inc., 50 Hampshire St., 5th Floor, Cambridge, MA 02139, USA. https://doi.org/10.1016/C2018-0-02349-X
31. Husen A, Siddiqi KS (2014) Phytosynthesis of nanoparticles: concept, controversy and application. Nanoscale Res Lett 9(229):1–24. https://doi.org/10.1186/1556-276X-9-229
32. Husen A, Siddiqi KS (2014) Carbon and fullerene nanomaterials in plant system. J Nanobiotechnol 12(16):1–10
33. Husen A, Siddiqi KS (2023) Advances in Smart Nanomaterials and Their Applications. Elsevier Inc., 50 Hampshire St., 5th Floor, Cambridge, MA 02139, USA. https://doi.org/10.1016/C2021-0-02202-1
34. Husen A (2010) Growth characteristics, physiological and metabolic responses of teak (*Tectona grandis* Linn. f.) clones differing in rejuvenation capacity subjected to drought stress. Silvae Genetica 59:124–136
35. Husen A, Iqbal M, Aref MI (2014) Growth, water status and leaf characteristics of *Brassica carinata* under drought stress and rehydration conditions. Braz J Botany 37:217–227
36. Hughes J, Hepworth C, Dutton C, Dunn JA, Hunt L, Stephens J, Waugh R, Cameron DD, Gray JE (2017) Reducing stomatal density in barley improves drought tolerance without impacting on yield. Plant Physiol 174(2):776–787
37. Ilyas M, Nisar M, Khan N, Hazrat A, Khan AH, Hayat K, Fahad S, Khan A, Ullah A (2021) Drought tolerance strategies in plants: a mechanistic approach. J Plant Growth Regul 40:926–944
38. Jalil SU, Zahera M, Khan MS, Ansari MI. Biochemical synthesis of gold nanoparticles from leaf protein of Nicotiana tabacum L. cv. xanthi and their physiological, developmental, and ROS scavenging responses on tobacco plant under stress conditions. IET Nanobiotechnol 13(1):23–9
39. Jiang M, Dai S, Wang B, Xie Z, Li J, Wang L, Li S, Tan Y, Tian B, Shu Q, Huang J (2021) Gold nanoparticles synthesized using melatonin suppress cadmium uptake and alleviate its toxicity in rice. Environ Sci Nano 8(4):1042–1056
40. Joshi R, Wani SH, Singh B, Bohra A, Dar ZA, Lone AA, Pareek A, Singla-Pareek SL (2016) Transcription factors and plants response to drought stress: current understanding and future directions. Front Plant Sci 14(7):1029

41. Kapoor D, Yadav S, Sharma MM, Sharma P (2023) Interaction between metal nanoparticles and PGPR on the plant growth and development. In: Nanomaterials and nanocomposites exposures to plants: response, interaction, phytotoxicity and defense mechanisms, pp 327–351. Springer Nature Singapore, Singapore
42. Karandish F, Šimůnek J (2017Nov) Two-dimensional modeling of nitrogen and water dynamics for various N-managed water-saving irrigation strategies using HYDRUS. Agric Water Manag 1(193):174–190
43. Kaur G, Asthir B (2017) Molecular responses to drought stress in plants. Biol Plant 61:201–209
44. Khan MB, Hussain M, Raza A, Farooq S, Jabran K (2015) Seed priming with CaCl2 and ridge planting for improved drought resistance in maize. Turk J Agric For 39(2):193–203
45. Khan SA, Shahid S, Lee CS (2020) Green synthesis of gold and silver nanoparticles using leaf extract of *Clerodendrum inerme*; characterization, antimicrobial, and antioxidant activities. Biomolecules 10(6):835
46. Konapala G, Mishra AK, Wada Y, Mann ME (2020) Climate change will affect global water availability through compounding changes in seasonal precipitation and evaporation. Nat Commun 11(1):3044
47. Kosma DK, Bourdenx B, Bernard A, Parsons EP, Lu S, Joubes J, Jenks MA (2009) The impact of water deficiency on leaf cuticle lipids of Arabidopsis. Plant Physiol 151(4):1918–1929
48. Kulus D, Tymoszuk A (2021) Gold nanoparticles affect the cryopreservation efficiency of in vitro-derived shoot tips of bleeding heart. Plant Cell Tissue Organ Culture (PCTOC) 146(2):297–311
49. Loyal A, Pahuja SK, Rani N, Pooja, Srivastava RK, Sharma P (2023a) Titanium oxide nanoparticles: plant response, interaction, phytotoxicity, and defence mechanisms. In: Nanomaterials and nanocomposites exposures to plants: response, interaction, phytotoxicity and defense mechanisms, pp 263–284. Springer Nature Singapore, Singapore
50. Loyal A, Pahuja SK, Sharma P, Malik A, Srivastava RK, Mehta S (2023b) Potential environmental and human health implications of nanomaterials used in sustainable agriculture and soil improvement. In: engineered nanomaterials for sustainable agricultural production, soil improvement and stress management, pp 387–412. Academic Press
51. Manna M, Thakur T, Chirom O, Mandlik R, Deshmukh R, Salvi P (2021) Transcription factors as key molecular target to strengthen the drought stress tolerance in plants. Physiol Plant 172(2):847–868
52. Misra V, Solomon S, Mall AK, Prajapati CP, Hashem A, Abd_Allah EF, Ansari MI (2020) Morphological assessment of water stressed sugarcane: a comparison of waterlogged and drought affected crop. Saudi J Biol Sci 27(5):1228–36
53. Mukarram M, Choudhary S, Kurjak D, Petek A, Khan MM (2021) Drought: sensing, signalling, effects and tolerance in higher plants. Physiol Plant 172(2):1291–1300
54. Pareek A, Sopory SK, Bohnert HJ (2009) Abiotic stress adaptation in plants. Springer, Dordrecht, The Netherlands
55. Rani K, Rani A, Sharma P, Dahiya A, Punia H, Kumar S, Sheoran S, Banerjee A (2022) Legumes for agroecosystem services and sustainability. In: Advances in legumes for sustainable intensification, pp 363–380. Academic Press
56. Rani K, Sharma P, Kumar S, Wati L, Kumar R, Gurjar DS, Kumar D, Kumar R (2019) Legumes for sustainable soil and crop management. Sustain Manage Soil Environ 2019:193–215
57. Rasheed A, Li H, Tahir MM, Mahmood A, Nawaz M, Shah AN, Aslam MT, Negm S, Moustafa M, Hassan MU, Wu Z (2022Nov) The role of nanoparticles in plant biochemical, physiological, and molecular responses under drought stress: a review. Front Plant Sci 24(13):976179
58. Ray DK, West PC, Clark M, Gerber JS, Prishchepov AV, Chatterjee S (2019) Climate change has likely already affected global food production. PLoS ONE 14(5):e0217148
59. Sangwan S, Sharma P, Wati L, Mehta S (2023) Effect of chitosan nanoparticles on growth and physiology of crop plants. In: Engineered nanomaterials for sustainable agricultural production, soil improvement and stress management, pp 99–123. Academic Press

60. Sharma MM, Kapoor D, Rohilla R, Sharma P (2023a) Nanomaterials and their toxicity to beneficial soil microbiota and fungi associated plants rhizosphere. In: nanomaterials and nanocomposites exposures to plants: response, interaction, phytotoxicity and defense mechanisms, pp 353–380. Springer Nature Singapore, Singapore
61. Sharma MM, Kapoor D, Rohilla R, Sharma P (2023) Nanomaterials and their toxicity to beneficial soil microbiota and fungi associated plants rhizosphere. In: Nanomaterials and nanocomposites exposures to plants: response, interaction, phytotoxicity and defense mechanisms, pp 353–380. Springer Nature Singapore, Singapore
62. Sharma MM, Ramekar RV, Park NI, Choi IY, Choi SK, Park KC (2021) Characterization of transcription factor genes related to cold tolerance in Brassica napus. Genomics Informatics 19(4)
63. Sharma MM, Sharma P, Kapoor D, Beniwal P, Mehta S (2021c) Phytomicrobiome community: an agrarian perspective towards resilient agriculture. Plant performance under environmental stress: hormones, biostimulants and sustainable plant growth management. 2021c:493–534
64. Sharma P, Meyyazhagan A, Easwaran M, Sharma MM, Mehta S, Pandey V, Liu WC, Kamyab H, Balasubramanian B, Baskaran R, Klemeš JJ (2022) Hydrogen sulfide: a new warrior in assisting seed germination during adverse environmental conditions. Plant Growth Regul 98(3):401–420
65. Sharma P, Pandey V, Sharma MM, Patra A, Singh B, Mehta S, Husen A (2021) A review on biosensors and nanosensors application in agroecosystems. Nanoscale Res Lett 16:1–24
66. Sharma P, Patra A, Singh B, Mehta S (2023e) Microbial rejuvenation of soils for sustainable agriculture. In: Sustainable Agriculture Reviews 60: Microbial Processes in Agriculture, pp 293–323. Springer Nature Switzerland, Cham
67. Sharma P, Sangwan S, Kaur H, Patra A, Anamika, Mehta S (2023d) Diversity and evolution of nitrogen fixing bacteria. Sustain Agric Rev 60: Microbial Processes in Agriculture, 95–120
68. Sharma P, Sangwan S, Kaur H, Patra A, Mehta S (2023c) Microbial remediation of agricultural residues. In: Sustainable agriculture reviews 60: microbial processes in agriculture, pp 325–358). Cham: Springer Nature Switzerland.
69. Sharma P, Sangwan S, Kumari A, Singh S, Kaur H. Impact of climate change on soil microorganisms regulating nutrient transformation. InPlant Stress Mitigators: Action and Application 2022b May 6 (pp. 145–172). Singapore: Springer Nature Singapore.
70. Sharma P, Sangwan S, Mehta S (2023b) Emerging role of phosphate nanoparticles in agriculture practices. In: Engineered nanomaterials for sustainable agricultural production, soil improvement and stress management, pp 71–97. Academic Press
71. Sharma P, Sharma MM, Anamika, Kapoor D, Rani K, Singh D, Barkodia M (2020b) Role of microbes for attaining enhanced food crop production. Microb Biotechnol: Basic Res Appl 55–78
72. Sharma P, Sharma MM, Malik A, Vashisth M, Singh D, Kumar R, Singh B, Patra A, Mehta S, Pandey V (2021b) Rhizosphere, rhizosphere biology, and rhizospheric engineering. In: Plant growth-promoting microbes for sustainable biotic and abiotic stress management, pp 577–624. Springer International Publishing, Cham
73. Sharma P, Sharma MM, Patra A, Vashisth M, Mehta S, Singh B, Tiwari M, Pandey V (2020a) The role of key transcription factors for cold tolerance in plants. In: Transcription factors for abiotic stress tolerance in plants, pp. 123–152). Academic Press
74. Shekhawat GS, Mahawar L, Rajput P, Rajput VD, Minkina T, Singh RK (2021) Role of engineered carbon nanoparticles (CNPs) in promoting growth and metabolism of *Vigna radiata* (L.) Wilczek: Insights into the biochemical and physiological responses. Plants 10(7):1317
75. Singh A, Sharma P, Kumari A, Kumar R, Pathak DV (2019) Management of root-knot nematode in different crops using microorganisms. Plant biotic interactions: state of the art 85–99
76. Singh D, Laxmi A (2015) Transcriptional regulation of drought response: a tortuous network of transcriptional factors. Front Plant Sci 29(6):895
77. Singh S, Sangwan S, Sharma P, Devi P, Moond M (2021) Nanotechnology for sustainable agriculture: An emerging perspective. J Nanosci Nanotechnol 21(6):3453–65.Nayana AR, Joseph BJ, Jose A, Radhakrishnan EK (2020) Nanotechnological advances with PGPR applications. Sustain Agriculture Reviews 41: Nanotechnology for Plant Growth and Development 163–80

78. Song Y, Zheng C, Basnet R, Li S, Chen J, Jiang M (2022Dec) Astaxanthin synthesized gold nanoparticles enhance salt stress tolerance in rice by enhancing tetrapyrrole biosynthesis and scavenging reactive oxygen species in vitro. Plant Stress. 1(6):100122
79. Sreelakshmi B, Induja S, Adarsh PP, Rahul HL, Arya SM, Aswana S, Haripriya R, Aswathy BR, Manoj PK, Vishnudasan D (2021Jan) Drought stress amelioration in plants using green synthesised iron oxide nanoparticles. Mater Today: Proc 1(41):723–727
80. Sultan B, Defrance D, Iizumi T (2019) Evidence of crop production losses in West Africa due to historical global warming in two crop models. Sci Rep 9(1):12834
81. Taiz L, Zeiger E, Møller IM, Murphy A. Plant physiology and development. Sinauer Associates Incorporated 2015
82. Vakilian KA (2019) Gold nanoparticles-based biosensor can detect drought stress in tomato by ultrasensitive and specific determination of miRNAs. Plant Physiol Biochem 1(145):195–204
83. Van Nguyen D, Nguyen HM, Le NT, Nguyen KH, Nguyen HT, Le HM, Nguyen AT, Dinh NT, Hoang SA, Van Ha C (2022) Copper nanoparticle application enhances plant growth and grain yield in maize under drought stress conditions. J Plant Growth Regul 41(1):364–375
84. Venzhik Y, Deryabin A, Popov V, Dykman L, Moshkov I (2022) Gold nanoparticles as adaptogens increazing the freezing tolerance of wheat seedlings. Environ Sci Pollut Res 29(36):55235–55249
85. Venzhik Y, Deryabin A, Popov V, Dykman L, Moshkov I (2022) The effects of gold nanoparticles on redox status and freezing tolerance of wheat seedlings. Acta Physiol Plant 44(11):113
86. Venzhik YV, Deryabin AN, Popov VN, Dykman LA, Titov AF, Moshkov IE (2022) Influence of gold nanoparticles on the tolerance of wheat to low temperature. In: Doklady Biochemistry and Biophysics, Vol 502, No 1, pp 5–9. Pleiades Publishing, Moscow
87. Wahab A, Batool F, Muhammad M, Zaman W, Mikhlef RM, Naeem M (2023) Current knowledge, research progress, and future prospects of phyto-synthesized nanoparticles interactions with food crops under induced drought stress. Sustainability 15(20):14792
88. Wahid I, Rani P, Kumari S, Ahmad R, Hussain SJ, Alamri S, Tripathy N, Khan MI (2022) Biosynthesized gold nanoparticles maintained nitrogen metabolism, nitric oxide synthesis, ions balance, and stabilizes the defense systems to improve salt stress tolerance in wheat. Chemosphere 1(287):132142
89. Warner K, Afifi T (2014) Where the rain falls: Evidence from 8 countries on how vulnerable households use migration to manage the risk of rainfall variability and food insecurity. Climate Dev 6(1):1–7
90. Werner C, Correia O, Beyschlag W (1999) Two different strategies of *Mediterranean macchia* plants to avoid photoinhibitory damage by excessive radiation levels during summer drought. Acta Oecologica 20(1):15–23
91. Xiao S, Liu L, Zhang Y, Sun H, Zhang K, Bai Z, Dong H, Li C (2020) Fine root and root hair morphology of cotton under drought stress revealed with RhizoPot. J Agron Crop Sci 206(6):679–693
92. Yahaya MA, Shimelis H (2022) Drought stress in sorghum: Mitigation strategies, breeding methods and technologies—A review. J Agron Crop Sci 208(2):127–142
93. Yang X, Lu M, Wang Y, Wang Y, Liu Z, Chen S (2021) Response mechanism of plants to drought stress. Horticulturae 7(3):50
94. Zargar SM, Gupta N, Nazir M, Mahajan R, Malik FA, Sofi NR, Shikari AB, Salgotra RK (2017) Impact of drought on photosynthesis: molecular perspective. Plant Gene 1(11):154–159

Mitigation of Salinity Stress in Plants Using Gold Nanoparticles

Muhittin Kulak, Mustafa Guven Kaysim, and Canan Gulmez Samsa

Abstract Salt stress is one of the most critical threats to agroecosystems, ranking first due to its negative consequences in crop or non-crop plant species. The agricultural lands are being salinized day by day. The high level of salinization in soils is manifested as stunted plant growth and reduced crop productivity. In order to bolster plant performance, a number of attempts have been done. Nano-structured materials are among the most promising candidates for buffering the severity of stress among the attempts of the last decade. Herewith in the chapter, we have had a discussion and review of the very new novel "gold nanoparticles" and their potential use against salt stress. According to the literature survey, there were very limited reports about the interactive effects of golden nanoparticles and salt stress. The limited reports suggested that gold nanoparticles can be considered novel growth promoting materials due to their regulatory roles in redox status, stress-related attributes, nitrogen metabolism, ion balance, biosynthesis of nitric oxide synthesis, and tetrapyrrole. However, such endpoints can be regarded as the "basic theme" of almost nanoparticles, but the motor theme of gold nanoparticles is still not fully understood. The chapter was concluded with a set of open questions for possible future research.

Keywords Nano-engineered materials · Stress response · Abiotic stress · Nanotechnology

---

All authors contributed equally to the chapter

M. Kulak (✉)
Department of Herbal and Animal Production, Vocational School of Technical Sciences, Igdir University, Igdir 76000, Türkiye
e-mail: muhyttynx@gmail.com

M. G. Kaysim
Department of Field Crops, Faculty of Agriculture, Igdir University, Igdir 76000, Türkiye

C. G. Samsa
Department of Pharmacy Services, Tuzluca Vocational High School, Igdir University, Igdir 76000, Türkiye

A. Husen (ed.), *Plant Response to Gold Nanoparticles*, Smart Nanomaterials Technology, https://doi.org/10.1007/978-981-97-7174-5_11

1 Introduction

1.1 Salt Stress and Its Devastating Effects on Plant Health and Crop Productivity

Due to their sessile entity, plants are constantly challenged by stress factors. Plants have evolved a number of adaptations related to the growth-defense trade-off to ensure their sustainability. The cost of defense is reflected in the stunted or incomplete plant performance. Among the abiotic stresses, as evidenced by over 30,000 documents, salt stress is one of the most studied. There is a growing interest in understanding the mechanisms of salt stress using integrated approaches of physiological, biochemical, and molecular tools. The knowledge of the mechanism is being extended day by day but a new finding makes the system more complicated than ever imagined. We can specify the complexity of mechanism by quoting words of two respected scientists. Of pioneers of plant proteomics, in his two articles [36, 48], Prof. Dr. Jesús Valentín Jorrín Novo talks about the complexity of the plant system. Prof. Jorrin-Novo quotes that "....*No. Proteomics, and more concretely plant proteomics, is in its infancy, at the descriptive stage, with the proteomes observed being just the tip of the iceberg. We are assembling the pieces of a puzzle that will help us to understand how the cell is built and how it works. We are striving to see light at the end of the very long tunnel that links genotype and phenotype that, however, is still too dark. Every proteomics experiment shows us that life is more complex than we have ever imagined, while research continues to be reductionist and simple*".

In a very similar manner, Prof. Dr. Eric Lander (Principal Investigator of the International Human Genome Project) underlined the complexity of the organism after the completion of the Human Genome Project by quoting "*Having the gene map in our possession is like having a passenger plane and its operating manual spread out on the floor of a large hangar, broken down into its smallest parts. Making sense of the information in the gene map and understanding what goes right and wrong in health and disease will be like putting those parts together by looking at the instruction manual and reassembling a plane*".

With regard to the proposed hypothesis on salt stress and its severity, soil salinity is likely to increase in parallel with changes in global climate. Number of reports related to salt stress and its effects on plant health support the hypothesis considered. Agricultural land is being salinized at a rate of 2 million ha per year, according to [35], which also supports the hypothesis. Up to our best literature survey, we can classify plant physiologists into two categories. The first group has addressed their interest in understanding and expanding knowledge of the action mechanism. In this respect, genomic, proteomic, and transcriptomic approaches are among the molecular tools that are being used.

In some cases, mutant variants are used to determine the precise role of a particular molecule. The second group of plant physiologists (mostly agronomist, horticulturist) design and develop methods/strategies to improve plant health and consequently increase productivity. Out of the management, soil amendments [1], irrigation management [49], selection of salt-tolerant crops [9, 10], crop rotation and diversification [13, 47], improved drainage systems [57], fertilizer management [22], genetic engineering [16, 68] are of the strategies employed. However, in the last decade, nano-structured materials have been put into action to increase the tolerance spectra of plants against biotic and abiotic stress factors. Herein, we have had a discussion and review of the very new novel "gold nanoparticles" and their potential use against salt stress. The first sections of the chapter were associated with the synthesis and characterization of the gold nanoparticles and the use of nanoparticles in plants was discussed in the following sections.

1.2 A Brief Overview of the Synthesis of Nanoparticles

Nanoparticles (NPs) (1–100 nm) have garnered significant attention worldwide due to their inspirational features such as adaptability to structural changes, manipulability, and rapid production on a nanoscale [15]. Nanobiotechnology, gaining a unique characteristic with the inclusion of biological systems, has facilitated the combination of a vast diversity of new materials. Nanomaterials have offered opportunities for extensive use in various fields due to their superior chemical, physical, magnetic, and optical properties, as well as electrical conductivity, thermal performance, and catalytic efficiency [28]. Among them, metallic nanoparticles (MNPs) have become highly popular due to their small sizes, high monodispersity, morphologies, and catalytic characteristics [21]. Due to these characteristics, it has become worth applying in multidisciplinary fields such as electronics [61], textiles [24], environmental protection [39], agriculture [31, 32, 44, 64], food (Couto and Almeida et al. 2022,[58]), biosensing [7] biomedical [34], microbiology [8] and cosmetics [46]. The synthesis of NPs with noble metals (Au, Ag, Pt, and Pd) has created a diversity of application areas, thanks to their high surface area, monodispersity, stability, biocompatibility, and non-toxic nature, coupled with advantages in chemical, optical, electrical, mechanical, thermal properties [20, 62, 69]. Gold nanoparticles (AuNPs), which have attracted the most attention from healthcare professionals and scientists in recent years, have significant potential in the fields of textiles [56], energy [17], environment and bioremediation [41], bioimaging [54], biosensors [25], agriculture [50], and medicine [27] due to their unique properties. Since AuNPs can be easily conjugated to proteins, dyes, drugs, antibodies, enzymes, and nucleic acids, they possess great potential diagnostic tracers, drug delivery, photothermal and radio-therapy, gene therapy, tumor detector [51].

In the synthesis of AuNPs, similar to other nanoparticle synthesis methods, chemical, physical, and green synthesis methods are preferred. Each method for the synthesis of AuNPs has its own advantages and disadvantages and NPs are synthesized in different sizes and physicochemical properties depending on the selected method and application requirements [30]. There, the most commonly favored method is the Turkevich Method, which relies on the reduction of gold ions by the reducing agent sodium citrate, leading to the formation of nanoparticles [38].

Another method is the Brust-Schiffrin Method, where gold salts are dissolved in an organic solvent along with a phosphine compound and a solvent, facilitating the synthesis of nanoparticles [63]. In the citrate reduction method, gold ions are typically reduced by the reducing agent sodium citrate, initiating nanoparticle synthesis by forming nuclei [4]. On the other hand, the most popular method of Au-NP in recent times is the "green synthesis" method, which has relatively few studies [52].

NPs synthesized by traditional methods have toxicity, expense, long reaction times, and reaction steps, so they have fallen behind environmentally friendly, cheap, and low-energy green synthesis methods [45]. The green synthesis stages roughly include the preparation of the biomaterial, addition of gold salts, control of reaction conditions, separation of nanoparticles, and characterization [37]. Recently, green synthesis procedures have been developed by harnessing plant extracts, leading to the production of highly efficient products with significant potential in various fields (Table 1).

While the beneficial or toxic effects of nanoparticles in various fields are still a subject of debate in the literature [14, 51, 59], the toxic potential of AuNPs, which are increasingly utilized in the health and biomedical sector, should be taken into consideration. Thus, in vitro and in vivo experiments have demonstrated that AuNPs induce oxidative damage in cell lines and tissues, leading to primary toxicity based on NP properties. Therefore, it is recommended to well-characterize the physicochemical changes before and after treatment with the biological environment [51].

1.3 What Has Been Done with Regard to the Use of Gold Nanoparticles in Plants that Suffer from Salt Stress?

To the best of our literature search, we have identified a few documents related to the use of gold nanoparticles to mitigate the adverse effects of salt stress. These documents have been in circulation for only a few years. Alongside this report, we have had in-depth discussions about the thematic content and key findings of the documents. Of the reports, Ferrari et al. [26] have assayed the effects of gold nanoparticles in *Arabidopsis thaliana* ecotype Columbia 0 (Col-0). The authors pointed out the lack of a comprehensive investigation of the fate of AuNPs and their effects when they are released into plant growth media. In this regard, a physiological, transcriptomic, and proteomic integrated study was handled by the authors. For the primary assays, 0 to 20 mg/ L was applied. After a dose-dependent activity manner, 10 mg/L dose was found to be the best according to the physiological attributes. For that reason, 10 mg/L dose was used for subsequent assays. Concerning transcriptomic analysis, the analysis was based on two different stages including short and long exposure, 6 h and 7 days, respectively. Out of the differentially expressed genes; following 6 h of golden nanoparticle treatment, 121 genes were up-regulated whereas 530 genes were down-regulated. In the case of long term exposure (7 day of AuNP treatment), 3 genes were up-regulated while 3 genes were down-regulated. The authors specified

Table 1 Production of AuNPs by green synthesis method

AuNP	Methods	Size	Shape	Characterization	Applications	Author (s)
Ephedra-AuNPs	Green synthesis	1.3- 15.6	Spherical and triangular	UV–vis, FTIR, HRTEM, EDX, XPS, XRD, and ζ-potential	Antipyretic activity,antiasthma activity and antimicrobial activity	[6]
Cyclodextrin-capped AuNPs	Green synthesis	~1 μm average diameter	Sphere-like	FTIR and 1 H NMR, TEM, EDS, XPS, and XRD	Electrocatalyst	[42]
AuNPs using *licorice* root extract	Green synthesis	2.647–16.25	Spherical	UV–Vis, FTIR, XRD, EDX, TEM, SEM, DLS and HPLC	Antibacterial, antifungal activity	[5]
AuNPs using *Physalis minima*	Green synthesis	15	Spherical	UV–vis, XRD, FTIR, SEM, EDAX, HR-TEM, DLS, and ζ-potantial	Antioxidant, antidiabetic, and antibacterial activity	[53]
Flower-liked AuNP@graphene quantum dots	Green synthesis	47	Spherical	UV–vis, TEM, and EDX	Detection of colorants in the food industry and prohibited dye contaminants in aquaculture	[23]
Phytochemical-stabilized AuNPs	Green synthesis	20	Spherical	TEM, UV/Vis, FT-IR spectroscopy, and ζ-potential	Biocompatibility and antioxidative activity	[40]
AuNPs	Green synthesis	5–11	Spherical	UV–Vis, TEM and HRTEM	Antibacterial activity	[55]
AuNPs using upland cress	Green synthesis	10.7 ± 2.2	Spherical	V-vis, ζ-potential, SEM, EDX, AFM, ATR-FTIR, and XRD	Catalysis, pharmaceutics, and biomedicine	[33]

(continued)

Table 1 (continued)

AuNP	Methods	Size	Shape	Characterization	Applications	Author (s)
AuNPs using *T. arjuna* fruit extract	Green synthesis	20–50	Spherical	UV–vis, FTIR, XRD, AFM, EDX, TEM, DLS, and ζ-potential	Enhancing seed germination activity of *Gloriosa superba*	[29]
AuNPs using mix of *Olea europaea* fruit extract and *Acacia nilotica* husk extract	Green synthesis	44.96	Spherical	UV–vis, FTIR, TEM, SEM, EDS	Antibacterial and anticancer activity	[11]
Au NPs using *Abelmoschus esculentus (L.)* pulp extract	Green synthesis	~ 14	Spherical	UV–vis, FTIR, XRD, EDX	Anticancer and antimicrobial activity	[43]
AuNPs using *Rosa canina* fruit extract	Green synthesis	20–30	Pseudo-spherical	UV–Vis, FT-IR, SEM, TEM, EDX, elemental mapping and XRD	Efficient catalyst in the reductive degradation of toxic water pollutant organic dyes	[3]
AuNPs using *Chenopodium formosanum* shell extract	Green synthesis	8 ± 6	Spherical	EDS, HRTEM, SAED, XRD, FTIR	Antibacterial activity	[18]
Saudi Origanum vulgare L. extract mediated AuNPs	Green synthesis	41–65	Spherical	UV–Vis, ζ-potential, TEM, EDS, FTIR, XRD	Defending against angiogenesis, apoptosis, and proinflammatory cytokines in age-related macular degeneration disease	[67]
AuNPs using *Sphaeranthus indicus* leaf extract	Green synthesis	25	Spherical	UV–Vis, FT-IR XRD, SEM, and TEM	Developing pollen germination media and plant tissue culture	[12]

that the short-term exposure caused the down-regulation of the majority of genes involved in disease resistance, defense response, response to oxidative stress, and metal response. The proteomic findings revealed that oxidative stress-linked proteins were mainly downregulated after exposure to AuNP treatments. Specifically, the translation levels of DGR1 (DUF642 L-GalL-responsive gene 1, At1g80240) and GSTF6 (Glutathione S-transferase F6, At1g02930) were genes/proteins that were reproducibly and robustly regulated upon exposure to AuNPs.

Song et al. [60] synthesized astaxanthin gold nanoparticles (Ast-AuNPs) and then assayed against salt stress in rice. It was reported that 100 mg/L Ast-AuNPs effectively alleviated the effects of salt stress (50 mM NaCl). In salt stress-submitted rice plants, exposure to Ast-AuNPs triggered the biosynthesis of tetrapyrrole and decreased the stress-induced oxidative stress. Corresponding to increases in activities of the antioxidant enzymes, viz. peroxidase, catalase, and superoxide dismutase, a significant reduction in those values of free proline, malondialdehyde, and hydrogen peroxide was observed. The authors further investigated the expression levels of ROS-scavenging genes (OsCATA, OsCATB, OsPRX65, OsPRX89, OsCu/ZnSOD1, and OsCu/ZnSOD2). In accordance with the parameters considered, a-2.19 to 9.92 fold increases have been reported after exposure of Ast-AuNPs.

Wahid et al. [66] showed the efficiency of the green synthesized AuNPs in wheat plants suffering from salt stress by estimating a series of physiological and biochemical analysis. Accordingly, AuNPs significantly regulated/maintained the Na+/K+ ratio and chlorophyll concentration. Furthermore, the accumulation of oxidative stress end-points reactive oxygen (ROS) and nitrogen species (RNS) were buffered and subsequently the ROS/RNS equilibrium was successfully maintained by AuNPs. Finally, the endpoints were translated into induced salt stress tolerance in wheat plants (Fig. 1). In addition to the very scarce documents of salt stress, gold nanoparticles exhibited similar behavior when interacting with cold stress. For instance, [65] demonstrated the effects of gold nanoparticles (diameter 15 nm, concentrations of 5, 10, 20, and 50 $\mu g\ ml^{-1}$) on the redox status and freezing tolerance of wheat seedlings. The nanoparticles were applied from root. The concentration of 20 $\mu g\ ml^{-1}$ was the most effective in combating the freezing at −3 °C for 24 h. The authors suggested that golden nanoparticles can increase lipid peroxidation and reactive oxygen species, which in turn initiates a cascade of responses that enhance stress tolerance (Fig. 1).

2 Conclusion and the Way Ahead

As expected and reported, the biotic or abiotic stress causes an oxidative ROS burst, which in turn brings about molecular damage to DNA, oxidation of amino acids, and lipid peroxidation. Such changes in the primary metabolism of the plants are reflected in the stunted growth of the plants and the reduced productivity of the crops. Among stress sources, salt stress ranks first because of its devastating effects on agroecosystems. In recent decades, the use of nanoparticles to mitigate salt stress has become very common. In this chapter, we have reviewed the documents on AuNPs-salt stress

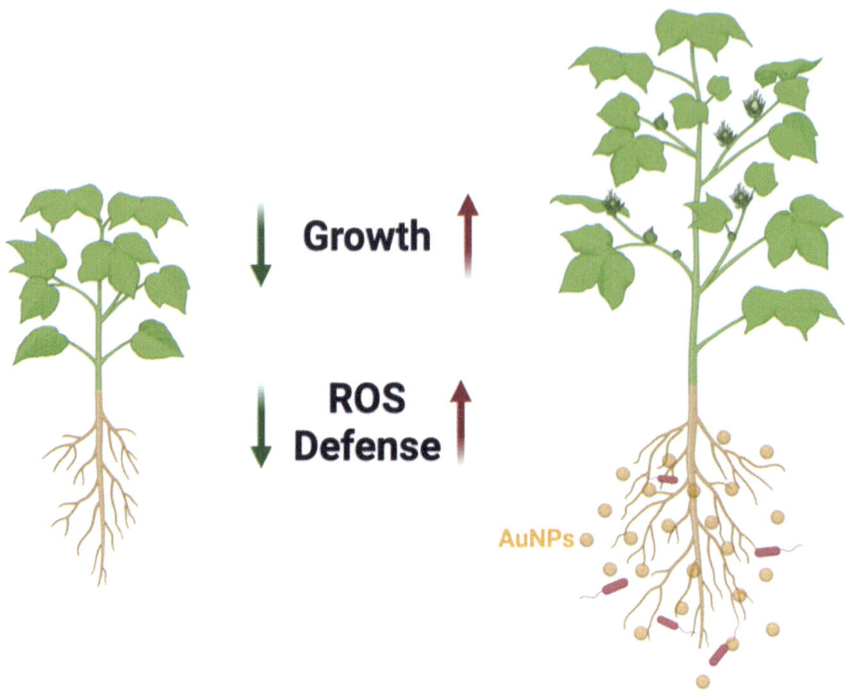

Fig. 1 Hypothetical scheme of the role of the AuNPs in the shift of the trade-off between growth and defense

interaction. However, the documents were very scarce. The general outputs in plant systems after exposure of AuNPs were as follows: gold nanoparticles, as in the case of other growth-promoting molecules, might be candidates in coping with the salt stress via regulating the antioxidant enzymes, stress-linked attributes, nitrogen metabolism, ions balance, biosynthesis of nitric oxide synthesis and tetrapyrrole. However, the reported data can be considered as the "basic theme" of almost all nanoparticles but the motor theme of gold nanoparticles is still not completely elucidated. In this context, there are many concerns regarding the use of gold nanoparticles to mitigate the damage of stress factors and to increase the productivity of plants under stress or stress-free conditions. The discussion could be conducted in a broad framework, including issues related to "mechanism of action, uptake, and distribution, toxicity, and biocompatibility, integration with other stressors, field application and scale-up, economic viability, and environmental impact". Up to the best of our literature review, we have found that the reports on the interactive effects of gold nanoparticles and salt stress are still in their early stages. For instance; (1) what will be the efficacy of gold nanoparticles in salt-stressed plants when other possible stressors (i.e. drought or heat stress or combined stress) occur? (2) What is the switch point between eu-stress or distress of the nanoparticles? (3) What kind of epigenetic regulations are involved with the exposure of the nanoparticles? (4) To what extent are the acquired responses/adaptations recorded in the memory of the plants?

References

1. Ali S, Rizwan M, Qayyum MF, Ok YS, Ibrahim M, Riaz M, Shahzad AN et al (2017) Biochar soil amendment on alleviation of drought and salt stress in plants: a critical review. Environ Sci Pollut Res 24:12700–12712
2. Aliakbarpour F, Vaziri A, Mohseni M (2020) Green synthesis of Au-Ag nanoparticles using *Mentha piperita* and effects of Au-Ag alloy nanoparticles on the growth of *Mentha piperita* under salinity stress. J Plant Process Funct 9:48
3. Alikhani N, Hekmati M, Karmakar B, Veisi H (2022) Green synthesis of gold nanoparticles (Au NPs) using *Rosa canina* fruit extractand evaluation of its catalytic activity in the degradation of organic dye pollutants of water. Inorg Chem Commun 139:109351
4. Al-Johani H, Abou-Hamad E, Jedidi A, Widdifield CM, Viger-Gravel J, Sangaru SS, Basset JM (2017) The structure and binding mode of citrate in the stabilization of gold nanoparticles. Nat Chem 9(9):890–895
5. Al-Radadi NS (2021) Facile one-step green synthesis of gold nanoparticles (AuNp) using licorice root extract: antimicrobial and anticancer study against HepG2 cell line. Arab J Chem 14(2):102956
6. Al-Radadi NS (2023) Ephedra mediated green synthesis of gold nanoparticles (AuNPs) and evaluation of its antioxidant, antipyretic, anti-asthmatic, and antimicrobial properties. Arab J Chem 16(1):104353
7. Arya A, Husen A (2023) Smart nanomaterials in biosensing applications, In: Husen A, Siddiqi KS (Ed) Advances in smart nanomaterials and their applications. Elsevier Inc., 50 Hampshire St., 5th Floor, Cambridge, MA 02139, USA, pp 207–231
8. Arya A, Tyagi PK, Kumar S, Husen A (2023) Nanomaterials and their application in microbiology disciplines. In: Husen A, Siddiqi KS (Ed) Advances in smart nanomaterials and their applications. Elsevier Inc., 50 Hampshire St., 5th Floor, Cambridge, MA 02139, USA, pp 175–194
9. Ashraf M, Munns R (2022) Evolution of approaches to increase the salt tolerance of crops. Crit Rev Plant Sci 41(2):128–160
10. Ashraf M (2004) Some important physiological selection criteria for salt tolerance in plants. Flora-Morphol Distribut Function Ecol Plants 199(5):361–376
11. Awad MA, Eisa NE, Virk P, Hendi AA, Ortashi KM, Mahgoub AS, Eissa FZ (2019) Green synthesis of gold nanoparticles: preparation, characterization, cytotoxicity, and anti-bacterial activities. Mater Lett 256:126608
12. Balalakshmi C, Gopinath K, Govindarajan M, Lokesh R, Arumugam A, Alharbi NS, Benelli G (2017) Green synthesis of gold nanoparticles using a cheap *Sphaeranthus indicus* extract: impact on plant cells and the aquatic crustacean *Artemia nauplii*. J Photochem Photobiol, B 173:598–605
13. Bani A, Daghari I, Hatira A, Chaabane A, Daghari H (2021) Sustainable management of a cropping system under salt stress conditions (Korba, Cap-Bon, Tunisia). Environ Sci Pollut Res 28:46469–46476
14. Bessa MJ, Brandão F, Viana M, Gomes JF, Monfort E, Cassee FR, Teixeira JP (2020) Nanoparticle exposure and hazard in the ceramic industry: an overview of potential sources, toxicity and health effects. Environ Res 184:109297
15. Bhushan I, Singh VK, Tripathi DK (Eds.) (2020) Nanomaterials and environmental biotechnology. Springer, Cham, p 342
16. Bressan RA, Bohnert HJ, Hasegawa PM (2008) Genetic engineering for salinity stress tolerance. Adv Plant Biochem Mol Biol 1:347–384

17. Chen C, Midelet C, Bhuckory S, Hildebrandt N, Werts MH (2018) Nanosurface energy transfer from long-lifetime terbium donors to gold nanoparticles. J Phys Chem C 122(30):17566–17574
18. Chen MN, Chan CF, Huang SL, Lin YS (2019) Green biosynthesis of gold nanoparticles using *Chenopodium formosanum* shell extract and analysis of the particles antibacterial properties. J Sci Food Agric 99(7):3693–3702
19. Couto C, Almeida A (2022) Metallic nanoparticles in the food sector: a mini-review. Foods 11(3):402
20. Craciun AM, Focsan M, Magyari K, Vulpoi A, Pap Z (2017) Surface plasmon resonance or biocompatibility-key properties for determining the applicability of noble metal nanoparticles. Materials 10(7):836
21. Dikshit PK, Kumar J, Das AK, Sadhu S, Sharma S, Singh S, Kim BS (2021) Green synthesis of metallic nanoparticles: applications and limitations. Catalysts 11(8):902
22. Djaman K, Mel V, Boye A, Diop L, Manneh B, El-Namaky R, Futakuchi K et al (2020) Rice genotype and fertilizer management for improving rice productivity under saline soil conditions. Paddy Water Environ 18:43–57
23. Fei X, Liu Z, Li Y, Yang G, Su C, Zhong H, Guo Z et al (2017) One-pot green synthesis of flower-liked Au NP@GQDs nanocomposites for surface-enhanced raman scattering. J Alloy Compd 72:1084–1090
24. Fernandes M, Padrão J, Ribeiro AI, Fernandes RD, Melro L, Nicolau T, Zille A et al (2022) Polysaccharides and metal nanoparticles for functional textiles: a review. Nanomaterials 12(6):1006
25. Ferrari E (2023) Gold nanoparticle-based plasmonic biosensors. Biosensors 13(3):411
26. Ferrari E, Barbero F, Busquets-Fité M, Franz-Wachtel M, Köhler HR, Puntes V, Kemmerling B (2021) Growth-promoting gold nanoparticles decrease stress responses in Arabidopsis seedlings. Nanomaterials 11(12):3161
27. Giljohann DA, Seferos DS, Daniel WL, Massich MD, Patel PC, Mirkin CA (2020) Gold nanoparticles for biology and medicine. Spherical Nucleic Acids 55–90
28. Gopakumar DA, Pai AR, Pasquini D, Ben LSY, Abdul Khalil HPS, Thomas S (2019) Nanomaterials—state of art, new challenges, and opportunities. Nanoscale Mater Water Purif 1–24
29. Gopinath K, Gowri S, Karthika V, Arumugam A (2014) Green synthesis of gold nanoparticles from fruit extract of *Terminalia arjuna*, for the enhanced seed germination activity of *Gloriosa superba*. J Nanostructure Chem 4:1–11
30. Hammami I, Alabdallah NM (2021) Gold nanoparticles: Synthesis properties and applications. J King Saud Univ-Sci 33(7):101560
31. Husen A (2022) Engineered nanomaterials for sustainable agricultural production, soil improvement and stress management. Elsevier Inc., 50 Hampshire St., 5th Floor, Cambridge, MA 02139, USA. https://doi.org/10.1016/C2021-0-00054-7
32. Husen A (2023) Nanomaterials and nanocomposites exposures to plants (response, interaction, phytotoxicity and defense mechanisms). Springer Nature Singapore Pte Ltd., 152 Beach Road, #21-01/04 Gateway East, Singapore 189721, Singapore. https://doi.org/10.1007/978-981-99-2419-6
33. Hutchinson N, Wu Y, Wang Y, Kanungo M, DeBruine A, Kroll E, Zhang W et al (2021) Green synthesis of gold nanoparticles using upland cress and their biochemical characterization and assessment. Nanomaterials 12(1):28
34. Illath K, Wankhar S, Mohan L, Nagai M, Santra TS (2021) Metallic nanoparticles for biomedical applications. Nanomater Their Biomed Appl 29–81
35. Ivushkin K, Bartholomeus H, Bregt AK, Pulatov A, Kempen B, De Sousa L (2019) Global mapping of soil salinity change. Remote Sens Environ 231:111260
36. Jorrin-Novo JV (2020) What is new in (plant) proteomics methods and protocols: the 2015–2019 Quinquennium. Plant Proteomics: Methods Protoc 1–10
37. Khan S, Sher Khan R, Khalid A, Gul M, Brekhna Wadood A, Ullah R et al. (2024) Biomedical and agricultural applications of gold nanoparticles (AuNPs): a comprehensive review. Z Für Phys Chem

38. Kimling J, Maier M, Okenve B, Kotaidis V, Ballot H, Plech A (2006) Turkevich method for gold nanoparticle synthesis revisited. J Phys Chem B 110(32):15700–15707
39. Latif A, Sheng D, Sun K, Si Y, Azeem M, Abbas A, Bilal M (2020) Remediation of heavy metals polluted environment using Fe-based nanoparticles: mechanisms, influencing factors, and environmental implications. Environ Pollut 264:114728
40. Lee J, Kim HY, Zhou H, Hwang S, Koh K, Han DW, Lee J (2011) Green synthesis of phytochemical-stabilized Au nanoparticles under ambient conditions and their biocompatibility and antioxidative activity. J Mater Chem 21(35):13316–13326
41. López-Miranda JL, Esparza R, Rosas G, Pérez R, Estévez-González M (2019) Catalytic and antibacterial properties of gold nanoparticles synthesized by a green approach for bioremediation applications. 3 Biotech 9:1–9
42. Mariappan S, Mutharani B, Kavitha T, Sarojin P, Chiu FC, Ranganathan P (2023) Green synthesis of cyclodextrin-capped AuNPs decorated on polystyrene microspheres for a furazolidone-responsive electrode. Colloids Surf A 664:131171
43. Mollick MMR, Bhowmick B, Mondal D, Maity D, Rana D, Dash SK, Chattopadhyay D et al. (2014) Anticancer (in vitro) and antimicrobial effect of gold nanoparticles synthesized using *Abelmoschus esculentus* (L.) pulp extract via a green route. RSC Adv 4(71):37838–37848
44. Ndaba B, Roopnarain A, Haripriya RAMA, Maaza M (2022) Biosynthesized metallic nanoparticles as fertilizers: an emerging precision agriculture strategy. J Integr Agric 21(5):1225–1242
45. Nguyen NPU, Dang NT, Doan L, Nguyen TTH (2023) Synthesis of silver nanoparticles: from conventional to 'modern'methods—a review. Processes 11(9):2617
46. Pan S, Goudoulas TB, Jeevanandam J, Tan KX, Chowdhury S, Danquah MK (2021) Therapeutic applications of metal and metal-oxide nanoparticles: Dermato-cosmetic perspectives. Front Bioeng Biotechnol 9:724499
47. Qadir M, Tubeileh A, Akhtar J, Larbi A, Minhas PS, Khan MA (2008) Productivity enhancement of salt-affected environments through crop diversification. Land Degrad Dev 19(4):429–453
48. Rey MD, Castillejo MÁ, Sánchez-Lucas R, Guerrero-Sanchez VM, López-Hidalgo C, Romero-Rodríguez C, Jorrín-Novo JV et al. (2019) Proteomics, holm oak (*Quercus ilex* L.) and other recalcitrant and orphan forest tree species: how do they see each other? Int J Mol Sci 20(3):692
49. Rubio JS, Rubio F, Martínez V, García-Sánchez F (2010) Amelioration of salt stress by irrigation management in pepper plants grown in coconut coir dust. Agric Water Manag 97(10):1695–1702
50. Sahu B, Kurrey R, Deb MK, Khalkho BR, Manikpuri S (2023) Recognition of malathion pesticides in agricultural samples by using α-CD functionalized gold nanoparticles as a colorimetric sensor. Talanta 259:124526
51. Sani A, Cao C, Cui D (2021) Toxicity of gold nanoparticles (AuNPs): a review. Biochem Biophys Rep 26:100991
52. Santhosh PB, Genova J, Chamati H (2022) Green synthesis of gold nanoparticles: an eco-friendly approach. Chemistry 4(2):345–369
53. Sekar V, Al-Ansari MM, Narenkumar J, Al-Humaid L, Arunkumar P, Santhanam A (2022) Synthesis of gold nanoparticles (AuNPs) with improved anti-diabetic, antioxidant and anti-microbial activity from *Physalis minima.* J King Saud Univ-Sci 34(6):102197
54. Si P, Razmi N, Nur O, Solanki S, Pandey CM, Gupta RK, de la Zerda A (2021) Gold nanomaterials for optical biosensing and bioimaging. Nanoscale Advances 3(10):2679–2698
55. Siegel J, Kolářová K, Vosmanská V, Rimpelová S, Leitner J, Švorčík V (2013) Antibacterial properties of green-synthesized noble metal nanoparticles. Mater Lett 113:59–62
56. Silva IO, Ladchumananandasivam R, Nascimento JHO, Silva KKO, Oliveira FR, Souto AP, Zille A (2019) Multifunctional chitosan/gold nanoparticles coatings for biomedical textiles. Nanomaterials 9(8):1064
57. Singh G, Lal K (2018) Review and case studies on biodrainage: an alternative drainage system to manage waterlogging and salinity. Irrig Drain 67:51–64
58. Singh N, Gaur S, Chawla S, Singh S, Husen A (2023) Use of smart nanomaterials in food packaging. In: Husen A, Siddiqi KS (Ed) Advances in smart nanomaterials and their applications. Elsevier Inc., 50 Hampshire St., 5th Floor, Cambridge, MA 02139, USA, pp 223–245

59. Sonawane JM (2023) Bionanomaterials to avoid the environmental hazards of conventional nanoparticles. In: Synthesis of bionanomaterials for biomedical applications, Elsevier, pp 37–56
60. Song Y, Zheng C, Basnet R, Li S, Chen J, Jiang M (2022) Astaxanthin synthesized gold nanoparticles enhance salt stress tolerance in rice by enhancing tetrapyrrole biosynthesis and scavenging reactive oxygen species in vitro. Plant Stress 6:100122
61. Tan HW, An J, Chua CK, Tran T (2019) Metallic nanoparticle inks for 3D printing of electronics. Advanced Electronic Materials 5(5):1800831
62. Toropov N & Vartanyan T (2019) 1.03 noble metal nanoparticles: synthesis and optical properties. Comprehen Nanosci Nanotechnol 61
63. Uehara A, Booth SG, Chang SY, Schroeder SL, Imai T, Hashimoto T, Dryfe RA (2015) Electrochemical insight into the Brust-Schiffrin synthesis of Au nanoparticles. J Am Chem Soc 137(48):15135–15144
64. Vanti G, Belur S, Husen A (2023) Use of nanomaterials in agricultural sectors, *In:* Husen A, Siddiqi KS (Ed) Advances in Smart Nanomaterials and their Applications. Elsevier Inc., 50 Hampshire St., 5th Floor, Cambridge, MA 02139, USA, pp. 445–467
65. Venzhik Y, Deryabin A, Popov V, Dykman L, Moshkov I (2022) The effects of gold nanoparticles on redox status and freezing tolerance of wheat seedlings. Acta Physiol Plant 44(11):113
66. Wahid I, Rani P, Kumari S, Ahmad R, Hussain SJ, Alamri S, Khan MIR et al (2022) Biosynthesized gold nanoparticles maintained nitrogen metabolism, nitric oxide synthesis, ions balance, and stabilizes the defense systems to improve salt stress tolerance in wheat. Chemosphere 287:132142
67. Wang J, Wang Y, Li Q (2020) Synthesis of AuNPs using plant polyphenols and their potential treatment for age-related macular degeneration. J Drug Deliv Sci Technol 55:101377
68. Yang X, Liang Z, Wen X, Lu C (2008) Genetic engineering of the biosynthesis of glycinebetaine leads to increased tolerance of photosynthesis to salt stress in transgenic tobacco plants. Plant Mol Biol 66:73–86
69. Zhu C, Du D, Eychmüller A, Lin Y (2015) Engineering ordered and nonordered porous noble metal nanostructures: synthesis, assembly, and their applications in electrochemistry. Chem Rev 115(16):8896–8943

Stress Responses of Wetlands and Aquatic Plants to Gold Nanoparticles

Bhakti Naik, Ishita Oza, Ujwala Jadhav, and Hina Alim

Abstract Concerns over the potential toxicity of projected nanoparticles (NPs) when released into the environment, especially their effects on marine natural elements, have been raised by the growing usage of nanomaterials (NMs) in numerous organizations. The detrimental effects of metallic nanoparticles on marine life, including microalgae, seem to be related to their composition and physical characteristics as well as their ability to lead in conditions that are conducive to deterioration, accumulation, and agglomeration in both land and water. On a fundamental level, the composition displays a wide range of destructive outcomes in light of variations in the techniques applied in microalgae bioassays. Because of the gold nanoparticles (AuNPs), these assortments can cause differentiations of up to five significant degrees in EC50 values. In order to focus on their environmental significance, the message also emphasizes the significance of early cell thickness and the necessity of maintaining an awareness of test circumstances that are as closely aligned with regular conditions as feasible. The fate and toxicity of gold compound nanoparticles on microalgae, which are creatures that thrive in both water and land, are the main topics of this book chapter. The overview also considers the design of bioassays and the direction of future research in the field of nano ecotoxicology. Electron microscopy is also used to examine the impact of AuNP size on species-unequivocal take-up, revealing that the absorption of gold nanoparticles from suspension is associated with root microfibril thickness. The revelations highlight the need to ponder various components, both biotic and abiotic, while anticipating nanomaterial bioavailability to maritime plants. It concludes with a brief analysis of recent advancements in the manufacturing of Au bimetallic NPs by green (green development mediated) methods.

Keywords Gold nanoparticles · Stress response · Aquatic plants · Wetland · Nanoecotoxicology

B. Naik · I. Oza · U. Jadhav · H. Alim (✉)
Department of Life Sciences, University of Mumbai, Santacruz (East), Mumbai 400098, India
e-mail: hina.alim@lifesciences.mu.ac.in

A. Husen (ed.), *Plant Response to Gold Nanoparticles*, Smart Nanomaterials Technology, https://doi.org/10.1007/978-981-97-7174-5_12

1 Introduction

Infinitesimal particles have existed on Earth since its start, where ordinary cycles like volcanic activity and fluid activities have ordinarily made manometer-scale particles [84]. Regardless, non-ordinary planned nanoparticles have emerged in the environment in view of human activities. The area of nanotechnology has experienced basic advancement since the mid 21st 100 years, including the creation, use, and mix of materials under 1 mm, as portrayed by "nanotechnology" [103]. Present-day expressing much of the time insinuates these materials as nanoparticles (NPs), which are materials with something like one angle more unassuming than 100 nm, according to rules like ASTM (2006) and Klaine et al. [56]. NPs track down numerous applications such as in paints, magnificence care items, bioremediation, antibacterial trained professionals, against fouling subject matter experts, antagonistic to sickness drugs, aquaculture, water therapy, air cleansing, food packaging, biosensing, clothing, energy limit things, toothpaste, plastics, security of archeological stones and so on [5, 11, 24, 34, 47–52, 55, 57, 70, 71, 113]. Interests in nanotechnologies extended around 10 billion bucks in 2005 [84], and it is by survey that about 66,000 measurement loads of NPs are conveyed into surface waters consistently. This example should be augmented in the following couple of years [107]. The speedy improvement of the nanotechnology business has raised stresses over the fate and possible risks of NPs, particularly in the environment, including surface waters and oceans, which go about as a conclusive receptor for waste NPs. Some gigantic degree genuinely should intensify creators making NPs arranged near marine or estuarine delivery centers (Klaine et al. 2008). No matter what this is, our understanding of the effects and harmfulness of various NPs on land and water proficient circumstances, especially marine life, remains insufficient [8, 122].

Nanoparticles show unique credits related to their little size [56, 57]. It is evaluated that 30–40% of the particles in each NP are appropriated on a shallow level, achieving high reactivity even in substances that are routinely non-open at greater extensions, introducing potential and surprising prosperity and regular risks [60, 64, 84]. This reactivity is furthermore suitable to metallic NPs like gold mixture NPs (AuNPs). These metallic NPs have found late applications in various fields, and the central delayed consequences of hurtfulness tests on land and water-proficient living things are starting to appear in the sensible composition. For example, there are 213 circulations for AuNPs' toxicity in marine circumstances, 301 in maritime circumstances, with 107 in land and water proficient circumstances. Colloidal normal gold has been used for a seriously lengthy timespan in development treatment and is as of now used in various applications, including versatile driving inks, impulses, threatening to develop creams, paints, sun-situated cell planning, memory contraptions, water cleaning, catalysis, defilement control, and illness treatment and finding. Like AuNPs, concluding the particular proportion of AuNPs conveyed, consumed, and wasted consistently is convoluted, but a harsh evaluation for consumed AuNPs is around 3 metric tons every year. In this chapter, one directs an expansive evaluation of the predetermination, lead, and noxiousness of AuNPs on photosynthetic marine microbiota, consolidating a relationship with freshwater creature classes. One also can explore a progressing methodology for the green biosynthesis of AuNPs, including plant-mediated processes, including green development and microalgae. Likewise, one offers a couple of shutting thoughts on openings in our knowledge and suggest districts for future assessment in this field.

2 Fate of Gold NP in Environment

The fate and behavior of gold nanoparticles (NPs) in the environment are influenced by agglomeration and collection processes. Agglomeration, which is the desolation of particles, is typically the fate for most NPs in maritime environments. Homo-collection is common in bioassays with high NP concentrations, but it doesn't occur in bioassays with low NP concentrations. Hetero-completeness of NPs with various particles present in the media occurs similarly. The settling of NPs in land and water-resistant media can vary depending on factors like collection with suspended matter. However, the combination of NPs doesn't necessarily diminish their harmful effects on maritime natural substances. Ingestion by slow eaters can further increase harm, and its association with harm depends on living organisms and conditions.

Various techniques have been used to understand the size and transport of AuNPs and related particles in bioassays, including diffusive spectroscopy, molecule explicit terminals, ultrafiltration, ICP-MS, spectroscopy, dynamic light scattering, X-Pillar spectroscopy, cyclic voltammetry, stream field stream fractionation (FIFF), separation electron microscopy (SEM), transmission electron microscopy (TEM), wet analyzing electron microscopy (WetSEM), energy-dispersive X-radiates assessment (EDAX or EDX), atomic power microscopy, confocal microscopy, stream cytometry, and neutron activation assessment (NAA). Mesocosm tests have been conducted to better understand the behavior and predetermination of AuNPs in estuarine food organizations. Studies have shown that biofilms play a crucial role in the presence of AuNPs, with a small portion ending up in biofilms. These studies highlight the complexity of NP directly in maritime conditions and the potential for NPs to enter food organizations. While biomagnification processes have not been represented as Au, NPs travel through trophic levels. Researchers continue to study the harm and storage of NPs in the environment using mesocosms, microcosms, and multispecies bioassays.

3 Gold Nanoparticles Toxicity on Microalgae

Metallic nanoparticles pose a significant threat to land and water-efficient ecosystems, as they are the foundation of these ecosystems. In multicellular animals, nanoparticles can cause damage at block levels, similar to breath or skin. However, for unicellular organisms like microalgae, NP-activated damage impacts the entire individual. Metallic NPs and particles have been linked to an increase in reactive oxygen species (ROS) in microalgae, which can be detected by examining the oxidative stress on the board. The harmfulness of metallic NPs to microalgae and other marine animals appears to be lower than that of free metallic particles, but this lower destructiveness doesn't generally affect each marine living thing. Bioassays assessing free particles and metallic NPs in microalgae and other marine living things are not entirely the same. An equilibrium model allows for the assessment of free particles in liquid media, but microalgae surfaces show a high compound proclivity for cations. In group social orders used in bioassays, free particles are quickly absorbed onto cell surfaces, extracellular microalgae exudates, or held within the intracellular space, potentially causing the relentless appearance of broken down metal from NPs.

The low dissolution of most NPs in all development media, reinforced by released substances through green development, presents another challenge in microalgae harm bioassays. The use of sub-lethal surfactant surfactants as carriers should not be excused, as they can help detect disaggregated NPs even in marine environments. Additionally, the potential for damage frameworks to go beyond compound or nuclear interactions and incorporate real joint efforts is another variable to address in metallic NPs toxicity. An extension in cell weight due to NP adsorption could oust microalgae from the phototic zone, subsequently influencing photosynthesis.

4 Toxicity of Gold Nanoparticles (AuNPs)

Gold has long been considered low-toxicity for living natural substances, but when reduced to the nanoscale, it shows extended reactivity due to its fascinating atomic consistency. Continuous assessments using lipid bilayers as model cell layers have shown that AuNPs can interact with lipid bilayers, causing cell damage. Small AuNPs (1.4 nm) have been linked to genotoxicity through their association with DNA grooves. Tannic destructive coated AuNPs show higher dispersal in fake layers than PVP-covered ones, with more unassuming AuNPs showing more essential reactivity than larger ones. Cationic NPs entered cells through scattering, while anionic NPs likely entered through endocytosis. Perreault et al. (2012a) coordinated specific studies on wild-type and odd-type strains of *C. reinhardtii*, investigating the harmful arrangement of AuNPs covered with glycodendrimers at extremely low NP centers. They found culture assortment in wild-type strains, possibly due to the relationship of NPs with cell walls.

The surface study of AuNPs is considered a fundamental factor in their harmfulness. Renault et al. [105] focused on the harmfulness of 10 nm distance across AuNPs on S. subspicatus using a high starting cell thickness, finding a decrease in the fundamental number of cells even at the most negligible AuNP center. Issues associated with biomass arise when coordinating total tests, as low cell densities don't make sufficient biomass for quantitative metal assessment. Moreno-Garrido et al. [82] examined the harmfulness of separated Au and citrate-covered AuNPs on five marine microalgae, finding EC50 values for separated Au for various species.

5 Nanoparticle Harmfulness Data, Bioassays, and Regulatory Scene

The specifying of preliminary designs for bioassays including microalgae and nanoparticles (NPs) remains a creating area of examination. Close to the start of the 21st hundred years, OECD (Relationship for Financial Co-movement and Improvement) rules for land and water proficient harmfulness testing didn't address nanomaterials and their original credits [89]. It was solely in the second 10 years of the century that the OECD introduced reports unequivocally watching out for the prosperity and destructiveness testing of NMs (nanomaterials) [88, 90–96]. Regardless, these reports didn't loosen up to examinations of marine circumstances.

Nanoparticles present incites in hurtfulness testing due to their probable reactivity with various parts inside the land and water proficient circumstances, whether they are designed or regularly occurring. This integrates nutritive courses of action, humic or fulvic acids, amino acids, proteins, sugars, normal matter [23], and salts. The high unambiguous surface locale and reactivity of NPs are key components in these affiliations [4, 101, 124]. The shortfall of standardized NP bioassays for any living being, including microalgae, is a significant issue requiring objective [42]. Prominently, critical assortments can be tracked down in the continuous composition concerning uncovered EC50 values for unequivocal NPs in microalgae [77].

All through the long haul, ecotoxicology guidelines have habitually waited behind legitimate movements. In the area of ecotoxicology concerning nanomaterials (nanoecotoxicology), it is basic for scientists and executives to collaborate, particularly taking into account the quick augmentation of the nanotechnology business. Nanoparticles should be seen as needed substances [7, 40]. A watchful and moderate procedure is central to making an effort not to repeat past mistakes. A start to finish study of the state of biological guidelines associated with the regulatory control of nanomaterials can be found in a report of Milieu Ltd, AMEC, and Specific School of Denmark, 2011. In any case, this report genuinely doesn't address the clever peculiarities of NPs ecotoxicology in estuarine, marine, or oceanic circumstances. In like manner, another change by Kühnel and Nickel [58] of the OECD ace gathering on ecotoxicology and environmental predetermination of NMs moreover neglects this essential normal setting. Subsequently, scientists should heighten their undertakings to appreciate, assess, and model the normal hazards of NPs in marine circumstances to give executives the data expected to make informed decisions around here.

6 Knowledge Gaps and Future Research Directions in Nanoecotoxicology

Nanoecotoxicology is an inherently interdisciplinary field, mentioning collaboration among regular specialists, researchers, and physicists. The outstanding properties of NPs, which stretch the boundaries of quantum-level reactions, require organized work to comprehend the approach to acting, predetermination, and environmental harmfulness. Considering the ongoing gathering of data concerning microalgae, particularly marine microalgae, and the noxiousness of metallic nanoparticles, unequivocally gold (AuNPs) a couple of data openings persevere through that warrant further assessment and investigation.

i. Standardized Bioassays and Insightful Models

One crushing need is the improvement of standardized bioassays highlighted working with the relationship of harmfulness results. It is fundamental to spread out insightful models using a total informational index of hurtfulness responses and the physico-engineered depiction of nanoparticles. These models will add to a more significant cognizance of the potential dangers introduced by NPs. Besides, research attempts should focus on further developing steady noxiousness bioassays, bioaccumulation tests to examine the trophic trade of NPs, and surveys including mesocosms and multi-species bioassays.

ii. Specific Lead in Marine Circumstances

The approach to acting of AuNPs in marine circumstances generally shifts from their lead in freshwater conditions. Hence, it is improper to extrapolate destructiveness results from freshwater life structures to marine animals. To address this, standard techniques and assessment drives should encompass both freshwater and marine circumstances, including estuarine settings, perceiving their outstanding characteristics. Unequivocal complement should be placed on microorganisms staying in marine leftovers, with explicit care in regards to benthic microalgae like benthic diatoms. Stream research on macroalgae's response to NPs stays insufficient and loaded with weaknesses, requiring more complete assessments.

iii. Genotoxicity Investigation

Despite the apparent potential for genotoxicity in unambiguous NPs, for instance, AuNPs, confined assessments have been directed to overview genotoxic influences using microalgae as test animals [95]. The progression of genomics, as well as other 'omics' approaches including proteins or synthetics, is prepared to uncover new encounters into microalgae noxiousness and genotoxicity. Future assessment in this space is crucial to totally survey the genotoxic capacity of NPs.

iv. Interchanges with Various Poisons

A beguiling and generally disregarded element of nano ecotoxicology pivots around the logical interchanges of NPs with various toxins and their joined destructiveness to maritime animals. While it is for the most part recognized that NPs rapidly absolute in freshwater and marine circumstances, progressing assessments have shown the way that low, non-lethal levels of surfactants can stay aware of NP dispersing, even in seawater. The antagonistic or synergistic effects of tainting blends under normally significant conditions warrant further assessment, as such associations would have colossal consequences for maritime natural frameworks.

7 Eco-Nanotoxicology: Investigating the Normal Impact of Nanomaterials

The saying "eco nanotoxicology" is founded from the mix of "nanotoxicology" and "ecotoxicology." It tends to be a particular sensible discipline that bright lights on understanding the normal effects of nanomaterials on normal territory. Given the wide creation and utilization of planned nanomaterials, as well as their reasonable presence in the environment, different assessments have been directed to survey the destructiveness related to nanomaterials. One rehashing issue in the space of eco-nanotoxicology is the deficit of state-authorized testing shows, which has provoked amazing conversation inside the coherent writing concerning the regular impact of nanomaterials. [6] have suggested that the joint revelations associated with the normal perils introduced by nanomaterials should outfit regulatory associations with a sensation of reassurance. Their thinking is laid out in the discernment that a couple of assessments use test conditions with very high nanomaterial centers to provoke any recognizable response. While harmfulness examinations truly ought to be driven under normally relevant transparency centers, it is moreover essential to make datasets that consolidate extraordinary openings, sometimes requiring higher obsessions. Such serious datasets fill a twofold need: choosing no recognizable effect obsessions and laying out the foundation for future nanomaterial rules. Coincidentally, the ongoing collection of composing is stacked with weaknesses concerning the predetermination of nanomaterials in the environment and their natural repercussions.

Maurier-Jones et al. (2013) perceived fundamental openings in eco-nanotoxicity research, particularly in the space of coherent science, encompassing the areas of disclosure, speciation, and change of nanomaterials. These assessment openings feature the necessity for exhaustive eco-nanotoxicity data. Autotrophic species, similar to plants, emerge as a vital place of combination for settling unanswered requests inside the area of eco-nanotoxicology. Taking into account that plants involve the focal pieces of dominance hierarchies and show close correspondences with the incorporating air, water, and soil, they are likely going to encounter nanomaterials. This not only makes plants proper opportunities for looking at nanomaterial take-up and destructiveness but likewise for investigating potential nanomaterial trophic trade processes. Notwithstanding the way that surveys including natural plants have been driven, there is an extraordinary lack of investigation on associations between maritime plants and nanomaterials, except for several examinations.

8 Possible Mechanisms of Nanomaterial Uptake in Plants

[85] have insisted that the fundamental framework essential for cell mischief and noxiousness impelled by nanoparticles is the period of responsive oxygen species (ROS). This period of ROS is attributed to the high surface locale-to-volume extent of

nanoparticles. As particles decrease in size, the degree of surface iotas experiences an extraordinary turn of events. This contrary relationship through and through raises the reactivity of these particles, with the smallest particles showing the most expressed reactivity.

Exceptionally, planned nanomaterials don't be ensured ought to be integrated inside plant cells to set off a response. Amassed nanomaterials on plant roots can influence pressure-driven conductivity and reduce water take-up [3]. At this point, the recognized technique for action concerning how nanomaterials speak with plants is cell take-up and entrance, yet the specific part of take-up remains insufficiently described (Lin S et al. 2009,[15, 54, 59, 63]). Nanomaterials likely follow a similar maintenance way as clamminess and enhancements through plant roots [63]. Nanomaterials can enter plant roots and be moved inside plants through three possible courses. The chief course incorporates entry into the cell wall pores. These pores are portrayed as the joined microfibril network that is the cell wall structure. The size and thickness of these openings or openings between microfibrils contrast depending upon the plant species. These pore structures have been comprehensively assessed in the composition and are by and large perceived to fall inside the 5–10 nm range, for specific reports regardless, exhibiting breadths of up to 50 nm [9, 14–16, 76, 80]. The ensuing vehicle course incorporates apoplastic moves. The apoplast wraps the cell wall and the intercellular space between the cell wall and the plasma layer. This pathway licenses nanomaterials to enter the plant yet holds them back from crossing the plasma layer. Lin et al. 2008 certified the presence of ZnO nanomaterial on the root surface of ryegrass using electron microscopy. The scholars then, at that point, used transmission electron microscopy to confirm that ZnO nanoparticles were adsorbed inside the root apoplast and the cytoplasm. To show up at the cytoplasm, these particles likely crossed the plasma layer and entered the symplast region of the plant. The symplastic pathway, portrayed as the intracellular locale of the plant, involves the third passage and transport pathway. If nanomaterials enter the symplast, they live inside the plasma layer, allowing them to move intracellularly through channels known as plasmodesmata that interface close by cells. Plasmodesmata are customarily around 40 nm in broadness. Though advancing exploration is beginning to determine a couple of requests with respect to nanomaterial coordinated effort and transport inside higher vascular plants, it stays essential to extrapolate the pathways of take-up, development, and bioconcentration. This is especially colossal given that plants go about as a fundamental course of receptiveness to nanomaterials for higher species, including individuals [80, 106].

9 Toxicity Towards Algae and Higher Plants

How we could decipher the potential risks introduced by nanomaterials to higher plant species stays confined [66, 80, 106]. The ongoing gathering of composing features the presence of colossal openings in getting a handle on the predetermination, harmfulness, change, and potential perils related to nanomaterials and their collaborations with plants [109],Helpful et al. 2008; Klaine et al. [56, 57]. Over the past four to five years, there has been a creating focus on the participation of nanomaterials with photosynthetic living things, particularly higher plants, consolidating focuses associated with their vehicle, take-up, and hurtfulness responses.

Planned nanomaterials, beginning from human activities, present a bet in light of their copious responsive districts, high surface locale, and flexibility [75, 109]. In 2008, Helpful et al. 2008 underlined that a critical piece of the ecotoxicological data accumulated for nanomaterials fundamentally connected with standard freshwater living creatures, achieving a deficiency of data concerning tiny organic entities, natural species, marine species, and higher plants. Despite resulting advancement in datasets, it is by and large apparent that there are still respectably several focuses on nanotoxicity with higher plants [99]. In 2008, Navarro et al. [84] coordinated an assessment concerning the potential perils introduced by planned nanomaterials to green development, plants, and parasites. Due to the close relationship between plants and green development in their close by environment, these fundamental creators are helpless against the results of receptiveness to nanomaterials [22, 84].

Different endpoints are available for assessing the phytotoxic responses of plants following receptiveness to anticipated harms. These endpoints consolidate genotoxicity and cytotoxicity, as well as seed germination and seedling improvement. A couple of assessments have reviewed both turns of events and cyto/genotoxicity endpoints in higher plants. For instance. Strangely, *Lactuca sativa* introduced to comparable particles showed root prolongation prevention. This addresses how responses to nanomaterial transparency can vary among different species. In a concentrate by Ghosh et al. [31], the inherited response of plants introduced to nanomaterials was explored. *Allium cepa* and *Nicotiana tabacum* were introduced to 100 nm TiO_2 at combinations of 2–10 mM. The makers noted DNA hurt in *A. cepa*, particularly at the 4 mM center, as shown by a comet look at. Though the mitotic rundown didn't show a gigantic effect, chromosomal mutilations as anaphase-telephase ranges and micronuclei were recognized. In *N. tabacum*, the most raised DNA hurt was seen at a centralization of 2 mM.

Concerning this, both prompt and roaming effects of nanomaterials have been taken note. [45] noted size-subordinate hurtfulness of TiO_2 on the green development *Desmodesmus subspicatus*, with additional humble particles inciting higher harmfulness. This gathers that the size and surface district of the nanomaterial expect an essential part [45]. Roaming responses have in like manner been documented. In 2007, Franklin et al. [25] investigated the destructiveness effects of nanoparticulate ZnO, mass ZnO, and $ZnCl_2$ on the freshwater green development *Pseudokirchneriella subcapitata*. The harmfulness response was credited to the separated zinc segment rather than the nanoparticle itself. Lee et al. [62] on *Pseudokirchneriella subcapitata* introduced to ZnO and TiO_2 nanoparticles, inhibitory effects on improvement were seen in view of the photoreactive thought of these nanomaterials. Regardless, pre-light didn't impact the improvement rate. The harmfulness of nanomaterials was expected by the presence of free metal particles, highlighting the atypical effects of nanomaterials, including the breaking down and appearance of noxious metal particles [61, 84]. In another survey, Lee et al. [61] declared developmental destructiveness in *Arabidopsis thaliana* when introduced to metal oxide nanomaterials, expressly ZnO. This study revealed a critical reduction in seed germination, with simply 6% germination seen at 400 mg/L ZnO (indistinguishable from 14.6 mg/L dissolvable Zn). Besides, Zn demolished root extending by 75% diverged from control drugs [61].

The implications of the natural appearance of nanomaterials (NMs) remain questionable. Continuous papers have highlighted the broad data openings concerning the predetermination and consequences of NMs in the environment, which forestall the quantitative examination of biological bet and the strong organization of NMs in land and water proficient conditions. Research attempts highlighted understanding the appearance of NMs in the environment wraps a far-reaching range, going from transparency showing to life cycle assessments to predict surface water centers [35, 65]. Additionally, concentrates on research on the impact of NMs on crop plants and terrestrial vegetation, similar to wheat, squash, pumpkin, tobacco, and rice. In any case, careful appraisals of NMs' joint efforts with maritime plants are at this point limited [33, 84].

Plants expect a fundamental part in getting over the environment and the biosphere in both natural and maritime organic frameworks. This association point requires further examination of the course, predetermination, and anticipated effects of NMs. Continuous study papers composed by have recorded The correspondences among NMs and natural plants have been recorded by Rico et al. [106] and Miralles et al. [80]. Typically studied endpoints for plant-nanoparticle interchanges consolidate germination and root/shoot prolongation, which give huge datasets. Regardless, there is a prerequisite for extra complete assessments that jump into the major parts. Rico et al. [106] highlights the deficiency of information concerning NM relationship with plants and the various factors influencing their bioavailability. These factors integrate NM size, shape, the course of take-up, mass NM game plan, and surface science. Miralles et al. [80] feature the requirement for overhauled encounters into the parts managing NM take-up and move inside plants, including thoughts of bioavailability.

Terrestrial plants could encounter NMs through various means, including barometrical declaration, unseemly evacuation of NM-containing substances, and the utilization of sewage slop to land, an apparent vault for NMs. Following whirlwind events, flood, crumbling, or the deficiency of feasible best organization practices, the maritime environment could transform into a vault for NMs. While a huge gathering of composing is committed to assessing the impact of NMs on terrestrial yield species, a part of the disclosures so far are pertinent to maritime plants. Different equivalents exist between the physiological patterns of maritime and terrestrial plants, wrapping vascular structures, photosynthetic plans, and plant organ systems (roots and leaves). Before long, land and water-proficient macrophytes can ingest supplements through both their establishments and leaves, rather than various natural plants [21]. This qualification in supplementing features the meaning of looking at the limits that influence NM maintenance in maritime vascular plants, as potential roots may not be the sole site for NM take-up. Studies have demonstrated the way that NMs can enter the underpinnings of various plant species, fundamentally restricted to pores present in the microfibril network plan of the telephone wall. These pores, found inside the microfibril layers, constrain the distance across plans fit for entering through to the telephone film.

All things considered, watching out for these data openings and seeking extra assessment in nano ecotoxicology is vital to impel how one could decipher the regular impact of nanoparticles. Agreeable undertakings across various legitimate disciplines are essential to open the probable benefits and moderate the potential perils related to nanomaterials in the ordinary territory.

10 Conclusions

In the recently referenced choices, one put forward a hypothesis suggesting that the part was essential for the stacking of AuNPs into plant roots is not set in stone by the scattering of AuNPs into the root cortex of maritime plants. This scattering cooperation appears, apparently, to be immovably associated with the advancement of water into the roots. Regardless, it's basic to observe that despite this root stacking, the centralizations of AuNPs in the shoot tissue didn't augment basically, exhibiting immaterial vehicles through the xylem. It's possible that the AuNPs became trapped in the root apoplast and couldn't move into the plant's vascular system on account of the presence of the Casparian strip. While it is speculatively functional for AuNPs to clearly enter the vasculature at the root tip, bypassing the Casparian strip and entering the symplast, this pathway is likely not going to address the greater part of AuNP advancement in maritime plants. This is maintained by past investigations showing that root harming doesn't further develop AuNP take-up. Obviously, the fundamental take-up of AuNPs is size-subordinate, with particles that are excessively tremendous to try and consider diffusing across cell walls staying adhered to the root surface. At the point when an AuNP shows up at the root apoplast, it binds to cell walls, perhaps in regions containing charged surface social events like sugars and polysaccharides These revelations suggest that AuNPs partner with land and water-proficient plants and total inside their tissues. This social event is impacted not simply by water quality factors like ionic strength, hardness, and pH but also by the presence of typical regular matter. The maintenance of AuNPs vacillates among different species and is subject to the presence of roots by virtue of *A. caroliniana*. Regardless of the maintenance assortments, huge combinations of gold were recognized in the plant tissues across all prescriptions.

Notwithstanding, there is a pressing requirement for extra assessments on the ecotoxicity and transportation of nanomaterials in maritime circumstances. Specifically, more thought should be composed towards understanding the parts of NM take-up, the components affecting their bioavailability, and the potential for trophic trade of AuNPs inside and perhaps out of the maritime climate. Also, the results show that extended evapotranspiration, close to the advancement of water inside the plant, prompts higher AuNP stacking in the roots, as demonstrated by tissue obsessions and electron micrographs. These revelations suggest that the speeds of evapotranspiration and water improvement expect a critical part in concluding the maintenance speed of AuNPs by maritime macrophytes into their root tissue. Prominently, there was no connection between shoot tissue centers and evapotranspiration rates, exhibiting that the development of AuNPs from the roots to the shoots is liberated from the root stacking rate. This suggests that scattering is the central purpose behind the improvement of AuNPs from the roots to the shoots at whatever point they are consumed into the root apoplast. All things considered, watching out for these data openings and seeking extra assessment in nano ecotoxicology is pressing to impel how we could decipher the normal impact of nanoparticles. Helpful undertakings across various consistent disciplines are central to open the possible benefits and moderate the potential risks related to nanomaterials in the normal living space.

References

1. Aruoja V, Dubourguier HC, Kasemets K, Kahru A (2009) Toxicity of nanoparticles of bibbibCuO, ZnO and TiO2 to microalgae *Pseudokirchneriella subcapitata*. Sci Total Environ 407:1461–1468
2. Aschberger K, Micheletti C, Sokull-Klüttgen B, Christensen FM (2011) Analysis of currently available data for characterising the risk of engineered nanomaterials to the environment and human health e lessons learned from four case studies. Environ Int 37(6):1143–1156
3. Asli S, Neumann PM (2009) Colloidal suspensions of clay or titanium dioxide nanoparticles can inhibit lead growth and transpiration via physical effects on root water transport. Plant, Cell Environ 32:577–584
4. Auffan M, Rose J, Wiesner MR, Bottero JY (2009) Chemical stability of metallic nanoparticles: a parameter controlling their potential cellular toxicity in vitro. Environ Pollut 157(4):1127–1133
5. Bachheti RK, Bachheti A, Husen A (2024) Metal and Metal-Oxide Based Nanomaterials (*Synthesis, Agricultural, Biomedical and Environmental Interventions*). Springer Nature Singapore Pte Ltd., 152 Beach Road, #21-01/04 Gateway East, Singapore 189721, Singapore
6. Batley GE, Kirby JK, McLaughlin MJ (2012) Fate and risks of nanomaterials in aquatic and terrestrial environments. Acc Chem Res 46:854–862
7. Baun A, Hartmann NB, Grieger KD, Hansen SF (2009) Setting the limits for engineered nanoparticles in European surface waters are current approaches, pp 1774–1781
8. Baun A, Hartmann NB, Grieger KD, Kusk KO (2008) Ecotoxicity of engineered nanoparticles to aquatic invertebrates: a brief review and recommendations for future toxicity testing. Ecotoxicology 17:387–395
9. Berestovsky GN, Ternovsky VI, Kataev AA (2001) Through pore diameter in the cell wall of *Chara corallina*. J Exp Bot 52:1173–1177
10. Bour A, Mouchet F, Silvestre J, Gauthier L, Pinelli E (2015) Environmentally relevant approaches to assess nanoparticles ecotoxicity: a review. J Hazard Mater 283:764–777
11. Boxall A, K Tiede, Q Chaudhry, R Aitken, A Jones, B Jefferson, J Lewis (2007) Current and future predicted exposure to engineered nanoparticles
12. Bozich JS, Lohse SE, Torelli MD, Murphy CJ, Hamers RJ, Klaper RD (2014) Surface chemistry, charge and ligand type impact the toxicity of gold nanoparticles to *Daphnia magna*. Environ Sci Nano 1:260
13. Caballero-Díaz E, Valcarcel M (2014). Toxicity of gold nanoparticles. In Barcelo D (Ed.) Gold Nanoparticles in Analytical Chemistry. Elsevier, Amster-dam, The Netherlands
14. Carpita N, Montezions D, Delmer DP (1979) Determination of the pore size of cell walls of living plant cells. Science 205:1144–1147

15. Chen R, Ratnikova TA, Stone MB, Lin S, Lard M, Huang G, Hudson JS, Ke PC (2010) Differential uptake of carbon nanoparticles by plant and mammalian cells. Small 6:612–617
16. Chesson A, Gardner PT, Wood TJ (1997) Cell wall porosity and available surface area of wheat straw and wheat grain fractions. J Sci Food Agric 75:289–295
17. Crespo E, Lozano P, Blasco J, Moreno-Garrido I (2013) Effect of copper, irgarol and atrazine on epiphytes attached to artificial devices for coastal ecotoxicology bioassays. Bull Environ Contam Toxicol 91:656–660
18. Crespo E, Lozano P, Blasco J, Moreno-Garrido I (2014) Epiphyte toxicity bioassay for coastal monitoring. Environ Monit Assess 186:4647–4654
19. Da Silva BFD, Perez S, Gardinalli P, Singhal RK, Mozeto AA, Barcel, oD (2011) Analytical chemistry of metallic nanoparticles in natural environments. Rev Trends Anal Chem 30(3):528–540
20. Dahoumane SA, Djediat C, Yepr emian C, Cout e A, Fi evet F, Coradin T, Brayner R (2012) Recycling and adaptation of *Klebsormidium flaccidum* microalgae for the sustained production of gold nanoparticles. Biotechnol Bioeng 109(1):284–288
21. Denny P (1972) Sites of nutrient absorption in aquatic macrophytes. J Ecol 60:81 829
22. Dietz KJ, Herth S (2011) Plant nanotoxicology. Trends Plant Sci 16:582–589
23. Egorova EM (2011) Interaction of silver nanoparticles with biological objects: antimicrobial properties and toxicity for the other living organisms. III Nanotechnology International Forum. J Phys Conf Ser 291:1–8
24. Essa AMM, Khallaf MK (2014) Biological nanosilver particles for the protection ofarchaeological stones against microbial colonization. Int Biodeterior Biodegrad 94:31–37
25. Franklin NM, Rogers NJ, Apte SC, Batley GE, Gadd GE, Casey PS (2007) Comparative toxicity of nanoparticulate ZnO, bulk ZnO, and ZnCl2 to a freshwater microalga (*Pseudokircheriella subcapitata*): the importance of particle solubility. Environ Sci Technol 41:8484–8490
26. Gambardella C, Gallus L, Gatti AM, Faimali M, Carbone S, Antisari LV, Falugi C, Ferrando S (2014) Toxicity and transfer of metal oxide nanoparticles from microalgae to sea urchin larvae. Chem Ecol 30:308–316
27. García-Negrete CA, Blasco J, Volland M, Rojas TC, Hampel M, Lapresta Fernandez A, Jimenez De Haro MC, Soto M, Fernandez A (2013) Behaviour of Au-citrate nanoparticles in seawater and accumulation in bivalves at environmentally relevant concentrations. Environ Pollut 174:134–141
28. Garner KL, Keller AA (2014). Emerging patterns for engineered nanomaterials in the environment: a review of fate and toxicity studies. J. Nanoparticle Res 16(8);art. no. 2503
29. Garnham GW, Codd GA, Gadd GM (1992) Kinetics of uptake and intracellular location of cobalt, manganese and zinc in the estuarine green alga *Chlorella salina*. Appl Microbiol Biotech 37:270–276
30. Gerber A, Bundschuch M, Klingelhofer D, Groneberg DA (2013) Gold nanoparticles: recent aspects for human toxicology. J Occup Med Toxicol 8:32–38
31. Ghosh S, Mashayekhi H, Bhowmik P, Xing B (2009) Colloidal stability of Al2O3 nanoparticles as affected by coating of structurally different humic acids. Langmuir 26:873–879
32. Gilroy KD, Neretina S, Sanders RW (2014) Behavior of gold nanoparticles in an experimental algal-zooplankton food chain. J Nanopart Res 16:2414
33. Glenn JB, White SA, Klaine SJ (2012) Interactions of gold nanoparticles with aquatic macrophytes are size and species dependent. Environ Toxicol Chem 31:194–201
34. Gong N, Shao K, Feng W, Lin Z, Liang C, Sun Y (2011) Biotoxicity of nickel oxide nanoparticles and bio-remediaton by microalgae *Chlorella vulgaris*
35. Gottschalk F, Sonderer T, Scholz RW, Nowack B (2009) Modeled environmental concentrations of engineered nanomaterials (TiO2, ZnO, Ag, CNT, Fullerenes) for different regions. Environ Sci Technol 43:9216–9222
36. Handy RD (2012) Environmental toxicity of engineered nanomaterials: focus on aquatic species. Toxicol Lett 211:S12–S13

37. Handy RD, Cornelis G, Fernandes T, Tsyusko O, Decho A, Sabo-Attwood T, Metcalfe C, Steevens JA, Klaine SJ, Koelmans AA, Horne N (2012) Ecotoxicity test methods for engineered nanomaterials: practical experiences and recommendations from the bench. Environ Toxicol Chem 31(1):15–31
38. Handy RD, Owen R, Valsami-Jones E (2008) The ecotoxicology of nanoparticles and nanomaterials: current status, knowledge gaps, challenges, and future needs. Ecotoxicology 17:315–325
39. Handy RD, Van der Brink N, Chapell M, Mühling M, Behra R, Dusinska M, Simpson P, Ahtiainen J, Jha AN, Seiter J, Bednar A, Kennedy A, Fernandes TF, Riediker M (2012) Practical considerations for conducting ecotoxicity test methods with manufactured nanomaterials: what have we learnt so far? Ecotoxicology 21:933–972
40. Hansen SF, Ganzleben C, Baun A (2011) Nanomaterials and European water framework directive. Eur J Law Technol 2(3):111
41. Hartmann NB (2011) Ecotoxicity of engineered nanoparticles to freshwater organisms (PhD thesis). Department of Environmental Engineering, Technical University of Denmark, p. 90
42. Hartmann NB, Engelbrekt C, Zhang J, Ulstrup J, Ole Kusk K, Baun A (2012) The challenges of testing metal and metal oxide nanoparticles in algal bioassays: titanium dioxide and gold nanoparticles as case studies. Nanotoxicology 1–13
43. Hou WC, Moghadam BY, Corredor C, Westerhoff P, Posner JD (2012) Distribution of functionalized gold nanoparticles between water and lipid bilayers as model cell membranes. Environ Sci Technol 46:1869–1876
44. Hou WC, Westerhoff P, Posner JD (2013) Review. Biological accumulation of engineered nanomaterials: a review of current knowledge. Environ Sci Process Impacts 15(1):103–122
45. Hund-Rinke K, Simon M (2006) Ecotoxic effect of photocatalytic active nanoparticles TiO2 on algae and daphnids. Environ Sci Pollut Res 13:225–232
46. Husen A (2022) Engineered nanomaterials for sustainable agricultural production, soil improvement and stress management. Elsevier Inc., 50 Hampshire St., 5th Floor, Cambridge, MA 02139, USA. https://doi.org/10.1016/C2021-0-00054-7
47. Husen A (2023a) Secondary metabolites based green synthesis of nanomaterials and their applications. Springer Nature Singapore Pte Ltd., 152 Beach Road, #21–01/04 Gateway East, Singapore 189721, Singapore. https://doi.org/10.1007/978-981-99-0927-8
48. Husen A (2023b) Nanomaterials from agricultural and horticultural products. Springer Nature Singapore Pte Ltd., 152 Beach Road, #21-01/04 Gateway East, Singapore 189721, Singapore. https://link.springer.com/book/9789819934348
49. Husen A (2023c) Nanomaterials and nanocomposites exposures to plants (response, interaction, phytotoxicity and defense mechanisms). Springer Nature Singapore Pte Ltd., 152 Beach Road, #21-01/04 Gateway East, Singapore 189721, Singapore. https://doi.org/10.1007/978-981-99-2419-6
50. Husen A, Bachheti RK, Bachheti A (2023) Current trends in green nano-emulsions (Food, Agriculture and Biomedical Sectors). Springer Nature Singapore Pte Ltd., 152 Beach Road, #21-01/04 Gateway East, Singapore 189721, Singapore
51. Husen A, Siddiqi KS (2023) Advances in Smart Nanomaterials and Their Applications. Elsevier Inc., 50 Hampshire St., 5th Floor, Cambridge, MA 02139, USA. https://doi.org/10.1016/C2021-0-02202-1
52. Jin-Chul K, Madhusudhan A, Husen A (2021) Smart Nanomaterials in Biomedical Applications. Springer Nature Switzerland AG, Gewerbestrasse 11, 6330 Cham, Switzerland. https://doi.org/10.1007/978-3-030-84262-8
53. Joubert Y, Pan JF, Buffet PE, Pilet P, Gilliland D, Valsami-Jones E, Mouneyrac C, Amiard-Triquet C (2013) Subcellular localization of gold nanoparticles in the estuarine bivalve *Scrobicularia plana* after exposure through the water. Gold Bull. 46(1):47–56
54. Judy JD, Unrine JM, Rao W, Wirick S, Bertsch PM (2012) Bioavailability of gold nanoparticles to plants: importance of particle size and surface coating. Environ Sci Technol 46:8467–8474
55. Kaegi R, Sinnet B, Zuleeg H, Hagendorfer E, Mueller R, Vonbank R, Boller M, Burkhardt M (2010) Release of silver nanoparticles from outdoor facades

56. Klaine SJ, Koelmans AA, Horne N, Carley S, Handy RD, Kapustka L, Nowack B, Von der Kammer F (2012) Paradigms to assess the environmental impacts of manufactured nanomaterials. Environ Toxicol Chem 31:1–14
57. Klaine SJ, Koelmans AA, Horne N, Carley S, Handy RD, Kaputska L, Nowack B, Von der Kammer F (2012) Paradigms to assess the environmental impact of manufactures nanomaterials. Environ Toxicol Chem 33(1):3–14
58. Kuhnel D, Nickel C (2014) The OECD expert meeting on ecotoxicology and environmental fate e towards the development of improved OECD guidelines for the testing of nanomaterials. Sci Total Environ 472:347353
59. Kurepa J, Paunesku T, Vogt S, Arora H, Rabatic BM, Lu J, Wanzer MB, Woloschak GE, Smalle JA (2010) Uptake and distribution of ultrasmall anatase TiO2 alizarin red S nanoconjugates in *Arabidopsis thaliana.* Nano Lett 10:2296–2302
60. Lapresta-Fernandez A, Fernandez A, Blasco J (2012) Nanoecotoxicity effects of engineered silver and gold nanoparticles in aquatic organisms. Review. Trends Anal Chem 32:40–59
61. Lee CW, Mahendra S, Zodrow K, Li D, Tsai YC, Braam J, Alvarez PJJ (2010) Developmental phytotoxicity of metal oxide nanoparticles to Arabidopsis thaliana. Environ Toxicol Chem 29:669–675
62. Lee WM, An YJ (2013) Effects of zinc oxide and titanium dioxide nanoparticles on green algae under visible, UVA, and UVB irradiations: no evidence of enhanced algal toxicity under UV pre-irradiation. Chemosphere 91:536–544
63. Lee WM, An YJ, Yoon H, Kweon HS (2008) Toxicity and bioavailability of copper nanoparticles to the terrestrial plants mung bean (*Phaseolus radiatus*) and wheat (*Triticum aestivum*): plant agar test for water-insoluble nanoparticles. Environ Toxicol Chem 27:1915–1921
64. Levard C, Hotze EM, Lowry GV, Brown GE (2012) Review e environmental transformations of silver nanoparticles: impact on stability and toxicity. Environ Sci Technol 46(13):69006914
65. Limbach LK, Bereiter R, Muller E, Krebs R, Galli R, Stark WJ (2008) Removal of oxide nanoparticles in a model wastewater treatment plant: influence of agglomeration and surfactants on clearing efficiency. Environ Sci Technol 42:5828–5833
66. Lin D, Xing B (2008) Root uptake and phytotoxicity of ZnO nanoparticles. Environ Sci Technol 42:5580–5585
67. Lin S, Reppert J, Hu Q, Hudson JS, Reid ML, Ratnikova TA, Rao AM, Luo H, Ke PC (2009) Uptake, translocation, and transmission of carbon nanomaterials in rice plants. Small 5:1128–1132
68. Ma S, Zhou K, Yang K, Lin D (2015) Heteroagglomeration of oxide nanoparticles with algal cells: effects of particle type, ionic strength and pH. Environ Sci Technol 49(2):932–939
69. Ma Y, Kuang L, He X, Bai W, Ding Y, Zhang Z, Zhao Y, Chai Z (2010) Effects of rare earth oxide nanoparticles on root elongation of plants. Chemosphere 78:273–279
70. Madhusudhan A, Purohit SD, Prasad R, Husen A (2024) Functional smart nanomaterials and their theranostics approaches. Springer Nature Singapore Pte Ltd., 152 Beach Road, #21-01/04 Gateway East, Singapore 189721, Singapore
71. Mandal AK, Ghorai S, Husen A (2023) Functionalized smart nanomaterials for point-of-care testing. Springer Nature Singapore Pte Ltd., 152 Beach Road, #21-01/04 Gateway East, Singapore 189721, Singapore
72. Manzo S, Miglietta ML, Rametta G, Buono S, Di Francia G (2013) Toxic effects of ZnO nanoparticles towards marine algae *Dunaliella tertiolecta.* Sci Total Environ 445–446:371–376
73. Matranga V, Corsi I (2012) Toxic effects of engineered nanoparticles in the marine environment: model organisms and molecular approaches. Mar Environ Res 76:32–40
74. Maurer-Jones MA, Gunsolus IL, Murphy CJ, Haynes CL (2013) Toxicity of engineered nanoparticles in the environment. Anal Chem 85:3036–3049
75. Maynard AD, Aitken RJ, Butz T, Colvin V, Donaldson K, Oberdörster G, Philbert MA, Ryan J, Seaton A, Stone V, Tinkle SS, Tran L, Walker NJ, Warheit DB (2006) Safe handling of nanotechnology. Nature 444:267–269

76. McCann MC, Wells B, Roberts K (1990) Direct visualization of cross-links in the primary plant cell wall. J Cell Sci 96:323–334
77. Menard A, Drobne D, Jemec A (2011). Ecotoxicity of nanosized TiO_2. Review of in vivo data. Environ Pollut 159(3):677–684
78. Milieu Ltd; AMEC Environment & Infrastructure UK Ltd & Technical University of Denmark (2011) Review of environmental legislation for the regulatory control of nanomaterials
79. Miller RJ, Lenihan HS, Muller EB, Tseng A, Hanna SK, Keller AA (2010) Impacts of metal oxide nanoparticles on marine phytoplankton. Environ Sci Technol 44:7329–7334
80. Miralles P, Church TL, Harris AT (2012) Toxicity, uptake, and translocation of engineered nanomaterials in vascular plants. Environ Sci Technol 46:9224–9239
81. Moreno-Garrido I, Lubian LM, Soares AMVM (2000) In fluence of cellular density on determination of EC50 in microalgal growth inhibition tests. Ecotoxicol Environ Saf 47:112–116
82. Moreno-Garrido I, Trombini C, García-Negrete C, Lapresta-Fernandez A, Fernandez A, Blasco J (2012) Toxicity bioassay for dissolved Au and Au-citrate nanoparticles on five marine microalgal species. In: III Simposio internacional en Ciencias del Mar y XVI Seminario Iberico de Química Marina, C adiz, Spain, 24th-26th January 2012
83. Mueller NC, Nowack B (2008) Exoposure modeling of engineered nanoparticles in the environment 42:4447–4453
84. Navarro E, Baun A, Behra R, Hartmann NB, Filser J, Miao AJ, Quigg A, Santschi PH, Sigg L (2008) Environmental behavior and ecotoxicity of engineered nanoparticles to algae, plants, and fungi. Ecotoxicology 17:372–386
85. Nel A, Xia T, Madler L, Li N (2006) Toxic potential of materials at the nanolevel. Science 311:622–627
86. Nowack B, Ranville JF, Diamond S, Gallego-Urrea JA, Metcalfe C, Rose J, Horne N, Koelmans AA, Klaine SJ (2012) Potential scenarios for nanomaterial release and subsequent alteration in the environment. Environ Toxicol Chem 31(1):50–59
87. Odzak, N, Kistler D, Behra R, Sigg L (2014). Dissolution of metal and metal oxide nanoparticles under natural freshwater conditions
88. OECD Environment Health and Safety Publications. Paris. OECD (2010) Guidance manual for the testing of manufactured nanomaterials : OECD's sponsorship programme; first revision. In: Series on the Safety of Manufactured Materials. No. 25. OECD, Paris
89. OECD (2000) Guidance documentation on aquatic toxicity testing of difficult substances and mixtures. In: Series on Testing and Assessment, No. 23
90. OECD (2012) Guidance to sample preparation and dosimetry for the safety testing of manufactured nanomaterials. In: Series on the Safety of Manufactured Materials. No. 36. OECD, Paris
91. OECD (2013a) Co-operation on Risk Assessment: Priorisation of Important Issues on Risk Assessment of Manufactured Nanomaterials e Final Report. In: Series on the Safety of Manufactured Materials. No. 38. OECD, Paris
92. OECD (2013b) Environmentally Sustainable Use of Manufactured Nanomaterials. In: Series on the Safety of Manufactured Materials. No. 39. OECD, Paris
93. OECD (2014a) Addendum to ecotoxicology and environmental fate of manufactured nanomaterials: test guidelines. In: Series on the Safety of Manufactured Materials. No. 40. OECD, Paris
94. OECD (2014b) Addendum to the Report of the OECD Expert Meeting on the Physical Chemical Properties of Manufactured Nanomaterials and Test Guidelines: Presentations Given at the Workshop. In: Series on the Safety of Manufactured Materials. No. 41
95. OECD (2014d) Genotoxicity of manufactured nanomaterials: report of the OECD expert meeting. In: Series on the Safety of Manufactured Materials. No. 43. OECD, Paris
96. OECD, Paris. OECD (2014c). Ecotoxicology and environmental fate of manufactured nanomaterials: test guidelines. In Series on the Safety of Manufactured Materials. No. 40. OECD, Paris

97. Pan JF, Buffet PE, Poirier L, AmiardTriquet C, Gilliand D, Joubert Y, Pilet P, Guibbolini M, De Faverney CR, Romeo M, Valsami-Jones E, Mouneyrac C (2012) Size dependent bioaccumulation and ecotoxicity of gold nanoparticles in an endobenthic invertebrate: the tellinid clam Scrobicularia plana. Environ Pollut 168:37–43
98. Peng X, Palma S, Fisher NS, Wong SS (2011) Effect of morphology of ZnO nanostructures on their toxicity to marine algae. Aquat Toxicol 102:186–196
99. Peralta-Videa JR, Zhao L, Lopez-Moreno ML, de la Rosa G, Hong J, Gardea-Torresdey JL (2011) Nanomaterials and the environment: a review for the biennium 2008–2010. J Haz Mater. 186:1–15
100. Perez S, Moreno-Garrido I, Capit an-Valley LF, Lapresta-Fernandez, A, Lubian, LM, Blasco J (2014) Toxicity of silver and glod-silver alloy nanoparticles in
101. Perreault F, Pedroso Melegari S, Fussinatto CF, Bogdan N, Morin M, Popovic R, Matias WG (2012b) Toxicity of PAMAM-coated gold nanoparticles in different unicellular models
102. Quik JTK, Velzeboer I, Wouterse M, Koelmans AA, van de Meent D (2014) Heteroaggregation and sedimentation rates for nanomaterials in natural waters. Water Res 48(1):269–279
103. Rai M, Duran N (eds) (2011) Metal Nanoparticles in Microbiology. Springer-Verlag, Berlin e Heidelberg, p 305
104. Ramakritinan CM, Kaarunya E, Shankar S, Kumaraguru AK (2013) Antibacterial effects of Ag, Au and bimetallic (Ag-Au) Nanoparticles synthesized from red algae. Solid State Phenom 210:211–230
105. Renault S, Baudrimont M, Mesmer-Dudons N, Gonzalez P, Mornet S, Brisson A (2008) Impacts of gold nanoparticle exposure on two freshwater species: a phytoplanktonic alga (*Scenedesmus subspicatus*) and a benthic bivalve (*Corbicula fluminea*). Gold Bull. 41(2):116–126
106. Rico CM, Majumdar S, Duarte-Gardea M, Peralta-Videa JR, GardeaTorresdey JL (2011) Interaction of nanoparticles with edible plants and their possible implications in the food chain. J Agric Food Chem 59:3485–3498
107. Savolainen K, Bakman U, Brouwer D, Fadeel B, Fernandes T, Kuhlbusch T, Landsiedel R, Lynch I, Pylkkanenm L (2013). 2015e2025. Towards safe and sustainable nanomaterials and nanotechnology innovations. Finnish Institute of Ocupational Health, Helsinki. SCENIHR (Scientific Committee on Emerging and Newly Identified Health Risk), 2014. Nanosilver: Safety, Health and Environmental Effects and Role in Antimicrobial Resistance. European Commission, Luxembourg
108. Schrimer K, Behra R, Sigg L, Suter MJ-F (2013) Ecotoxicological aspects of nanomaterials in the aquatic environment. In: Luther W, Zweck A (Eds.) Safety Aspects of Engineered Nanomaterials. Pan Stanford Publishing
109. Scown TM, van Aerle R, Tyler CR (2010) Review: do engineered nanoparticles pose a significant threat to the aquatic environment? Crit Rev Toxicol 40:653–670
110. Sigg L, Behra R, Groh K, Isaacson C, Odzak N, Piccapietra F, Rohder L, Schug H, Yue Y, Schrimer K (2014) Chemical aspects of nanoparticle ecotoxicology. Chimia 68:806–811
111. Stone V (2009) (Project's coordinator). Engineered nanoparticles: review of health and environmental safety (ENRHES). Project Final Report. Edinburgh Napier University, p 426
112. Su S, Chen B, He M, Xiao Z, Hu B (2014) A novel strategy for sequential analysis of gold nanoparticles and gold ions in water samples by combining magnetic solid phase extraction with inductively coupled plasma mass spectrometry. J Anal Atomic Spectrom 29(3):444–453
113. Taghiyari HR, Morrell JF, Husen A (2023) Emerging nanomaterials (*Opportunities and Challenges in Forestry Sectors*). Springer Nature Switzerland AG, Gewerbestrasse 11, 6330 Cham, Switzerland. https://doi.org/10.1007/978-3-031-17378-3
114. The marine and freshawater microalgae. In: Abstracts of the IV Congress of Marine Sciences. Las Palmas de Gran Canaria, June 11th to 13th, 2014. Perreault, F., Bogdan, N., Morin, M., Claverie, J., Popovic, R., 2012a. Interaction of gold nanoglycodendrimers with algal cells (*Chlamydomonas reinhardtii*) and their effect on physiological processes. Nanotoxicol 6(2):109–120

115. Tiede K, Boxall ABA, Tear SP, Lewis J, David H, Hassellov M (2008) Detection and characterization of engineered nanoparticles in food and the environment. Food Addit Contam Part A Chem Anal Control, Expo Risk Assess 25(7):795–821
116. Tiede K, Hassellov M, Breitbarth E, Chaudhry Q, Boxall ABA (2009) Consid- € erations for environmental fate and ecotoxicity testing to support environmental risk assessments for engineered nanoparticles. J Chromatogr A 1216(3):503–509
117. Tiede K, Tear SP, David H, Boxall ABA (2009) Imaging of engineered nanoparticles and their aggregates under fully liquid conditions in environmental matrices. Water Res 43(13):3335–3343
118. Torkamani S, Wani SN, Tang YJ, Sureshkumar R (2010) Plasmon-enhanced microalgal growth in miniphotobioreactors. Appl Phys Lett 97(4):art. no. 043703
119. Van Hoecke De Schamphelaere KAC, Ali Z, Zhang F, Elsaesser A, RiveraGil P, Parak WJ, Smaggh G, Howard CV, Janssen CR (2011) Ecotoxicity and uptake of plymer coated gold nanoparticles. Nanotoxicology 1–11
120. Velzeboer I, Quik JTK, Van de Meent D, Koelmans AA (2014) Rapid settling of nanoparticles due to heteroaggregation with suspended sediment. Environ Toxicol Chem 33(8):1766–1773
121. Von Moos N, Slaveykova VI (2014) Oxidative stress induced by inorganic nanoparticles in bacteria and aquatic microalgae estate of the art and knowledge gaps. Nanotoxicology 8(6):605–630
122. Walters CR, Pool EJ, Somerset VS (2014) Ecotoxicity of silver nanomaterials in the aquatic environment: a review of literature and gaps in nano-toxicological research. J Environ Sci Health Part A 49:1588–1601
123. Yah CS (2013). The toxicity of gold nanoparticles in relation to their physicochemical properties. Biomed Res 24(3):400e413
124. Zhang H, Smith JA, Oyanedel-Craver V (2012) The effect of natural water conditions on the anti-bacterial performance and stability of silver nanoparticles capped with different polymers. Water Res 46(3):691–699

Effects of Au-NPs on Wetland Plant and Soil Systems

Noureddine Chaachouay, Lahcen Zidane, and Azamal Husen

Abstract Nanomaterials offer significant prospects in agriculture due to their distinctive physicochemical characteristics. The use of gold nanoparticles in biological systems has lately been recognized. However, more knowledge is needed on their potential impacts on plant growth and development. Understanding nanoparticle accumulation, translocation, growth response, and stress regulation in plant systems needs to be improved. Exposure of plants to gold and gold nanoparticles has been shown to elicit both beneficial and detrimental effects. Their development and yield differ among different species. Comprehending the physiological, morphological, and genotoxic alterations that occur during the interaction between Au-NPs and sol plants is crucial for efficiently applying Au-NPs in soil and wetland plants. Therefore, it is anticipated that Au-NPs, among the most extensively utilized engineered NPs in various consumer products, will infiltrate natural ecosystems via multiple pathways, including soil and plants. After reviewing the existing literature, conducting future investigations using a multidisciplinary integrated approach is highly recommended. This research should focus on understanding the fate, transformation, accumulation, and toxicity potential of Au-NPs in soil–plant systems. Additionally, it should investigate the overall impact of Au-NPs on environmental and human health.

Keywords Agriculture · Gold nanoparticles · Wetland plants · Soil systems · Translocation · Uptake

N. Chaachouay (✉)
Agri-Food and Health Laboratory (AFHL), Faculty of Sciences and Techniques of Settat, Hassan First University, Address: Po Box 382, 26000 Settat, Morocco
e-mail: nour.chay@gmail.com

L. Zidane
Animal Productions and Agro-Industry Laboratory, Department of Biology, Faculty of Sciences, Ibn Tofail University, B.P. 133 14000, Kenitra, Morocco

A. Husen
Wolaita Sodo University, PO Box 138, Wolaita, Ethiopia

Department of Biotechnology, Graphic Era (Deemed to Be University), Dehradun, Uttarakhand 248002, India

A. Husen (ed.), *Plant Response to Gold Nanoparticles*, Smart Nanomaterials Technology, https://doi.org/10.1007/978-981-97-7174-5_13

1 Introduction

The growing use of gold nanoparticles (Au-NPs) in many domains leads to their inadvertent discharge into terrestrial, aquatic, and atmospheric ecosystems [38]. Plants play a crucial role in all ecosystems, and when Au-NPs interact with plants, they are absorbed and stored in the plant's biomass. This interaction determines how Au-NPs are distributed and transported in the environment. Recently, there has been a surge in research emphasizing the possible impacts of Au-NPs on plants [17, 42, 49].

Au-NPs adhere to various surfaces of plants and are then absorbed via tiny apertures in the plant at the micrometer and nanometer size. ANPs may enter the plant body by many routes, and the pace at which they are absorbed depends on the size, shape, concentration, and surface charge of the Au-NPs [9]. The aerial component enhances airborne Au-NPs' interaction with the shoots' surfaces. Additionally, the fraction of designed Au-NPs discharged into the environment may get scattered by wind and subsequently reach the foliage of plants. Plants engage with atmospheric Au-NPs via their leaves, namely the leaf stomata and hydathodes, which are crucial for absorbing Au-NPs through the foliage [39]. The stomatal channel exhibits high capacitance due to its vast size exclusion limit exceeding 10 nm and rapid transport velocity. Nevertheless, the presence of permeability fluctuation renders this pathway very unpredictable. The particles' dimensions and concentration significantly influence nanoparticle absorption and dispersion in the plant system. Small hydrophobic Au-NPs may be internalized into nonpolar regions of the cuticle with nonpolar and polar entry routes [6]. Larger particles that lack a protective cuticle, such as stomata, hydathodes, and the stigma of flowers, are absorbed via some parts of the plant's surface [27].

The soil Au-NPs interact with the root system of plants, and the process of adsorption allows the Au-NPs to be incorporated into the cell wall and subsequently taken up by the cells [19]. Most primary roots possess a suberized exodermis and endodermis, which hinders the apoplastic passage of solutes and water from the soil to the central cylinder. Nevertheless, the apoplastic bypass might occur via the newly developed lateral roots that penetrate the cortex [36]. Therefore, the Au-NPs might penetrate the xylem via the cortex and central cylinder at the site where lateral roots are formed. Thus, Au-NPs can enter the plant system and lead to favorable or undesirable changes [55]. Further, on the given facts, the fate of engineered gold nanoparticles in the ecosystem and food chain integrity is of great concern. Plant cells having a negative surface charge facilitate the movement of negatively charged surface chemicals into the apoplast. Hence, introducing Au-NPs into ecosystems gives rise to substantial apprehensions about their safety and potential environmental harm [7, 57].

Given that plants play a crucial role in the environment as the main level of the food chain, it is essential to thoroughly comprehend the effects of Au-NPs on plants to evaluate their toxicity [28]. Furthermore, the computation of the ecological ramifications of Au-NPs necessitates the collaboration of a diverse group of specialists from various scientific disciplines: physicists analyze their structures, environmental engineers quantify the number of substances discharged into the environment, chemists, biologists, and toxicologists assess any adverse or harmful effects they may have [24, 47]. As a result, given the limited efforts and investments regarding the potential environmental consequences of this current Au-NPs boom, it is crucial to concentrate on this particular facet of nanotechnology. This chapter provides a comprehensive understanding of the effects of Au-NPs on wetland plant and soil systems.

2 Sources of Au-NPs in Wetlands

2.1 Anthropogenic Sources

The presence of Au-NPs in wetlands may be ascribed to various causes, including natural and human-induced factors (Fig. 16.1). Human activities are a significant method for introducing Au-NPs into wetland ecosystems [31]. Wastewater discharges, especially from enterprises producing and using nanomaterials or goods containing gold, are an essential anthropogenic source [42]. These discharges may include Au-NPs as byproducts or effluents from activities utilizing gold nanoparticles. They may reach wetlands via sewage systems or runoff, affecting both these ecosystems' aquatic and terrestrial components [23]. Furthermore, industrial operations such as gold mining and processing have the potential to release Au-NPs into the environment in addition to wastewater discharges. Gold mining activities can potentially discharge Au-NPs into nearby water bodies, especially those linked to wetlands [5]. This may occur when sediments containing gold particles erode and are carried away by runoff. Besides, unintended spills and inadequate disposal methods in industrial environments might lead to the emission of Au-NPs into the surroundings, augmenting the likelihood of their occurrence in wetlands. For instance, Au-NPs may serve as a means to eliminate contaminants from water, although they can also damage aquatic life. The presence of Au-NPs in wetlands resulting from human activities raises concerns about their possible biological impacts on wetland vegetation, soil, and the overall ecosystem [15, 59]. Comprehending and controlling these origins is vital for safeguarding the authenticity of wetland ecosystems and reducing possible ecological consequences.

2.2 Natural Sources

The primary source of Au-NPs in wetlands is natural, resulting from geological processes, namely the erosion of gold materials. Gold is an element that occurs naturally and may be found in different minerals [16]. This can result in the release of Au-NPs into the environment due to weathering. Hydrological and geological processes, such as erosion, transportation, and deposition, substantially impact the movement of gold particles from mineral deposits and their transport to wetland

Fig. 16.1 Sources of Au-NPs in wetlands

environments [32]. Over long periods, these natural processes may add to the buildup of Au-NPs in wetlands, integrating them into the natural background levels of the domain [25]. Furthermore, wetlands may naturally produce Au-NPs. Some wetlands may have sediments or soils that possess elevated quantities of gold, maybe due to the past accumulation of gold-laden substances or geological structures. When water comes into contact with these sediments and soils, it could release Au-NPs into the water. Moreover, wetlands often change water levels and redox conditions, which might impact the ecosystem's movement and distribution of Au-NPs.

3 Uptake and Accumulation in Wetland Plants

The process of uptake and accumulation of Au-NPs is a complex and ever-changing occurrence that may be influenced by several factors, such as the circumstances of exposure, the chemical characteristics of Au-NPs (such as particle size, surface charge, aggregation state, hydrophilic/hydrophobic properties, and the adsorption of biomolecules/proteins), and the specific kind of crop being studied (Fig. 16.2). The concentration and charge of the Au-NPs play a crucial role in their absorption by roots and their movement to aboveground plant parts, such as leaves and stems [6, 8, 33].

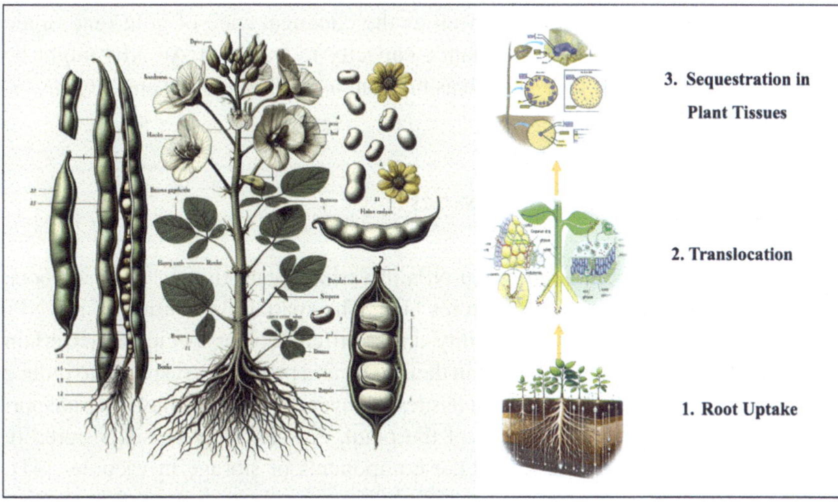

Fig. 16.2 The process of uptake and accumulation of Au-NPs in Wetland Plants

3.1 Root Uptake

The mechanism of Au-NP uptake in wetland plants begins with root absorption. Plant roots may uptake Au-NPs in the soil via several processes [30]. Passive diffusion is a significant process whereby tiny Au-NPs may enter plant roots by traversing the gaps between cell membranes in the roots. The involvement of active uptake processes is also likely since specialized transporters or channels in the membranes of root cells facilitate the absorption of Au-NPs [12]. Differences in root architecture, physiology, and the existence of specialized transporters may lead to variations in the efficiency of root absorption across plant species [20]. Moreover, the presence of Au-NPs in the nearby soil and the physicochemical characteristics of the nanoparticles, such as their dimensions and surface charge, might impact the absorption of these particles by plant roots.

3.2 Translocation

After being absorbed by the roots, Au-NPs may be transferred throughout the plant via translocation. Translocation may occur via the plant's vascular system, including the xylem and phloem [27]. Many variables influence the plant's translocation of Au-NPs. Plant species, such as Au-NPs, vary in their ability to translocate substances [6, 30]. Some species are adept at transporting Au-NPs from the roots to the shoots, while others tend to concentrate them in the root system [33, 44]. The pace and amount of translocation may be influenced by factors such as the dimensions and morphology of the nanoparticles, as well as the chemical state of gold (e.g., ionic or nanoparticulate). Moreover, the plant's capacity to transport Au-NPs might be affected by environmental factors such as nutrient availability and water stress.

3.3 Sequestration in Plant Tissues

Sequestration entails accumulating Au-NPs in various plant tissues, such as roots, stems, leaves, and reproductive structures [12]. The spatial distribution of Au-NPs inside these tissues may exhibit variability across different plant species [30]. Certain plants selectively accumulate Au-NPs in their roots as a protective response to reduce the possible harm in their aboveground tissues. Conversely, some plants can transport Au-NPs to the aboveground sections of the plant. Au-NPs may be sequestered in plant tissues by either binding to cellular components or storage in vacuoles [41]. The precise methods of sequestration may differ and may be impacted by factors such as plant species, growing circumstances, and the properties of the Au-NPs.

The processes are greatly influenced by the type of plants and the environment in which they develop [10]. Plant species differ in their capacities to absorb, transport, and store Au-NPs, leading to distinct patterns of Au-NP accumulation in plant tissues [6]. Growth factors, such as soil qualities, nutrient availability, and environmental stresses, may influence these processes. For instance, soils rich in nutrients might modify the absorption and movement of gold Au-NPs when plants adjust to changing nutritional circumstances [4]. Furthermore, environmental factors such as dryness or metal toxicity could influence the processes of Au-NPs absorption and accumulation.

4 Effects on Wetland Plant Physiology

The advent of enhanced technology using gold nanoparticles has significant potential for future applications across many industries. Moreover, the interaction between these nanoparticles and biological systems raises concerns over potential toxicological consequences attributed to their distinctive physiochemical features [26, 45]. While gold is not a necessary element for plant development, the likelihood of soil contamination with gold increases as it is applied more often, leading to increasing levels of gold exposure for plants. Gaining insight into the interaction between nanoparticles (NPs) and plants is crucial for assessing their hazardous properties. A study investigated the impact of surface charge on the absorption and dispersion of gold nanoparticles in four distinct plant species: rice, radish, pumpkin, and ryegrass [30, 53]. Seedlings were subjected to hydroponic exposure to an NP solution for five days. It was noted that the surface charge and the specific plant species influenced the absorption and dispersion of NPs. Positively charged NPs were absorbed by the roots of the plants, while negatively charged NPs were effectively transported to the shoots of the plants [53, 56].

Ryegrass and radish roots exhibited more significant NPs than rice and pumpkin roots. The impact of nanoparticle characteristics on the absorption and movement inside an organism shows a substantial correlation between NPs and cellular biomolecules [29]. The absorption of nanoparticles is also influenced by their size, as demonstrated by the observation that gold NP aggregates were found in the cytoplasm of tobacco roots when exposed to 3.5 nm gold nanospheres but not when exposed to 18 nm Au-NPs [27, 63].

Additionally, seedling toxicity experiments were conducted utilizing *Arabidopsis* plants treated with $KAuCl_4$ at varying concentrations. It was noted that an elevated concentration of $KAuCl_4$ impeded the growth of roots, while germination remained unaffected by any of the gold concentrations [48]. Besides, the genetic and physiological reactions of Arabidopsis to Au-NPs were examined [34]. A 75% reduction in the root lengths of seedlings cultivated on nutrient agar media supplemented with 100 ppm Au-NPs was documented in the report [1]. Oxidized gold was detected in both soil and stems, while reduced gold in nanoparticles was exclusively detected in root tissues.

An investigation was conducted to examine the growth profile and yield of Indian mustard when treated with five distinct doses of Au-NPs by foliar spray in a field setting [52]. According to this study, the germination percentage of Indian mustard seeds was enhanced when treated with 25 ppm Au-NPs compared to the control seeds. This improvement may be attributed to the increased permeability of the seed capsules, which allowed for more entry of water and di-oxygen into the cells. The treatment of Au-NPs resulted in an observed increase in the number of leaves per plant, with the most significant increase reported when treated with ten ppm Au-NPs [18]. Furthermore, it was noted that the application of NPs did not increase the average leaf area. Treated plants exhibited an enhanced growth profile characterized by increased plant height, average stem diameter, and number of branches per plant. The treatment with ten ppm Au-NPs resulted in the highest observed increase [2]. A higher average yield was seen when the number of pods per plant rose under the NPs treatment.

The addition of Ag-NPs into the nutritional medium resulted in a drop in seed germination, a slowdown in the production of nodules (due to a decrease in the population of the symbiotic *bacteria Rhizobium leguminosarum*), a reduction in shoot growth, and a drop in root length in *Vicia faba* L. Ag-NPs did not hurt seed germination in *Solanum lycopersicum* L. and *Raphanus sativus* L. under hydroponic culture [13, 14]. However, it did lead to decreased root and shoot length and a modest reduction in photosynthetic activity. The viability of the microalga *Skeletonema costatum* Grev. was reduced, and its chlorophyll content decreased due to an abundance of reactive oxygen species from exposure to Ag-NPs [60].

Several studies have shown the interaction between gold nanoparticles and tobacco and wheat plants [22, 46, 61]. The researchers conducted an experiment where tobacco and wheat seedlings were exposed to gold nanomaterials of varying sizes (10, 30, and 50 nm) coated with either tannate or citrate. The seedlings were treated hydroponically for three days for wheat and seven days for tobacco [58]. The experiment aimed to investigate how nanoparticles' size and surface chemistries affect their bioavailability to plants [21]. This research elucidated the function of plant cell wall pores in the absorption of nanoparticles. It examined how soil components affect the inherent characteristics of nanomaterials and their subsequent movement into plants.

5 Impact on Wetland Soil Systems

Soil is the primary recipient of nanoparticles. Studying how nanoparticles behave in soil and assessing their potential risks in agricultural soil ecosystems and other environmental situations is now very relevant. Soil is a naturally occurring substance with a high concentration of natural nanoparticles, which exist as individual particles and clusters or groups of particles [35]. Intentionally introducing nanoparticles into the soil may have a substantial impact due to their high resistance to degradation and ability to accumulate in the soil [43]. Au-NPs are a specific category of artificially created nanomaterials that have garnered significant interest in several disciplines, such as environmental research [51]. When Au-NPs are introduced into wetland soil ecosystems, they could impact soil features such as structure, microbial populations, and nutrient cycling.

Au-NPs may alter soil structure; these nanoparticles can engage with soil particles and organic materials, which might modify the arrangement and permeability of the soil [54]. The adhesion of Au-NPs to soil constituents may induce alterations in the soil's ability to retain water and facilitate aeration [40]. Consequently, these modifications can impact the development of plants and the general well-being of the wetland environment. In addition, the presence of Au-NPs may influence the soil's hydraulic conductivity, thereby causing alterations in the movement of water within the wetland [5]. This could affect soil erosion and the transportation of silt.

Furthermore, Au-NPs may have ramifications for microbial populations in marsh soil. Microorganisms are essential in the processes of nutrient cycling and decomposition. Au-NPs could disturb microbial communities by interacting with microbial cell membranes and influencing microbial physiology [43, 54]. Alterations in microbial composition and activity may impact the decomposition of organic matter, the availability of nutrients, and the overall processes of nutrient cycling in the soil of wetlands [13]. Consequently, this may affect the character of the plant community and the availability of nutrients for wetland plants.

In addition, Au-NPs may influence the nitrogen cycling process in marsh soil. They can adsorb nutrients, such as phosphorus, and modify their bioavailability [42]. These effects may have substantial implications for the absorption of nutrients by plants and the overall productivity of wetlands. Moreover, the existence of Au-NPs might influence the movement and accessibility of heavy metals in the soil, possibly resulting in either heightened or reduced toxicity to plants and soil organisms (Murali et al., 2022b). Comprehending the processes by which Au-NPs engage with soil constituents and their consequences on the circulation of nutrients is crucial for evaluating their possible ecological repercussions in wetland habitats.

6 Ecological Consequences

Au-NPs in wetland environments may have broader ecological implications beyond individual plants and soils, including plant communities, food chains, and ecosystem services. The significance of these implications lies in that wetlands are biologically crucial ecosystems that provide a range of essential functions and services. The incorporation of Au-NPs may modify wetland plant communities via many mechanisms. Varying plant species may exhibit distinct responses to the presence of Au-NPs, resulting in changes in species composition [62]. Certain plant species may demonstrate enhanced resistance to Au-NPs, whilst others may be more susceptible, resulting in alterations in competitive dynamics [3]. In addition, if specific plant species have a high capacity for collecting Au-NPs, they might grow dominant and outcompete local species. Changes in plant communities may have a series of repercussions on other parts of the ecosystem. Modifications in plant communities may impact herbivores and, subsequently, higher trophic levels in wetland food webs [37]. Changes in plant availability and quality may affect herbivorous insects and other creatures. Furthermore, the buildup of Au-NPs in plants may result in the consumption of these nanoparticles by herbivores, thereby impacting their physiology and overall health [52]. These alterations can propagate across the food chain, affecting predators and decomposers and possibly causing disturbances in trophic relationships.

Wetlands provide several ecological services, such as water filtration, flood control, and habitat creation. The existence of Au-NPs might influence these services. Changes in plant communities may impact the wetland's capacity to filter and cleanse water, possibly leading to a decline in water quality [21]. Alterations in the arrangement of plant life may also affect the wetland's ability to reduce the intensity of flooding. Moreover, modifications in plant communities might influence the accessibility of animal habitats, thereby influencing the local biodiversity and the wetland's function as a habitat for various species [50]. It is crucial to acknowledge that the ecological effects of Au-NPs in wetlands are now being actively studied. The precise repercussions might differ based on several aspects, including the concentration, size, and coating of Au-NPs and the distinct features of different wetland ecosystems. Hence, it is essential to conduct thorough investigations to understand the enduring consequences of Au-NPs on wetland ecosystems. These studies will provide valuable insights for developing management approaches that mitigate possible adverse effects and optimize wetland services and biodiversity conservation.

7 Mitigation and Future Research Directions

Comprehensive strategies are necessary to address the possible adverse impacts of Au-NPs in wetlands. Initially, it is essential to enforce regulatory measures to manage the emission of Au-NPs from industrial sources and wastewater discharges. Implementing these criteria may effectively restrict the entry of Au-NPs into wetland habitats [11]. It is crucial to establish monitoring programs to verify adherence to these requirements. Furthermore, it is essential to prioritize the development of conscientious nanoparticle design, explicitly targeting the alteration of the dimensions, configuration, or surface characteristics of Au-NPs to mitigate their harmful effects on living organisms and the environment. Collaboration between researchers and manufacturers may result in the creation of Au-NPs that have enhanced biocompatibility. Furthermore, the restoration of wetlands may play a crucial role in alleviating the impacts caused by Au-NPs. Establishing buffer zones between possible Au-NPs sources and wetlands, together with advocating for native wetland vegetation, might effectively facilitate the filtration and containment of nanoparticles before they enter these vulnerable ecosystems. Restoring wetlands may improve their ability to withstand environmental pressures, which might help reduce the ecological impact of Au-NPs [42].

Conducting more studies is crucial to understand better the long-term effects and destiny of Au-NPs in wetland ecosystems. Research should include investigations on the bioavailability, accumulation, transformations, and toxicological impacts of Au-NPs on wetland plants and animals. Monitoring wetlands exposed to Au-NP may provide helpful information into these nanoparticles' long-term presence and ecological effects in these environments (Avellan et al., 2019b). Furthermore, it is essential to investigate the possibility of Au-NPs engaging in intricate interactions with wetland soils and sediments, as this might impact their behavior and environmental destiny. To mitigate the environmental concerns linked to Au-NPs and protect the integrity of wetland ecosystems, we may use a combination of regulatory measures, responsible nanoparticle design, ecosystem restoration, and extensive investigation.

8 Conclusion

The significance of technological developments in nanotechnology has dramatically increased. Nevertheless, it has resulted in detrimental consequences for the ecology. Scientists have discovered that nanoparticles may enhance plant development and regulate pollutants. However, these minuscule particles pose a significant risk to the environment and the populations of microorganisms in the rhizosphere, owing to their ability to accumulate. To fully understand the influence of NPs on biological interactions in soil, it is crucial to get relevant information on their dual activities. The impact of designed Ag-NPs on soil–plant systems is a significant issue because of their critical interaction with the environment. Furthermore, the literature reviewed in this paper indicates that our understanding of the modulation of Ag-NP-chemistry in soils and its importance for Ag-NPs and Ag/Ag^+ bioavailability and bio-accessibility to soil biota, predominantly plants, is still in its early stages. The prevailing reports on the interaction between Ag-NPs and soil and plants have consistently shown the adverse impacts of Ag-NPs. Furthermore, there needs to be more knowledge on the probable processes responsible for the uptake of Ag-NPs by plants and the harmful effects of Ag-NPs on plants. The transfer of Ag-NPs from seeds to the next generation of plants and their future destiny within food chains must be better understood. It is highly recommended that a multidisciplinary integrated approach be incorporated in future research to investigate the quantitative and qualitative aspects of the fate, transformation, accumulation, and toxicity potential of rapidly increasing Ag-NPs in soil–plant systems. This research should also examine the cumulative impact of Ag-NPs on environmental and human health.

References

1. Agrawal S, Kumar V, Kumar S, Shahi SK (2022) Plant development and crop protection using phytonanotechnology: a new window for sustainable agriculture. Chemosphere 299:134465. https://doi.org/10.1016/j.chemosphere.2022.134465
2. Akhtar MJ, Ahamed M, Alhadlaq HA, Kumar S, Alrokayan SA (2020) Mitochondrial dysfunction, autophagy stimulation and non-apoptotic cell death caused by nitric oxide-inducing Pt-coated Au nanoparticle in human lung carcinoma cells. Biochimica et Biophysica Acta (BBA)—General Subjects 1864(1):129452. https://doi.org/10.1016/j.bbagen.2019.129452
3. Alabresm A, Chandler SL, Benicewicz BC, Decho AW (2021) Nanotargeting of resistant infections with a special emphasis on the biofilm landscape. Bioconjug Chem 32(8):1411–1430. https://doi.org/10.1021/acs.bioconjchem.1c00116
4. Aqeel U, Aftab T, Khan MMA, Naeem M, Khan MN (2022) A comprehensive review of impacts of diverse nanoparticles on growth, development and physiological adjustments in plants under changing environment. Chemosphere 291:132672
5. Auvinen H, Gagnon V, Rousseau DPL, Du Laing G (2017) Fate of metallic engineered nanomaterials in constructed wetlands: Prospection and future research perspectives. Rev Environ Sci Bio/Technol 16(2):207–222. https://doi.org/10.1007/s11157-017-9427-0
6. Avellan A, Yun J, Zhang Y, Spielman-Sun E, Unrine JM, Thieme J, Li J, Lombi E, Bland G, Lowry GV (2019) Nanoparticle size and coating chemistry control foliar uptake pathways, translocation, and leaf-to-rhizosphere transport in wheat. ACS Nano 13(5):5291–5305. https://doi.org/10.1021/acsnano.8b09781
7. Bakshi MS (2020) Impact of nanomaterials on ecosystems: Mechanistic aspects in vivo. Environ Res 182:109099
8. Ballikaya P, Brunner I, Cocozza C, Grolimund D, Kaegi R, Murazzi ME, Schaub M, Schönbeck LC, Sinnet B, Cherubini P (2023) First evidence of nanoparticle uptake through leaves and roots in beech (*Fagus sylvatica* L.) and pine (*Pinus sylvestris* L.). Tree Physiol 43(2):262–276
9. Bhattacharjee R, Negi A, Bhattacharya B, Dey T, Mitra P, Preetam S, Kumar L, Kar S, Das SS, Iqbal D (2023) Nanotheranostics to target antibiotic-resistant bacteria : Strategies and applications. OpenNano 100138
10. Billings WD (1952) The environmental complex in relation to plant growth and distribution. Q Rev Biol 27(3):251–265. https://doi.org/10.1086/399022
11. Chauhan A, Basniwal RK, Gurnani M, Rath P, Ranjan A, Rajput VD, Jindal T, Bauer T, Zamulina I, Chernikova N (2022) Environmental emissions of nanoparticles. In: Rajput VD,

Verma KK, Sharma N, Minkina T (Éds.) The role of nanoparticles in plant nutrition under soil pollution, pp 245–279. Springer International Publishing. https://doi.org/10.1007/978-3-030-97389-6_11

12. De La Rosa G, Vázquez-Núñez E, Molina-Guerrero C, Serafín-Muñoz AH, Vera-Reyes I (2021) Interactions of nanomaterials and plants at the cellular level: Current knowledge and relevant gaps. Nanotechnol Environ Eng 6(1):7. https://doi.org/10.1007/s41204-020-00100-1
13. Dimkpa CO (2014) Can nanotechnology deliver the promised benefits without negatively impacting soil microbial life? J Basic Microbiol 54(9):889–904. https://doi.org/10.1002/jobm.201400298
14. Dykman LA, Shchyogolev SY (2017) Interactions of plants with noble metal nanoparticles. Agric Biol 52(1):13–24
15. Fernandes JP, Mucha AP, Francisco T, Gomes CR, Almeida CMR (2017) Silver nanoparticles uptake by salt marsh plants–Implications for phytoremediation processes and effects in microbial community dynamics. Mar Pollut Bull 119(1):176–183
16. Hough RM, Noble RRP, Reich M (2011) Natural gold nanoparticles. Ore Geol Rev 42(1):55–61
17. Husen A (2023) Nanomaterials and nanocomposites exposures to plants (Response, Interaction, Phytotoxicity and Defense Mechanisms). Springer Nature Singapore Pte Ltd., 152 Beach Road, #21-01/04 Gateway East, Singapore 189721, Singapore. https://doi.org/10.1007/978-981-99-2419-6
18. Jurkow R, Pokluda R, Sękara A, Kalisz A (2020) Impact of foliar application of some metal nanoparticles on antioxidant system in oakleaf lettuce seedlings. BMC Plant Biol 20(1):290. https://doi.org/10.1186/s12870-020-02490-5
19. Khan MR, Adam V, Rizvi TF, Zhang B, Ahamad F, Jośko I, Zhu Y, Yang M, Mao C (2019) Nanoparticle-plant interactions: two-way traffic. Small 15(37):1901794. https://doi.org/10.1002/smll.201901794
20. Kiba T, Krapp A (2016) Plant nitrogen acquisition under low availability : Regulation of uptake and root architecture. Plant Cell Physiol 57(4):707–714
21. Klaine SJ, Alvarez PJJ, Batley GE, Fernandes TF, Handy RD, Lyon DY, Mahendra S, McLaughlin MJ, Lead JR (2008) Nanomaterials in the environment: behavior, fate, bioavailability, and effects. Environ Toxicol Chem 27(9):1825–1851. https://doi.org/10.1897/08-090.1
22. Kongala SI, Nadendla SR, Mamidala P (2021) Internalization and induction of defense responses in tobacco by harpinPss conjugated gold nanoparticles as a foliar spray. Colloid Interface Sci Commun 43:100438
23. Krishnani KK, Boddu VM, Chadha NK, Chakraborty P, Kumar J, Krishna G, Pathak H (2022) Metallic and non-metallic nanoparticles from plant, animal, and fisheries wastes: Potential and valorization for application in agriculture. Environ Sci Pollut Res 29(54):81130–81165. https://doi.org/10.1007/s11356-022-23301-4
24. Lee KX, Shameli K, Yew YP, Teow S-Y, Jahangirian H, Rafiee-Moghaddam R, Webster T (2020) Recent developments in the facile bio-synthesis of gold nanoparticles (AuNPs) and their biomedical applications. Int J Nanomed 15:275–300. https://doi.org/10.2147/IJN.S233789
25. Lowry GV, Espinasse BP, Badireddy AR, Richardson CJ, Reinsch BC, Bryant LD, Bone AJ, Deonarine A, Chae S, Therezien M, Colman BP, Hsu-Kim H, Bernhardt ES, Matson CW, Wiesner MR (2012) Long-term transformation and fate of manufactured Ag nanoparticles in a simulated large scale freshwater emergent wetland. Environ Sci Technol 46(13):7027–7036. https://doi.org/10.1021/es204608d
26. Luyts K, Napierska D, Nemery B, Hoet PH (2013) How physico-chemical characteristics of nanoparticles cause their toxicity: Complex and unresolved interrelations. Environ Sci Process Impacts 15(1):23–38
27. Lv J, Christie P, Zhang S (2019) Uptake, translocation, and transformation of metal-based nanoparticles in plants: recent advances and methodological challenges. Environ Sci Nano 6(1):41–59
28. Ma C, White JC, Dhankher OP, Xing B (2015) Metal-based nanotoxicity and detoxification pathways in higher plants. Environ Sci Technol 49(12):7109–7122. https://doi.org/10.1021/acs.est.5b00685

29. Ma S, Lin D (2013) The biophysicochemical interactions at the interfaces between nanoparticles and aquatic organisms: Adsorption and internalization. Environ Sci Process Impacts 15(1):145–160
30. Malejko J, Godlewska-Ży\lkiewicz B, Vanek T, Landa P, Nath J, Dror I, Berkowitz B (2021) Uptake, translocation, weathering and speciation of gold nanoparticles in potato, radish, carrot and lettuce crops. J Hazard Mater 418:126219
31. Mo F, Li H, He Y, Zhou Q (2022) An updated review on Ag NP effects at organismal level: internalization, responses, and influencing factors. Rev Environ Contam Toxicol 260(1):17. https://doi.org/10.1007/s44169-022-00019-5
32. Moreno-Brush M, McLagan DS, Biester H (2020) Fate of mercury from artisanal and small-scale gold mining in tropical rivers: Hydrological and biogeochemical controls. A critical review. Critical Reviews in Environmental Science and Technology, 50(5):437-475. https://doi.org/10.1080/10643389.2019.1629793
33. Murali M, Gowtham HG, Singh SB, Shilpa N, Aiyaz M, Alomary MN, Alshamrani M, Salawi A, Almoshari Y, Ansari MA (2022) Fate, bioaccumulation and toxicity of engineered nanomaterials in plants: current challenges and future prospects. Sci Total Environ 811:152249
34. Nair R (2016) Effects of nanoparticles on plant growth and development. In: C Kole, DS Kumar, MV Khodakovskaya (Éds.). Plant Nanotechnology, p 95–118. Springer International Publishing. https://doi.org/10.1007/978-3-319-42154-4_5
35. Nowack B, Bucheli TD (2007) Occurrence, behavior and effects of nanoparticles in the environment. Environ Pollut 150(1):5–22
36. Ogorek LLP, de la Cruz Jiménez J, Visser EJ, Takahashi H, Nakazono M, Shabala S, Pedersen O (2023) Outer apoplastic barriers in roots: prospects for abiotic stress tolerance. Funct Plant Biol. https://www.publish.csiro.au/fp/FP23133
37. Ohgushi T (2005) Indirect interaction webs: herbivore-induced effects through trait change in plants. Annu Rev Ecol Evol Syst 36(1):81–105. https://doi.org/10.1146/annurev.ecolsys.36.091704.175523
38. Peng C, Zhang W, Gao H, Li Y, Tong X, Li K, Zhu X, Wang Y, Chen Y (2017) Behavior and potential impacts of metal-based engineered nanoparticles in aquatic environments. Nanomaterials 7(1):21
39. Perone A (2018) Monitoring environmental quality with trees-A multiscale approach towards the implementation of Nature-Based Solution in urban ecosystems. https://dspace.unitus.it/handle/2067/42730
40. Pradhan S, Mailapalli DR (2017) Interaction of engineered nanoparticles with the agri-environment. J Agric Food Chem 65(38):8279–8294. https://doi.org/10.1021/acs.jafc.7b02528
41. Qamar SUR, Ahmad JN (2021) Nanoparticles: Mechanism of biosynthesis using plant extracts, bacteria, fungi, and their applications. J Mol Liq 334:116040
42. Rai PK, Kumar V, Lee S, Raza N, Kim K-H, Ok YS, Tsang DC (2018) Nanoparticle-plant interaction: Implications in energy, environment, and agriculture. Environ Int 119:1–19
43. Rajput VD, Minkina T, Sushkova S, Tsitsuashvili V, Mandzhieva S, Gorovtsov A, Nevidomskyaya D, Gromakova N (2018) Effect of nanoparticles on crops and soil microbial communities. J Soils Sediments 18(6):2179–2187. https://doi.org/10.1007/s11368-017-1793-2
44. Rani S, Kumari N, Sharma V (2023) Uptake, translocation, transformation and physiological effects of nanoparticles in plants. Arch Agronomy Soil Sci 69(9):1579–1599. https://doi.org/10.1080/03650340.2022.2103549
45. Roy R, Kumar S, Tripathi A, Das M, Dwivedi PD (2014) Interactive threats of nanoparticles to the biological system. Immunol Lett 158(1–2):79–87
46. Sabo-Attwood T, Unrine JM, Stone JW, Murphy CJ, Ghoshroy S, Blom D, Bertsch PM, Newman LA (2012) Uptake, distribution and toxicity of gold nanoparticles in tobacco (*Nicotiana xanthi*) seedlings. Nanotoxicology 6(4):353–360. https://doi.org/10.3109/17435390.2011.579631
47. Sarfraz N, Khan I (2021) Plasmonic gold nanoparticles (AuNPs): properties, synthesis and their advanced energy, environmental and biomedical applications. Chem –Asian J 16(7):720–742. https://doi.org/10.1002/asia.202001202

48. Shamsul Harumain ZA (2016) Phytomining of precious metals from mine wastes [PhD Thesis, University of York]. https://etheses.whiterose.ac.uk/16001/
49. Siddiqi KS, Husen A (2016) Engineered gold nanoparticles and plant adaptation potential. Nanoscale Res Lett 11(400):1–10
50. Sievers M, Hale R, Parris KM, Swearer SE (2018) Impacts of human-induced environmental change in wetlands on aquatic animals. Biol Rev 93(1):529–554. https://doi.org/10.1111/brv.12358
51. Singh P, Kumari K, Vishvakrma VK, Mehrotra GK, Chandra R, Kumar D, Patel R, Shahare VV (2017) Metal NPs (Au, Ag, and Cu): synthesis, stabilization, and their role in green chemistry and drug delivery. In: Singh R, Kumar S (Éds.) Green technologies and environmental sustainability, pp 309–337. Springer International Publishing. https://doi.org/10.1007/978-3-319-50654-8_14
52. Singla R, Kumari A, Yadav SK (2019) Impact of nanomaterials on plant physiology and functions. In: Husen A, Iqbal M (Éds.) Nanomaterials and plant potential, p 349–377. Springer International Publishing. https://doi.org/10.1007/978-3-030-05569-1_14
53. Spielman-Sun E, Avellan A, Bland GD, Tappero RV, Acerbo AS, Unrine JM, Giraldo JP, Lowry GV (2019) Nanoparticle surface charge influences translocation and leaf distribution in vascular plants with contrasting anatomy. Environ Sci Nano 6(8):2508–2519
54. Stefani PVY, Carrillo-González R, del Carmen Angeles G-CM (2021) Interaction of metal nanoparticles–plants–microorganisms in agriculture and soil remediation. J Nanoparticle Res 23(9). https://search.proquest.com/openview/9ea3784fc626195cf67ab7bbd7b56ce4/1?pq-origsite=gscholar&cbl=1456352
55. Su Y, Ashworth V, Kim C, Adeleye AS, Rolshausen P, Roper C, White J, Jassby D (2019) Delivery, uptake, fate, and transport of engineered nanoparticles in plants: a critical review and data analysis. Environ Sci Nano 6(8):2311–2331
56. Sun H, Wang M, Wang J, Wang W (2022) Surface charge affects foliar uptake, transport and physiological effects of functionalized graphene quantum dots in plants. Sci Total Environ 812:151506
57. Surendranath A, Mohanan PV (2020) Impact of nanoparticles in balancing the ecosystem. Biointerface Res Appl Chem 11:10461–10481
58. Tirani MM, Haghjou MM, Ismaili A (2019) Hydroponic grown tobacco plants respond to zinc oxide nanoparticles and bulk exposures by morphological, physiological and anatomical adjustments. Funct Plant Biol 46(4):360–375
59. Tran T-K, Nguyen M-K, Lin C, Hoang T-D, Nguyen T-C, Lone AM, Khedulkar AP, Gaballah MS, Singh J, Chung WJ (2023) Review on fate, transport, toxicity and health risk of nanoparticles in natural ecosystems: emerging challenges in the modern age and solutions toward a sustainable environment. Sci Total Environ 169331
60. Trchounian A, Gabrielyan L, Mnatsakanyan N (2018) Nanoparticles of various transition metals and their applications as antimicrobial agents. Metal nanoparticles: properties, synthesis and applications 161–211
61. Venzhik YuV, Moshkov IE, Dykman LA (2021) Gold nanoparticles in plant physiology: principal effects and prospects of application. Russ J Plant Physiol 68(3):401–412. https://doi.org/10.1134/S1021443721020205
62. Verma SK, Das AK, Patel MK, Shah A, Kumar V, Gantait S (2018) Engineered nanomaterials for plant growth and development: a perspective analysis. Sci Total Environ 630:1413–1435
63. Zheng Y (2014) Impact of gold nanoparticles with different coatings on lettuce. Southern Illinois University at Carbondale. https://search.proquest.com/openview/ec200a21efd8daded7d597da00b376c5/1?pq-origsite=gscholar&cbl=18750

Uptake, Accumulation and Potential Toxic Effects of Gold Nanoparticles in Aquatic Plants: An Overview

Maha Krayem, Sami El Khatib, Pascal Labrousse, and Sanaa Khaled

Abstract Nanoparticles could be discharged into the aquatic ecosystem in many ways and cause serious problems to the aquatic fauna and flora. This chapter will critically discuss gold nanoparticles (AuNPs) in water ecosystems. More specifically, it gives emphasis on AuNPs bioavailability in surface waters, evaluates its chemical interactions with plants and their effects on plant growth and development especially on aquatic plants or macrophytes, as well as some final research recommendations and perspectives regarding this field. Scientific literature showed that AuNPs are more adsorbed on plants' surface areas rather than being assimilated inside tissues. In addition, Gold speciation could be affected by several physicochemical characteristics of the water including pH, water hardness, dissolved organic carbon, conductivity, and could be transformed from metallic form to other oxidative forms. Once inside, gold nanoparticles could exert toxicity symptoms on plants. However, most toxicity studies are done on terrestrial plants and algae as mentioned in this chapter. Therefore, more research on toxicity and bioaccumulation studies is needed in order to find more about the fate and behavior of AuNPs in aquatic ecosystem.

Keywords AuNPs · Bioavailability · Accumulation · Toxicity · Aquatic plants

M. Krayem (✉) · S. El Khatib · S. Khaled
LIU, Lebanese International University, Bekaa Campus, Al Khyara-West Bekaa, Lebanon
e-mail: maha.krayem@liu.edu.lb

P. Labrousse
Université de Limoges, PEIRENE EA 7500, Limoges, France

S. El Khatib
Center for Applied Mathematics and Bioinformatics (CAMB), Gulf University for Science and Technology, Mubarak Al-Abdullah, Kuwait

A. Husen (ed.), *Plant Response to Gold Nanoparticles*, Smart Nanomaterials Technology, https://doi.org/10.1007/978-981-97-7174-5_14

1 Introduction

Because of the industry and technology revolution, different kinds of contaminants (pharmaceutical, agricultural, and industrial molecules) are released into waters [45]. Recently, nanomaterials have been added to the list and caused serious problems to the aquatic fauna and flora. More than 65.000 metric tons of NPs are discharged into water surfaces each year whereas a major portion of NPs from air find their way also to the aquatic ecosystem [25]. Consequently, they could provoke health problems for humans. Nanomaterials have been widely used inside industry, medicine, agriculture and research [6, 12, 30–37, 40, 41–44]. They are very useful since they can be easily modified into different shapes and sizes [57]. Mechanical grinding (Top-down), or chemical synthesis (Bottom-up) approaches are two methods used normally in synthesizing nanomaterials [21, 57]. Top-down approaches consist of grinding bulk materials and reducing large-scale dimensions to the nanoscale such as the fabrication of nanochannels having a dimension within 1–100 nm range. The chemical approach (bottom-up) consists of using controlled chemical reactions to convert atoms into nanostructures having dimensions between 2 and 10 nm less than that obtained by the mechanical one. Typically, two methods of bottom-up approaches have been utilized. The first is gas phase method used to build carbon nanotubes (CNTs) and the Wet chemical method used in the fabrication of metallic nanoparticles [1, 57, 79]. In the latter, ionic metallic salts are reduced to a zero-valent state by electrochemical reactions, photo-reduction, or thermal treatment making the zero-valent metal ions able to interact and form crystal packages constituting the building blocks of the nanoparticle. This reduction is followed by adding a Capping agent to stop the growth of the particle, stabilize it, and change some of its properties such as solubility, shape, and others.... [1, 23]. For example, Gold nanoparticles (AuNPs) are mainly formed by wet synthesis via the reduction of tetrachloroaurate (III) followed by adding citrate as a capping agent [23, 80].

Nanomaterials are divided into three groups including: (i) Carbon nanomaterials, (ii) metal oxides, and (iii) metal nanoparticles. Carbon nanomaterials have been known since 1985 with the discovery of the 60-carbon atom hollow sphere. One derivative of C60 are carbon nantubes (CNTs). They have a broad range of applications including additives in plastics, bicycle frames, tennis racquets, baseball bats, and so on [56].

Metal oxides are chemical compounds that contain a metal atom combined with at least one oxygen molecule such as iron (III) oxide, commonly known as rust. Metal oxides can be found naturally in the earth's crust. Refinement of these minerals into fine powders has been used in many applications, such as pigments in pottery dating back thousands of years [14], food-grade additives and packaging [74], cosmetics and sunscreens [8] and even electronic devices and sensors in the twenty-first century [71].

Metallic nanoparticles are synthesized chemically through a bottom-up approach as said previously, by using a reducing agent and metal salts. Common reducing agents are sodium borohydride, citrate, and heat. Examples of metallic nanomaterials include gold and silver nanoparticles [23, 57, 59, 80]. AuNPs that will be reviewed in detail in this chapter are mainly used in biomedical imaging, as a contrasting agent [53], and as cellular delivery of drugs [11, 46]. A recent study [46] found doxorubicin conjugated with AuNPs coated with dextran was effective in reaching the nucleus of cancer cells and killing them.

Nanoparticles could be discharged into the aquatic ecosystem in different ways and cause serious problems to the aquatic fauna and flora [60].

Aquatic plants constitute the core of the present chapter since they are considered the primary producers of the aquatic food web. They are divided into three groups: submerged, floating, and emerged macrophytes [51]. In addition, the European EU Water Framework Directive has added the macrophytes (aquatic plants) as a bioindicator of water quality [2, 47, 67] and used their biomarkers in a variety of toxicity tests and remediation of metal-polluted sites [15–17, 22, 48–52].

Regarding nanoparticles, several studies have examined their impact on terrestrial plants [24] but little information exists on aquatic plants and AuNps especially. This chapter will critically discuss AuNPs in water ecosystems. More specifically, it will detail AuNPs bioavailability in surface waters, evaluate its chemical interactions with plants and their effects on plant growth and development especially on aquatic plants or macrophytes, as well as some final research recommendations and perspectives regarding this field.

2 Au-NPs: Definition, Applications and Synthesis

Gold is a yellowish soft metal widely distributed on Earth. It is chemically not reactive although it can be dissolved in mixtures of nitric and hydrochloric acids. It has been known since ancient times and was mostly worked with metal. This is because its high density allowed it to be concentrated in streambeds either in the form of large or small flakes. When finely divided, its color can be black, ruby or purple. It is the most malleable metal; it is alloyed in most applications to give it more strength. Gold is mainly used in jewelry, coinage, and electroplating material [19]. The elemental properties of gold are presented in Table 1.

Nanotechnology is a technology that enables scientists to deal with nano-metric-sized matter. The term nanometer was first proposed by Zsigmonday in 1925 and his work opened the door to the modern nanotechnology research led by Richard Feynman [29]. This technology was targeted because of the physio-chemical properties of material, however, it has featured a lot of advancement in the past years at the level of systems, materials, commerce, medicine, and pharmacy. Nanoparticles have a size that ranges between 1 and 100 nm, which is close to that of proteins. This makes nanoparticles suitable as biological tags, fluorescent biological labels, MRI contrast enhancement as well as many other applications in the medical field [70]. In pharmaceutical applications, nanotechnology highly impacted medical devices and drug delivery systems, and it also improved food industry and cosmetics applications at the level of production, shelf life, packages, and bioavailability [29].

Table 1 The elemental properties of gold. Retrieved from the Royal Society of Chemistry. (Rsc.org)

Atomic data	
Atomic number	79
Atomic weight	196.96657
Melting point	1,063° C (1,945 F)
Boiling point	2,966° C (5,371 F)
Specific gravity	19.3 at 20°C (68 F)
Electronic configuration	$[Xe]4f^{14}5d^{10}6s^{1}$
Atomic radius, non-bonded (Å)	2.14
Electron affinity (kJ mol^{-1})	222.749
Covalent radius (Å)	1.30
Electronegativity (Pauling scale)	2.4
Common oxidation states	5,4,3,2,1,−1

More attention was given to metallic nanoparticles during the last decades since they can form colloidal systems. They include gold, iron oxide, and silver, with different shapes and sizes and that enabled scientists to use them widely and in diverse biomedical and medical applications [23].

AuNPs have shown promising properties in nanotechnology applications. They can be shaped in different-sized chains, nanospheres, nanorods as well as nanoprisms [23]

Historically, and since the Middle Ages, AuNPs were used in medicine in order to cure many diseases such as heart, arthritis, epilepsy as well as syphilis. However, its use in the oxidation state caused serious health problems for humans. Recently, more research attention has been given for the use the use of AuNPs in current biomedicine applications [7].

Researchers focused on enhancing the physio-chemical properties of AuNPs for better use in cancerology field. AuNPs sow promising biocompatibility and drug delivery abilities because of its physical properties such as radioactivity, high X-ray absorption coefficient, and localized surface plasma resonance, as well as being able to attach to a variety of organic ligands or polymers [7].

Recently, many researchers documented the use of AuNPs to minimize heavy metals like Ni, Cu, and Al from water, soil, and food. Many attempts succeeded in synthesizing green AuNPs from plant extracts and lately, using lemon extract to prepare AuNPs and use it to decrease the nickel concentration in water presented very promising results for environmental applications [7].

In parallel research, AuNPs were used to detect and characterize the Polydiallyldimethylammonium chloride (poly-(DADMAC) that is globally used in many drinking water treatment plants as a flocculant to remove suspended solids from raw water. The reason behind this research was that residual poly-(DADMAC) is converted into a carcinogenic compound known as N-nitrosodimethylamine (NDMA) during the water treatment process [62].

2.1 Green Synthesis of AuNPs

According to Freitas de Freitas et al. [23], the methods of synthesis of AuNPs are classified into physical–chemical methods, physical and chemical methods. The physical–chemical methods are of three types. The first is degrading the bulk material into smaller particles to achieve the required scale, the second is using small precursors such as metallic salts or molecular seeds that nucleate and form the desired nanostructures while the third is the supercritical fluid technology. The physical methods are mainly photochemical processes, ionizing radiation, and microwave radiation. Finally, the chemical methods use strong/mild reducing and capping agents such as sodium borohydride, hydrazine, and citrate (Fig. 1) [23].

The fields of nanotechnology are evolving in a rapid way especially since their applications are widely spread, and scientists have raised many environmental problems and toxicity issues related to the precursor's material and degree of toxicity of the chemicals used in the chain of preparation especially that many of them remain free and non-reactive and since they consist a potential hazard to human health [77]. As a result, nanomaterial green approaches are increasing more and more as there

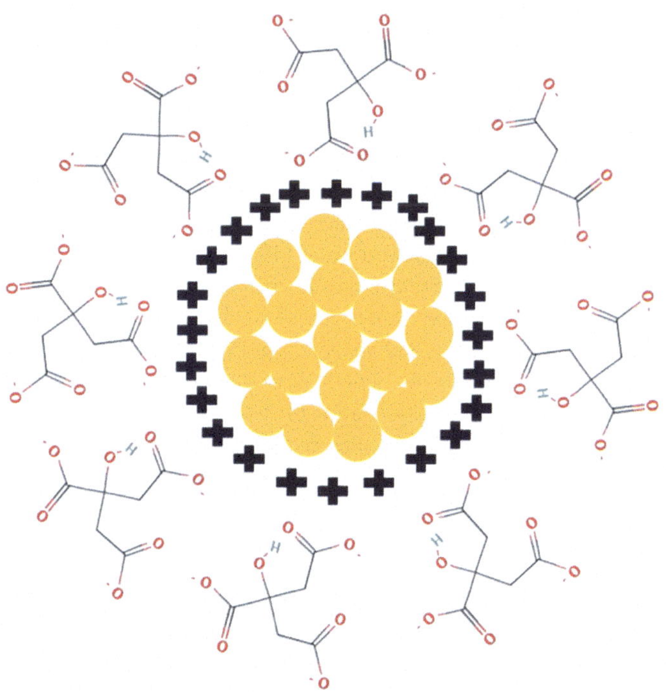

Fig. 1 Scheme of gold nanoparticles capped with citrate molecules. The yellow circles are the gold nanoparticles positively charged and attract citrate molecules

would be no need to use reducing agents in the synthesis process as well as reducing energy consumption.

Green synthesis methods use phytochemicals from bean extracts, low-molecular-weight proteins, carbohydrates from different sources, and antioxidants such as catechins from tea leaves that were used as reducing and capping agents in AuNPs. The phytochemicals in cumin stabilized by Gum arabic were also investigated as efficient generators of AuNPs. These protocols as well as others were proved to be environmentally friendly and biocompatible. Biological methods using bacteria and fungi also constitute another green approach in the synthesis of AuNPs especially since organic solvents are not used in these methods [23].

3 Plant/Au-NPs Interactions: Availability, Diffusion, and Accumulation

Due to the increasing diversification of the use of nanoparticles, it has long been anticipated that their emission routes into the environment would be multiple. Since most key products of nanoparticles are chemical substances, their emission and release into the environment is according to three routes: during production and use of the raw material, and release after disposal or waste management. These emissions can be directly released into the environment or indirectly via waste treatment facilities or landfills and have been estimated globally as follows 63–91% in landfills 8 to 28% in the soil, 7% in the water, and around 1.5% in the air [10]. Moreover, Anthropogenic activities unintentionally release different types of NP, such as Palladium emissions from traffic, iron-based NP in groundwater pipes, Nano pesticides of agricultural activities released into the underground water [10].

After the direct and indirect emission into the environment, NP will age with time either by reacting with other materials, or by aggregation and disaggregation reactions. Data must be collected and analyzed in order to study their current status, as well as estimate their real fate and whether laboratory control can be investigated more in order to get more realistic scenarios [10].

The potential toxicity of nanoparticles including Au-NPs on plants especially macrophytes is very scarce in terms of scientific literature. Literature was more focused on the impact of NMs on terrestrial plants, however, some minor studies were relevant to aquatic plants. Thus, it is very essential to invest more research in this domain [65].

Aquatic and terrestrial plants resemble by their vascular system, photosynthesis process, and their organs such as roots and leaves. However, aquatic macrophytes can absorb nutrients from both roots and leaves unlike terrestrial plants [54].

Before being absorbed, nanoparticles bioavailability is a complex mechanism that depends on plant species, its roots, particle size, its surface charge, water hardness, pH, the presence of chelators, and dissolved organic carbon [27, 28, 55, 76, 81]. For example, in a study done by [27], they found that the presence of dissolved organic carbon combined with 4- and 18-nm AuNPs in suspension decreased the particles aggregation and their absorption by three aquatic plants: *Azolla caroliniana* Willd, *Egeria densa* Planch., and *Myriophyllum simulans* Orch. In addition, organic acids such as humic and fulvic acids can affect the stability, charge, and surface chemistry of nanoparticles [5, 18, 63, 73]. Stankus et al. [73] found that AuNP surface chemistry in presence of organic acids is similar and particles become more stable. Thus, Diegoli et al. [18] suggested that citrate used as a capping agent in AuNps could be replaced by these organic acids. This could lead to more stable suspensions of particles and reduce their aggregation at different pH. Nason et al. [63] found that the higher molecular weight of humic acids, the higher the stability of citrate-coated AuNPs.

In another experiment done by Bergemann et al. [9], they studied the effect of seasonal changes and snail grazing on the bioavailability of AuNPs for *Egeria densa* species. They add weekly 44.8 μg/L of AuNP inside microcosms for 4 weeks and changed the light and temperature conditions simulating spring and fall conditions. They found that higher temperatures stabilize AuNPs inside water column by increasing the dissolution of organic carbon. In addition, Au was transferred from Egeria surface to the organic matter because of snail grazers.

Several studies and review papers have investigated the interaction between nanoparticles and terrestrial plants [24, 58, 68, 72]. From these reviews, a lack of data about nanoparticles interactions with plants was found. That's why research should be more invested in this domain related to nanoparticles in order to understand very well their speciation and the biotic and abiotic factors that affect their bioavailability especially in the aquatic ecosystem.

Once adsorbed, the mechanism of nanoparticle uptake and its penetration inside plants tissues is not fully known [24]. Nanoparticles may enter plants through their roots or leaves via stomata. From their roots, nanoparticles can be transported within plants via three possible routes. The cell wall pores are defined as the overlapping microfibrils network that is formed by the building of the cell wall having dimension sizes between 5 and 10 nm range and have even been reported up to 50 nm in diameter [13, 26]. The second route of transport is through apoplastic transfer defined as the cell wall and the intercellular space between the cell wall and the plasma membrane without crossing the latter [26, 75]. The third is the symplastic pathway, which is defined as the intracellular region of the plant. Here, nanoparticles have to pass through channels called plasmodesmata that connect adjacent cells 40 nm in diameter [26] (Fig. 2).

Once inside plant tissues, AuNPs are able to bioconcentrate and accumulate. As said previously, this bioaccumulation is not only affected by the physico-chemical properties of water but also the dissolved organic carbon. For example, [28] exposed three aquatic macrophytes: two submerged aquatic vascular plants (*Myriophyllum simulans Orch. and E. densa Planch.*), and a free-floating aquatic fern *(Azolla caroliniana* Willd.,), to 4- and 18-nm citrate-capped AuNPs at a concentration of 250 μg/L for up to 24 h (pH 7.1; TOC 8.56 mg/L; CaCO3 107 mg/L; conductivity 210 mS/cm). 4- and 18-nm AuNPs were adsorbed to the roots of the three species. Moreover, both particles were absorbed by *A. caroliniana*, whereas only the 4-nm was absorbed by *M. simulans*. Similarly, [4] studied the effect of adding AuNPs at

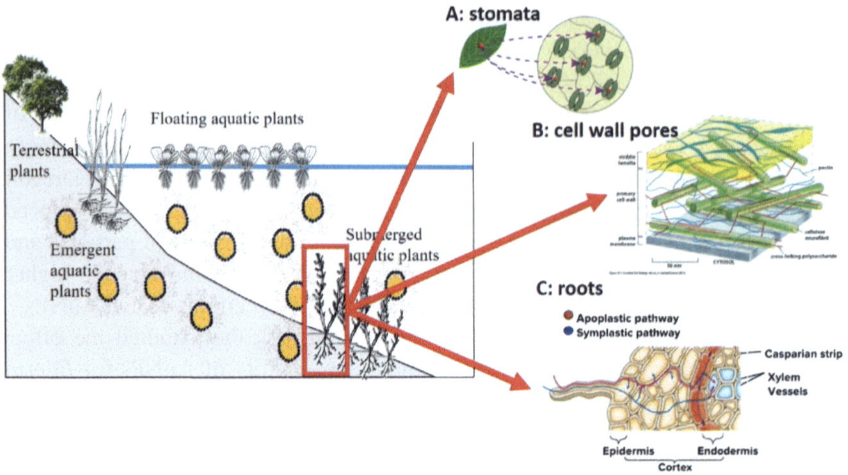

Fig. 2 Uptake and transport mechanism of gold nanoparticles in plants through their **a** stomata, **b** cell wall pores, and **c** their roots via apoplast pathway and symplast pathways. The yellow circles represent AuNPs

concentrations of $500 \pm 32 mgAu kg^{-1}$ (dry mass) in presence of the macrophyte *E. densa* for six months. They observed a decrease of 70% of the initial concentration added and relate this decrease to a mechanism of surface adsorption based on transmission electron microscope image results. In addition, they have observed that Au (zero valent) was biotransformed by the microbiome associated with the macrophyte to other oxidative forms complexed with cyanide ($43 \pm 8\%$), hydroxide ($38 \pm 4\%$), and thiol ($19 \pm 4\%$) ligands.

As a conclusion, the impact of the environmental discharge of AuNPs is still not fully known. This gap in knowledge prevents us from a good assessment of environmental risk of AuNPs in aquatic ecosystems. Thus, further investigation with regards to AuNPs, their bioavailability, uptake transport, and bioaccumulation inside plant tissues is needed.

4 Impact of Au-NPs on Aquatic Plants

One of the drawbacks of nanotechnology is the possible toxicity mechanisms of nanoparticles in humans and animals. A decade ago, Moore as well as Hund-Rinke and Simon pointed out the potential possibility of nanoparticles to cause harm to human beings by the formation of free radicals. Oxidative stress can cause several physiological changes inside bodies such as modifying hormones thus causing reproductive failure. In algae and aquatic plants, photosynthesis can be affected by the alteration of its pigment composition. Similarly, many researchers have reported the accumulation of nanoparticles in terrestrial plants causing changes in their biochemical and physiological functions [10].

The potential impacts of AuNPs on plants are still not fully known. In some studies, nanoparticles have shown a promoting effect on the growth and development of some plants and inhibitory on others [78]. Regarding phytoxicity, several endpoints could be used to measure the extent of the toxicity of nanoparticles on plants. They include genotoxicity and cytotoxicity endpoints, as well as seed germination and growth, free radical production, and enzymatic and biochemical endpoints.

Interaction of nanomaterials with photosynthetic organisms and in particular, higher plants, have grown over the last decade only [24]. Before 2008, no studies have been done on plants [64]. After that, studies with terrestrial plants have grown in number however, studies using aquatic plants are still sparse. In fact, studies of toxicity on terrestrial plants were found more than on aquatic plants [24, 72, 78]. The exposure of barley plants (*Hordeum vulgare which are terrestrial plants)* to 2–21 nm AuNPs at a concentration from 5 to 10 mg/L showed a yellowing and necrosis in their leaves, a dark brown color of their roots, and a decrease in fresh biomass after 21 days [20]. Similarly, necrotic lesions and biotoxicity were observed in the terrestrial tobacco plants *Nicotiana xanthi* after their exposure to 3.5 AuNPs at 48 mg/L as a concentration for 21 days [69].

The most common aquatic plants used in nanoparticles toxicity testing are *Lemna spp.* For example, a decline in *Lemna gibba* growth and an increase in oxidative damage was observed when exposed to 50 nm AgNPs at concentrations from 1 to 10 mg/L. These changes are correlated with the release of Ag + from AgNPs [66].

Regarding AuNPs toxicity in aquatic plants, a lack of studies was found. [55] studied the effect of 1000 μg/L of citrate (cit)- and branched polyethyleneimine (BPEI)-coated nAu (5, 20, and 40 nm) in 10% Hoagland's medium (10 HM) for 2, 7, and 14 days on *Salvinia minima* Bake. They observed that this concentration did not significantly affect their biomass and relative growth rate even after 14 days. However, [65] showed that phytoxicity of AuNPs (20 ± 5 nm) on *Ceratophyllum demesrsum* aquatic plant was observed at high concentration (3.64 mg.L-1) in which 50% of their shoots died after 28 days. One of the reasons maybe behind this lack of studies is related to the fact that gold is considered less toxic to organisms as mentioned by [61] in their review about silver and AuNPs toxicity to microalgae. In addition, they have found a high number of studies related to silver nanoparticles rather than AuNPs. Thus more research should be invested in this domain in order to better understand the toxic effects of AuNPs on aquatic macropyhtes.

5 Conclusion and Outlook

In conclusion, nanotoxicity and its related aspects need to be better investigated because the toxicity properties are different between bulk and nano-level especially since parameters such as size and charge, surface chemistry, and biological target differ between the applications as well as the time of exposure. Unfortunately, some studies showed no evidence of nanotoxicity unlike other studies. This is due to the lack of homogeneity between studies and this is the responsibility of researchers to homogenize the studied parameters [23]. Nanoaprticles including AuNPs are released into the environment and cause major health problems to organisms [3]. Since they are small in size less than 100 nm, they are able to diffuse in our body in different ways (Sajid et al. 2015). Similarly, Plants are also able to take nanoparticles from air, soil or water (Ebrahimbabaie et al. 2020; Mustafa and Komatsu, 2016) through their roots or shoots. Scientific literature showed that AuNPs are more adsorbed on plants' surface areas rather than being assimilated inside tissues. In addition, Gold speciation could be affected by different physico-chemical properties of water and could be transformed from metallic form to other oxidative forms. However, more detailed studies are therefore necessary to evaluate their fate and behavior in the food chain. Remarkably, a lack of studies was detected when searching scientific articles related to uptake, transport, and distribution of AuNPs inside plants, especially aquatic plants. More studies were found for terrestrial plants. Once inside, AuNPs could exert toxicity symptoms on plants. Similarly, almost most toxicity studies are done on terrestrial plants and algae as mentioned in this chapter. Therefore, more research toxicity and bioaccumulation studies in order to understand the fate and behavior of AuNPs should also be investigated. Macrophyte interaction to AuNPs is still a new topic with too few data and too many uncertainties. Thus, the development of genotocicity testing, proteins or enzymatic changes, physiology including photosynthesis will add new macrophyte toxicity data in the future.

Acknowledgments The first and last author have contributed equally in writing the manuscript. All the authors have reviewed it. Kind regards

References

1. Afolalu SA, Soetan SB, Ongbali SO, Abioye AA, Oni AS (2019) Morphological characterization and physio-chemical properties of nanoparticle—review. IOP Conf Ser Mater Sci Eng 640:012065. https://doi.org/10.1088/1757-899X/640/1/012065
2. Aguiar FC, Segurado P, Urbanič G, Cambra J, Chauvin C, Ciadamidaro S, Dörflinger G, Ferreira J, Germ M, Manolaki P, Minciardi MR, Munné A, Papastergiadou E, Ferreira MT (2014) Comparability of river quality assessment using macrophytes: a multi-step procedure to overcome biogeographical differences. Sci Total Environ 476–477:757–767. https://doi.org/10.1016/j.scitotenv.2013.10.021
3. Arif N, Yadav V, Singh S, Tripathi DK, Dubey NK, Chauhan DK, Giorgetti L (2018) Chapter 13—interaction of copper oxide nanoparticles with plants: uptake, accumulation, and

toxicity. In: Tripathi DK, Ahmad P, Sharma S, Chauhan DK, Dubey NK (Eds.) Nanomaterials in plants, algae, and microorganisms. Academic Press, pp 297–310. https://doi.org/10.1016/B978-0-12-811487-2.00013-X

4. Avellan A, Simonin M, McGivney E, Bossa N, Spielman-Sun E, Rocca JD, Bernhardt ES, Geitner NK, Unrine JM, Wiesner MR, Lowry GV (2018) Gold nanoparticle biodissolution by a freshwater macrophyte and its associated microbiome. Nat Nanotechnol 13:1072–1077. https://doi.org/10.1038/s41565-018-0231-y
5. Baalousha M, Sikder M, Poulin BA, Tfaily MM, Hess NJ (2022) Natural organic matter composition and nanomaterial surface coating determine the nature of platinum nanomaterial-natural organic matter corona. Sci Total Environ 806:150477. https://doi.org/10.1016/j.scitotenv.2021.150477
6. Bachheti RK, Bachheti A, Husen A (2024) Metal and metal-oxide based nanomaterials (*Synthesis, Agricultural, Biomedical and Environmental Interventions*). Springer Nature Singapore Pte Ltd., 152 Beach Road, #21-01/04 Gateway East, Singapore 189721, Singapore
7. Bai X, Wang Y, Song Z, Feng Y, Chen Y, Zhang D, Feng L (2020) The basic properties of gold nanoparticles and their applications in tumor diagnosis and treatment. Int J Mol Sci 21:2480. https://doi.org/10.3390/ijms21072480
8. Banerjee T, Samanta A (2022) Metal oxides for cosmetics and sunscreens. In: Metal oxides for biomedical and biosensor applications. Elsevier, pp 119–135. https://doi.org/10.1016/B978-0-12-823033-6.00002-8
9. Bergemann CM, Avellan A, Perrotta BG, Bernhardt ES, Simonin M (2023) Seasonal differences and grazing pressure alter the fate of gold nanoparticles in a microcosm experiment. Environ Sci Technol. https://doi.org/10.1021/acs.est.3c01839
10. Bundschuh M, Filser J, Lüderwald S, McKee MS, Metreveli G, Schaumann GE, Schulz R, Wagner S (2018) Nanoparticles in the environment: where do we come from, where do we go to? Environ Sci Eur 30:6. https://doi.org/10.1186/s12302-018-0132-6
11. Chandrakala V, Aruna V, Angajala G (2022) Review on metal nanoparticles as nanocarriers: current challenges and perspectives in drug delivery systems. Emergent Mater. 5:1593–1615. https://doi.org/10.1007/s42247-021-00335-x
12. Chawla S, Singh S, Husen A (2023) Smart nanomaterials targeting pathological hypoxia. Springer Nature Singapore Pte Ltd., 152 Beach Road, #21-01/04 Gateway East, Singapore 189721, Singapore ISBN: 978-981-99-1717-4. https://doi.org/10.1007/978-981-99-1718-1
13. Chen R, Ratnikova TA, Stone MB, Lin S, Lard M, Huang G, Hudson JS, Ke PC (2010) Differential uptake of carbon nanoparticles by plant and mammalian cells. Small 6:612–617. https://doi.org/10.1002/smll.200901911
14. Colomban P (2009) The use of metal nanoparticles to produce yellow, red and iridescent colour, from bronze age to present times in Lustre pottery and glass: solid state chemistry. Spectrosc Nanostruct J Nano Res 8:109–132. https://doi.org/10.4028/www.scientific.net/JNanoR.8.109
15. Delmail D (2014) Risk management of European inland waters using macrophyte biomonitoring. Front Environ Sci 2. https://doi.org/10.3389/fenvs.2014.00031
16. Delmail D, Labrousse P, Hourdin P, Larcher L, Moesch C, Botineau M (2011) Differential responses of *Myriophyllum alterniflorum* DC (Haloragaceae) organs to copper: physiological and developmental approaches. Hydrobiologia 664:95–105. https://doi.org/10.1007/s10750-010-0589-9
17. Delmail D, Labrousse P, Hourdin P, Larcher L, Moesch C, Botineau M (2011) Physiological, anatomical and phenotypical effects of a cadmium stress in different-aged chlorophyllian organs of *Myriophyllum alterniflorum* DC (Haloragaceae). Environ Exp Bot 72:174–181. https://doi.org/10.1016/j.envexpbot.2011.03.004
18. Diegoli S, Manciulea AL, Begum S, Jones IP, Lead JR, Preece JA (2008) Interaction between manufactured gold nanoparticles and naturally occurring organic macromolecules. Sci Total Environ 402:51–61. https://doi.org/10.1016/j.scitotenv.2008.04.023
19. Emsley J (2001) Nature's building blocks: an A-Z guide to the elements, 2nd edn. Oxford University Press, Oxford, New York

20. Feichtmeier NS, Walther P, Leopold K (2015) Uptake, effects, and regeneration of barley plants exposed to gold nanoparticles. Environ Sci Pollut Res 22:8549–8558. https://doi.org/10.1007/s11356-014-4015-0
21. Findik F (2021) Nanomaterials and their applications. Period Eng Nat Sci 9:62–75. https://doi.org/10.21533/pen.v9i3.1837
22. Fletcher J, Willby N, Oliver DM, Quilliam RS (2020) Phytoremediation using aquatic plants. In: Shmaefsky BR (Ed.) Phytoremediation: in-situ applications, concepts and strategies in plant sciences. Springer International Publishing, Cham, pp 205–260. https://doi.org/10.1007/978-3-030-00099-8_7
23. Freitas de Freitas L, Varca GHC, dos Santos Batista JG, Benévolo Lugão A (2018) An overview of the synthesis of gold nanoparticles using radiation technologies. Nanomaterials 8:939. https://doi.org/10.3390/nano8110939
24. Gao M, Chang J, Wang Z, Zhang H, Wang T (2023) Advances in transport and toxicity of nanoparticles in plants. J Nanobiotechnol 21:75. https://doi.org/10.1186/s12951-023-01830-5
25. Garner KL, Keller AA (2014) Emerging patterns for engineered nanomaterials in the environment: a review of fate and toxicity studies. J Nanoparticle Res 16:2503. https://doi.org/10.1007/s11051-014-2503-2
26. Glenn J (2013) An investigation into the interactions of gold nanoparticles with aquatic vascular macrophytes. Diss.
27. Glenn JB, Klaine SJ (2013) Abiotic and biotic factors that influence the bioavailability of gold nanoparticles to aquatic macrophytes. Environ Sci Technol 47:10223–10230. https://doi.org/10.1021/es4020508
28. Glenn JB, White SA, Klaine SJ (2012) Interactions of gold nanoparticles with freshwater aquatic macrophytes are size and species dependent: interactions of AuNPs with freshwater aquatic plants. Environ Toxicol Chem 31:194–201. https://doi.org/10.1002/etc.728
29. Hulla J, Sahu S, Hayes A (2015) Nanotechnology: history and future. Hum Exp Toxicol 34:1318–1321. https://doi.org/10.1177/0960327115603588
30. Husen A (2019) Natural product-based fabrication of zinc oxide nanoparticles and their application. In: Husen A, Iqbal M (Eds.) N*anomaterials* and plant potential. Springer International Publishing AG, Gewerbestrasse 11, 6330 Cham, pp 193–291 https://doi.org/10.1007/978-3-030-05569-1_7
31. Husen A (2020) Introduction and techniques in nanomaterials formulation. In: Husen A, Jawaid M (Eds.) Nanomaterials for agriculture and forestry applications. Elsevier Inc. 50 Hampshire St., 5th Floor, Cambridge, MA 02139, USA, pp 1–14. https://doi.org/10.1016/B978-0-12-817852-2.00001-9
32. Husen A (2022) Engineered nanomaterials for sustainable agricultural production, soil improvement and stress management. Elsevier Inc., 50 Hampshire St., 5th Floor, Cambridge, MA 02139, USA. https://doi.org/10.1016/C2021-0-00054-7
33. Husen A (2023a) Secondary metabolites based green synthesis of nanomaterials and their applications. Springer Nature Singapore Pte Ltd., 152 Beach Road, #21–01/04 Gateway East, Singapore 189721, Singapore. https://doi.org/10.1007/978-981-99-0927-8
34. Husen A (2023b) Nanomaterials from agricultural and horticultural products. Springer Nature Singapore Pte Ltd., 152 Beach Road, #21-01/04 Gateway East, Singapore 189721, Singapore. https://link.springer.com/book/9789819934348
35. Husen A (2023c) Nanomaterials and nanocomposites exposures to plants (Response, Interaction, Phytotoxicity and Defense Mechanisms). Springer Nature Singapore Pte Ltd., 152 Beach Road, #21-01/04 Gateway East, Singapore 189721, Singapor. https://doi.org/10.1007/978-981-99-2419-6
36. Husen A, Bachheti RK, Bachheti A (2023) Current trends in green nano-emulsions (Food, Agriculture and Biomedical Sectors). Springer Nature Singapore Pte Ltd., 152 Beach Road, #21-01/04 Gateway East, Singapore 189721, Singapore
37. Husen A, Iqbal M (2019) Nanomaterials and Plant Potential. Springer International Publishing AG, Gewerbestrasse 11, 6330 Cham, Switzerland. https://doi.org/10.1007/978-3-030-05569-1

38. Husen A, Jawaid M (2020) Nanomaterials for Agriculture and Forestry Applications. Elsevier Inc., 50 Hampshire St., 5th Floor, Cambridge, MA 02139, USA https://doi.org/10.1016/C2018-0-02349-X
39. Husen A, Siddiqi KS (2014) Phytosynthesis of nanoparticles: Concept, controversy and application. Nanoscale Res Lett 9(229):1–24. https://doi.org/10.1186/1556-276X-9-229
40. Husen A, Siddiqi KS (2023) Advances in smart nanomaterials and their applications. Elsevier Inc., 50 Hampshire St., 5th Floor, Cambridge, MA 02139, USA https://doi.org/10.1016/C2021-0-02202-1
41. Jin-Chul K, Madhusudhan A, Husen A (2021) Smart nanomaterials in biomedical applications. Springer Nature Switzerland AG, Gewerbestrasse 11, 6330 Cham, Switzerland. https://doi.org/10.1007/978-3-030-84262-8
42. Madhusudhan A, Purohit SD, Prasad R, Husen A (2024) Functional Smart Nanomaterials and Their Theranostics Approaches. Springer Nature Singapore Pte Ltd., 152 Beach Road, #21–01/04 Gateway East, Singapore 189721, Singapore
43. Mandal AK, Ghorai S, Husen A (2023) Functionalized smart nanomaterials for point-of-care testing. Springer Nature Singapore Pte Ltd., 152 Beach Road, #21-01/04 Gateway East, Singapore 189721, Singapore
44. Taghiyari HR, Morrell JF, Husen A (2023) Emerging Nanomaterials (*Opportunities and Challenges in Forestry Sectors*). Springer Nature Switzerland AG, Gewerbestrasse 11, 6330 Cham, Switzerland. https://doi.org/10.1007/978-3-031-17378-3
45. INERIS, 2014. Données technico-économiques sur les substances chimiques en France : cuivre, composés et alliages, DRC-14-136881-02236A, 91 p. (version mars 2015) [Technical and economic data on chemicals in France: copper, compounds and alloys, DRC-14-136881-02236A, 91 p. (March 2015 version)] [WWW Document]. URL //rsde.ineris.fr/ou http://www.ineris.fr/substances/fr/ (accessed 12.19.16)
46. Jang H, Ryoo S-R, Kostarelos K, Han SW, Min D-H (2013) The effective nuclear delivery of doxorubicin from dextran-coated gold nanoparticles larger than nuclear pores. Biomaterials 34:3503–3510. https://doi.org/10.1016/j.biomaterials.2013.01.076
47. Knight S, Hauxwell J (2009) Distribution and abundance of aquatic plants – human impacts. In: Likens GE (Ed.) Encyclopedia of inland waters. Academic Press, Oxford, pp 45–54. https://doi.org/10.1016/B978-012370626-3.00225-8
48. Krayem, Baydoun, M., Deluchat, V., Lenain, J.-F., Kazpard, V., Labrousse, P., 2016. Absorption and translocation of copper and arsenic in an aquatic macrophyte Myriophyllum alterniflorum DC. in oligotrophic and eutrophic conditions. Environ. Sci. Pollut. Res. 1–8. https://doi.org/10.1007/s11356-016-6289-x
49. Krayem M (2015) Etude des effets de l'arsenic et du cuivre sur un macrophyte aquatique, *Myriophyllum alterniflorum* D.C. : évaluation des biomarqueurs pour la détection précoce de pollution [Study of the effects of arsenic and copper on an aquatic macrophyte, Myriophyllum alterniflorum D.C.: evaluation of biomarkers for the early detection of pollution.]. Limoges
50. Krayem M, Deluchat V, Decou R, Labrousse P (2017) The arsenic threat in aquatic environments: from plant's point of view in Arsenic: risks of exposure, behavior in the environment, and toxicology. Nova Science Publishers Inc., Hauppauge, New York, Chemistry research and applications
51. Krayem M, Khatib SE, Hassan Y, Deluchat V, Labrousse P (2021) In search for potential biomarkers of copper stress in aquatic plants. Aquat Toxicol 239:105952. https://doi.org/10.1016/j.aquatox.2021.105952
52. Lewis M, Thursby G (2018) Aquatic plants: test species sensitivity and minimum data requirement evaluations for chemical risk assessments and aquatic life criteria development for the USA. Environ Pollut 238:270–280. https://doi.org/10.1016/j.envpol.2018.03.003
53. Luo D, Wang X, Burda C, Basilion JP (2021) Recent development of gold nanoparticles as contrast agents for cancer diagnosis. Cancers 13:1825. https://doi.org/10.3390/cancers13081825
54. Madsen TV, Cedergreen N (2002) Sources of nutrients to rooted submerged macrophytes growing in a nutrient-rich stream: nutrient uptake by submerged macrophytes. Freshw Biol 47:283–291. https://doi.org/10.1046/j.1365-2427.2002.00802.x

55. Mahaye N, Thwala M, Musee N (2021) Interactions of coated-gold engineered nanoparticles with aquatic higher plant *Salvinia minima* Baker. Nanomaterials 11:3178. https://doi.org/10.3390/nano11123178
56. Mathur RB, Singh BP, Pande S (2017) Carbon nanomaterials: synthesis, structure, properties and applications. CRC Press, Taylor & Francis Group, Boca Raton, FL
57. Mekuye B, Abera B (2023) Nanomaterials: an overview of synthesis, classification, characterization, and applications. Nano Sel 4:486–501. https://doi.org/10.1002/nano.202300038
58. Miralles P, Church TL, Harris AT (2012) Toxicity, uptake, and translocation of engineered nanomaterials in vascular plants. Environ Sci Technol 46:9224–9239. https://doi.org/10.1021/es202995d
59. Mody VV, Siwale R, Singh A, Mody HR (2010) Introduction to metallic nanoparticles. J Pharm Bioallied Sci 2:282–289. https://doi.org/10.4103/0975-7406.72127
60. Moloi MS, Lehutso RF, Erasmus M, Oberholster PJ, Thwala M (2021) Aquatic environment exposure and toxicity of engineered nanomaterials released from nano-enabled products: current status and data needs. Nanomaterials 11:2868. https://doi.org/10.3390/nano11112868
61. Moreno-Garrido I, Pérez S, Blasco J (2015) Toxicity of silver and gold nanoparticles on marine microalgae. Mar Environ Res 111:60–73. https://doi.org/10.1016/j.marenvres.2015.05.008
62. Mthembu SB, Akintayo DC, Moodley B, Gumbi BP (2023) Development of gold plasmonic nanoparticles for detection of polydiallyldimethylammonium chloride at Umgeni water treatment plants: an optimised study and case application. Heliyon 9:e17136. https://doi.org/10.1016/j.heliyon.2023.e17136
63. Nason JA, McDowell SA, Callahan TW (2012) Effects of natural organic matter type and concentration on the aggregation of citrate-stabilized gold nanoparticles. J Environ Monit 14:1885–1892. https://doi.org/10.1039/C2EM00005A
64. Navarro E, Baun A, Behra R, Hartmann NB, Filser J, Miao A-J, Quigg A, Santschi PH, Sigg L (2008) Environmental behavior and ecotoxicity of engineered nanoparticles to algae, plants, and fungi. Ecotoxicology 17:372–386. https://doi.org/10.1007/s10646-008-0214-0
65. Ostroumov SA, Poklonov VA, Kotelevtsev SV, Orlov SN (2014) Toxicity of gold nanoparticles for plants in experimental aquatic system. Mosc Univ Biol Sci Bull 69:108–112. https://doi.org/10.3103/S0096392514030080
66. Oukarroum A, Barhoumi L, Pirastru L, Dewez D (2013) Silver nanoparticle toxicity effect on growth and cellular viability of the aquatic plant *Lemna gibba*. Environ Toxicol Chem 32:902–907. https://doi.org/10.1002/etc.2131
67. Pereira SA, Trindade CRT, Albertoni EF, Palma-Silva C (2012) Aquatic macrophytes as indicators of water quality in subtropical shallow lakes, Southern Brazil. Acta Limnol Bras 24:52–63. https://doi.org/10.1590/S2179-975X2012005000026
68. Rico CM, Majumdar S, Duarte-Gardea M, Peralta-Videa JR, Gardea-Torresdey JL (2011) Interaction of nanoparticles with edible plants and their possible implications in the food chain. J Agric Food Chem 59:3485–3498. https://doi.org/10.1021/jf104517j
69. Sabo-Attwood T, Unrine JM, Stone JW, Murphy CJ, Ghoshroy S, Blom D, Bertsch PM, Newman LA (2012) Uptake, distribution and toxicity of gold nanoparticles in tobacco (*Nicotiana xanthi*) seedlings. Nanotoxicology 6:353–360. https://doi.org/10.3109/17435390.2011.579631
70. Salata O (2004) Applications of nanoparticles in biology and medicine. J Nanobiotechnol 2:3. https://doi.org/10.1186/1477-3155-2-3
71. Sharma A, Manocha D, Arora A, Kumar A, Srivastava CM, Rawat V, Kim H, Verma M, Rao GK (2022) Chapter 14—Applications of green nanomaterials in electronic and electrical industries. In: Shanker U, Hussain CM, Rani M (Eds.) Green nanomaterials for industrial applications, micro and nano technologies. Elsevier, pp 397–421. https://doi.org/10.1016/B978-0-12-823296-5.00016-2
72. Siddiqi KS, Husen A (2016) Engineered gold nanoparticles and plant adaptation potential. Nanoscale Res Lett 11:400. https://doi.org/10.1186/s11671-016-1607-2
73. Stankus DP, Lohse SE, Hutchison JE, Nason JA (2011) Interactions between natural organic matter and gold nanoparticles stabilized with different organic capping agents. Environ Sci Technol 45:3238–3244. https://doi.org/10.1021/es102603p

74. Stuparu-Cretu M, Braniste G, Necula G-A, Stanciu S, Stoica D, Stoica M (2023) Metal oxide nanoparticles in food packaging and their influence on human health. Foods 12:1882. https://doi.org/10.3390/foods12091882
75. Su Y, Ashworth V, Kim C, Adeleye AS, Rolshausen P, Roper C, White J, Jassby D (2019) Delivery, uptake, fate, and transport of engineered nanoparticles in plants: a critical review and data analysis. Environ Sci Nano 6:2311–2331. https://doi.org/10.1039/C9EN00461K
76. Thwala M, Klaine SJ, Musee N (2016) Interactions of metal-based engineered nanoparticles with aquatic higher plants: a review of the state of current knowledge: Engineered nanoparticle interactions with aquatic plants. Environ Toxicol Chem 35:1677–1694. https://doi.org/10.1002/etc.3364
77. Usman AI, Abdul Aziz A, Abu Noqta O (2018) Application of green synthesis of gold nanoparticles: a review. J Teknol 81. https://doi.org/10.11113/jt.v81.11409
78. Wang X, Xie H, Wang P, Yin H (2023) Nanoparticles in plants: uptake, transport and physiological activity in leaf and root. Materials 16:3097. https://doi.org/10.3390/ma16083097
79. Wang Y, Xia Y (2004) Bottom-up and top-down approaches to the synthesis of monodispersed spherical colloids of low melting-point metals. Nano Lett 4:2047–2050. https://doi.org/10.1021/nl048689j
80. Zhao P, Li N, Astruc D (2013) State of the art in gold nanoparticle synthesis. Coord Chem Rev 257:638–665. https://doi.org/10.1016/j.ccr.2012.09.002
81. Zhu Z-J, Wang H, Yan B, Zheng H, Jiang Y, Miranda OR, Rotello VM, Xing B, Vachet RW (2012) Effect of surface charge on the uptake and distribution of gold nanoparticles in four plant species. Environ Sci Technol 46:12391–12398. https://doi.org/10.1021/es301977w

Use of Gold Nanoparticles as a Nano-Weapon in Plant Disease Management

Murugan Karuvelan, Suganya Kaliyamoorthy, Suresh Dhanaraj, Ramachandran Chelliah, and Deog Hwan Oh

Abstract One of the world's most pressing issues at the moment is the diagnosis and treatment of plant diseases, which have a heavy financial cost for plant-based product sectors and farmers. The detrimental impact on plants in agriculture can primarily be attributed to the presence of plant pathogens, including viruses, bacteria, and fungi, and pollution stemming from nanomaterials, alongside other significant contributors to pollution. Currently, a diverse array of methodologies are employed for the detection of pathogens in plants. These encompass DNA-based techniques, alongside other identification and detection methods utilizing microscopy. Nevertheless, it is imperative to acknowledge that the implementation of these methodologies necessitates the utilization of intricate instruments and a considerable amount of time. In recent times, considerable efforts have been dedicated to the development and application of novel biosensing systems that rely on nanomaterials. These systems hold great promise for the early detection of diseases, offering enhanced specificity and sensitivity. The goal of this chapter is to explain the different biosensing methods used to find diseases, with a focus on those that use nanomaterials as the main sensing element. The implementation of a precise and time-efficient methodology for the identification of plant pathogens holds significant potential for mitigating losses incurred in the domains of agriculture and forestry. The present review endeavours to shed light on a diverse array of afflictions that plague plants, commonly referred to as plant diseases. The main goal is to explain what needs to be in place for a reliable, quick, and cost-effective way to test for these diseases. The other goal of this discussion is to look into new biosensing technologies that might help find diseases in field plants that are growing in forests early on before they do too much damage.

M. Karuvelan · S. Kaliyamoorthy · S. Dhanaraj
Department of Microbiology, Technology & Advanced Studies(VISTAS), Vels Institute of Science, Pallavaram, Chennai 603117, Tamilnadu, India

R. Chelliah (✉) · D. H. Oh
Department of Food Science and Biotechnology, College of Agriculture and Life Sciences, Kangwon National University, Chuncheon 24341, South Korea
e-mail: ramachandran865@gmail.com

R. Chelliah
Saveetha School of Engineering, (SIMATS) University, Sriperumbudur 600124, India

A. Husen (ed.), *Plant Response to Gold Nanoparticles*, Smart Nanomaterials Technology, https://doi.org/10.1007/978-981-97-7174-5_15

Keywords Gold nano-particles · Plant diseases · Plant pathogens bio-sensing · Forest diseases

1 Introduction

In recent years, there has been a growing body of literature discussing the pioneering utilization of nanomaterials across various domains within the realm of science and technology. The remarkable attributes possessed by nanomaterials have served as a catalyst for scientists to explore their immense potential in the realms of physical, chemical, and biological sciences [1–10]. Nanoscale dimensions confer distinct optical, electronic, physical, and mechanical properties upon nanomaterials, setting them apart from their bulk counterparts. The contemporary utilization of metal-based, non-metal, and metal-based oxide nanomaterials is pervasive, encompassing a diverse array of applications. These applications span various domains, including but not limited to drug delivery, medical devices and diagnostics, material research, optics, and imaging. Along with carbon nanotubes, graphene, and other metal-based nanoparticles, gold nanoparticles (AuNPs) have become an important group of nanomaterials because they can be used in a wide range of nanodevices and products [11]. AuNPs have many great qualities, such as a large surface area-to-volume ratio, the ability to work with biological systems, and the simplicity with which recognition probes like antibodies, aptamers, and peptides can be added to their surfaces. These qualities make them ideal for a lot of different detection platforms. The visualization of dispersion and aggregation states can be achieved by monitoring the properties of localized surface plasmon resonance (LSPR) in AuNPs. The observed variations in color, ranging from pink to violet to pale blue, can be attributed to the size of the AuNPs and the degree of particle aggregation [12]. The utilization of this particular concept has been observed across multiple diagnostic platforms, facilitating the colorimetric detection of diverse analyses.

In addition to the aforementioned application, AuNPs have been employed in the field of bioanalysis in the context of surface plasmon resonance (SPR) transduction. The underlying principle of this feature revolves around the modulation of the dielectric constant pertaining to the surface plasmons exhibited by the AuNPs. Consequently, the identification and quantification of analytes are facilitated by monitoring alterations in the angle, intensity, or phase of the reflected light [13]. Moreover, it has been observed that AuNPs possess the remarkable ability to facilitate electron transfer processes between the electrode surface and electroactive biological species. The utilization of this fundamental transfer mechanism has been widely employed in the field of redox enzyme biosensing. In such biosensing systems, the recognition probe, typically an enzyme, serves as a catalyst for the oxidation and reduction reactions of the analytes under investigation. The exploration of the application of AuNPs in various biosensors and detection modalities has been undertaken, taking into account the numerous facets associated with AuNPs [14]. The utilization of AuNPs in biosensors has garnered significant attention in the scientific community owing to their remarkable ability to enable colorimetric or optical detection. This unique characteristic of AuNPs has paved the way for numerous advancements in biosensing technologies. The field of biosensing disease-causing pathogens in plants through the utilization of nanomaterials has witnessed remarkable advancements, as documented in the literature [15]. The presence of diseases in plants constitutes a significant factor contributing to the decline in production and yield within the agricultural domain [16]. The prevention of disease manifestation and blowout has prompted a substantial reliance on chemical interventions, specifically pesticides encompassing fungicides and bactericides. This practice, while aiming to mitigate ecological challenges, inadvertently amplifies the financial burden associated with production inputs. The detrimental impact of both cultivation and post-cultivation (storage) diseases on crop plants has been widely acknowledged, resulting in significant economic losses for farmers and industries on a global scale. The current state of forestry species, particularly woody perennials, is a matter of concern. Forests and their counterparts, known as trees-outside-forests (TOFs), assume a central role in bolstering the resilience and adaptability of our ecosystem in the face of shifting global environmental circumstances. Moreover, they contribute significantly to the mitigation of climatic fluctuations and serve as repositories for carbon sequestration. Nevertheless, the escalating occurrence of weather anomalies and extremes, attributable to rapid and intensified human interventions, is progressively inflicting detrimental consequences upon forests. This is primarily due to the fact that critical thresholds have already been surpassed, exacerbating the adverse impacts [17].

The foremost cause of diminished productivity within forest ecosystems is attributed to the presence of tree diseases. It has been estimated that the cumulative impact of disease-induced mortality and growth impairment on timber resources is comparable in magnitude to the combined effects of all other stressors. Throughout the course of time, a multitude of significant pathogen outbreaks have transpired within arboreal environments, exerting a profound and deleterious impact on numerous species and ecosystems on a global scale [18]. The deleterious impact of diseases and pest infestations on the overall well-being of forest ecosystems, including trees situated beyond the confines of these wooded areas, is a matter of considerable concern. The concept of tree health encompasses a comprehensive understanding of the interplay between abiotic and biotic elements that exert an influence on the vitality and productivity of a tree. This influence manifests itself through a diverse array of symptoms and forms of impairment. The evaluation of forest health can be conducted through qualitative or quantitative means. Qualitative assessment involves the description and identification of symptoms or damage observed in the forest ecosystem. On the other hand, quantitative assessment involves the measurement and analysis of physiognomic characteristics of the forest [19]. Forest tree species worldwide exhibit a significant vulnerability to various diseases, as well as infestations by insect pests in agricultural crops, resulting in a wide range of ecological and economic detriments. The magnitude of harm inflicted by insects, pests, and diseases is contingent upon the developmental phase during which they infiltrate and the specific insect or pathogenic agent responsible for the infestation. Within the realm of insect pests, it is commonly observed that trees are susceptible to various forms of attack, primarily stemming from leaf eaters, defoliators, stem borers, sap suckers, gall formers, and seed infesters, among others. The etiology of tree diseases predominantly stems from the pathogenic activities of various microorganisms, namely fungi, bacteria, and viruses. The prevalent afflictions affecting arboreal organisms, primarily attributed to fungal pathogens, encompass a spectrum of maladies including rust, leaf blight, leaf spot, powdery mildew, wilt, and root rot, among others. Bacterial cankers, bacterial leaf scorch, necrosis, crown gall, and chlorosis have been identified as prevalent bacterial diseases affecting trees [20]. The predominant methodologies employed for the management of pathogens in arboreal organisms encompass rigorous pruning practices, the systematic removal of infected branches, and the implementation of insecticide/fungicide treatments to regulate the insect vector responsible for pathogen transmission. Both of these strategies may not possess economic and ecological viability, as the former necessitates substantial labor input, while the latter exerts detrimental impacts on our natural ecosystem and soil health.

The successful management and mitigation strategies pertaining to forest tree health issues are contingent upon the timely and precise identification thereof, with a particular emphasis on the prompt diagnosis and discernment of symptoms within the natural environment. Furthermore, it is imperative to acquire a comprehensive comprehension of the underlying biology of the forest disease in question. This entails gaining insights into its associated biological processes and mechanisms. Additionally, it is crucial to develop the ability to identify the stage of susceptibility that is most vulnerable to the disease's impact. Furthermore, understanding the peak duration of the disease's activity is of utmost importance in effectively combating its detrimental effects. Moreover, recognizing the reciprocal relationship between the disease and the environment is a critical aspect of devising strategies to mitigate its impact. By undertaking these key steps, forest disease problems can be effectively addressed, thereby minimizing economic losses [16, 21]. The imperative of fostering awareness and enhancing the capacity of forest fringe communities and other relevant stakeholders cannot be overstated when it comes to addressing the health-related challenges associated with forest ecosystems. The necessity for an expanded scope of visual acuity during the preliminary evaluation of forest well-being has been identified as a crucial factor in enhancing the prompt identification and effective mitigation of concerns [22].

In the ever-evolving field of forest management and silviculture, pest and disease control strategies have seen substantial advancements, introducing more complex control measures and necessitating the use of sophisticated tools for early disease detection. This is particularly crucial in developing and transitioning countries, where severe pest outbreaks in forest ecosystems can have far-reaching effects. Such outbreaks not only impact the national economy significantly, posing risks to regional welfare, but also deeply affect the livelihoods of local communities dependent on these forests. The consequences of these outbreaks are not limited to economic factors; they also threaten food security, highlighting the importance of understanding their broader socio-economic implications.

The development of methods to detect forest diseases and their causative agents requires a focus on achieving high sensitivity, specificity, and rapid results. In the realm of agriculture, the use of nanotechnology-based biosensors has proven exceptionally effective for various crops, demonstrating their potential in disease detection. Despite limited research on the application of nano-sensors for forest tree diseases, their implementation in large forest areas plagued by disease could significantly benefit growers, farmers, and forest departments. The adoption of these advanced techniques has the potential to not only improve productivity but also positively impact carbon fluxes within these ecosystems. The integration of nanotechnology in forest disease management represents a promising frontier, offering innovative solutions to enhance forest health and sustainability.

The detection methodologies employed in the field of disease detection have predominantly revolved around two primary approaches, namely direct and indirect detection methods. The utilization of traditional techniques, including microscopic examination of pathogens via culture and colony count, immunological detection assays, and molecular methods such as polymerase chain reaction, represents a collection of direct detection methodologies. These approaches aim to directly identify pathogens by targeting their molecular, morphological, and serological characteristics. The methods in question are widely regarded as the epitome of excellence, yet they present considerable difficulties, necessitating specialized expertise, dedicated equipment, meticulous sample preparation procedures, and typically entailing a longer duration for obtaining results and their subsequent interpretation [23]. Indirect methodologies have been developed to facilitate the detection of physiological or morphological alterations occurring in infected plants [16].

The methodologies employed in this study encompass a range of sophisticated techniques, including optical imaging, thermography, gas chromatography, and mass spectrometry. The aforementioned methodologies exhibit considerable efficacy in the identification of causal microorganisms, although each approach possesses its own set of constraints and advantages. The selection of an appropriate methodology is contingent upon several factors, including the intended objective, the disease's stage, the geographical location, the feasibility of experimental implementation, the accessibility of test samples, and the associated financial implications. The expeditious identification of plant diseases holds significant advantages in relation to the prompt implementation of measures to mitigate their dissemination and mitigate the adverse impact on crop productivity. The majority of both direct and indirect detection methodologies pose challenges in their applicability within on-site environments, necessitating additional time for the generation of results. The imperative for the advancement of expeditious and uncomplicated sensing platforms or devices has been pivotal in the realm of investigation pertaining to plant disease management.

The implementation of nanomaterial-based biosensors has become increasingly prevalent in a diverse array of applications encompassing environmental monitoring and medical diagnostics. The utilization of these technologies within the realm of plant sciences or agricultural practices has experienced a notable surge in recent times. A biosensor, in its typical configuration, consists of three fundamental components: a bio-recognition element, a transducer, and a signal interpretation system.

The utilization of gold nanoparticles and nanostructures has been widely employed in the process of immobilizing bio-recognition elements. This approach allows for the effective utilization of surface plasmon resonance effects in the transduction of signals. The utilization of nanoparticles has been discovered to enhance the efficacy of biosensing devices in relation to the limit of detection and sensitivity of the sensor, as indicated by previous studies [24, 25]. The investigation of various nanomaterials, including carbon nanotubes, graphene, and metal nanoparticles, has been undertaken in order to assess their potential impact on plant cells, primarily focusing on their toxicity effects. The utilization of gold nanoparticles (AuNPs) in sensing applications has been widely acknowledged due to their well-established mechanism and facile surface modification capabilities, which further enhance their biocompatibility. In this comprehensive review, the authors direct their attention toward an assortment of plant diseases that manifest within forested regions as well as domestic fields. The necessity for a temporally efficient, expeditious, and economically viable testing modality to identify field plants within forested environments and ascertain their disease status during the initial stages of pathogenic development is currently under investigation. The aforementioned review comprehensively elucidates the utilization and merits of nanomaterials in the realm of emerging biosensing technologies and the detection of plant diseases, as expounded upon in Fig. 1.

It provides a comprehensive and detailed diagrammatic representation that serves two primary purposes: firstly, it illustrates the various sources and pathways of nanomaterial pollution in the environment, and secondly, it showcases the innovative application of nanomaterials in the detection of plant pathogens. The first section of the figure details how nanomaterials, due to their extensive use in various industries, can enter and accumulate in different environmental compartments, such as

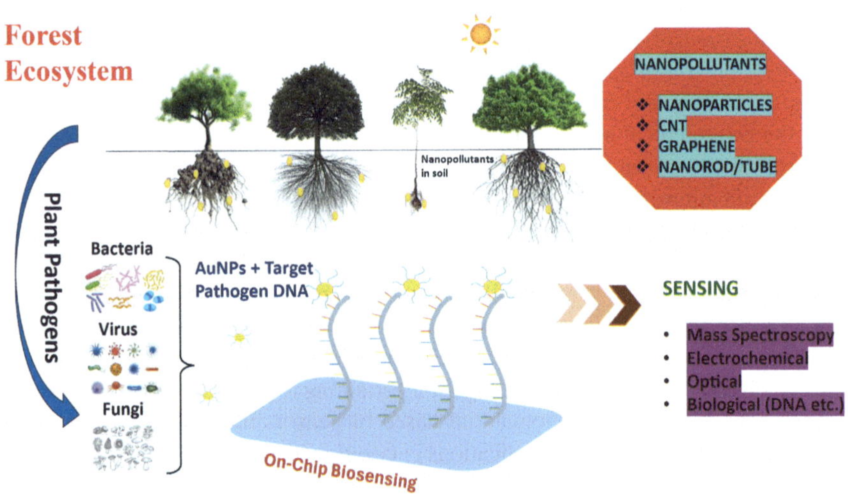

Fig. 1 Contrasts the challenges of nanomaterial pollution with their beneficial applications in agricultural pathogen detection

water bodies, soil, and air. It highlights the mechanisms through which these particles can cause pollution, including runoff from industrial sites, atmospheric deposition, and wastewater discharge. The figure also depicts the potential impact of such pollution on ecosystems and human health. In contrast, the second section of the figure focuses on the positive application of nanomaterials, particularly in the field of agriculture for the early detection of plant diseases. It illustrates the types of nanosensors used for identifying specific plant pathogens, showing how these devices can detect the presence of viruses, bacteria, or fungal infections in plants. The diagram explains the working principle of these sensors, which involves the interaction between nanomaterials and pathogen-specific biomolecules, leading to a detectable signal. This section emphasizes the benefits of using nanotechnology in agriculture, such as increased sensitivity, rapid detection, and the potential to prevent large-scale crop losses. Overall, Fig. 1 provides a contrasting view of nanomaterials, portraying both the challenges of environmental pollution and the opportunities for advanced pathogen detection in agriculture

2 Synthesis of Gold Nanoparticles

Nanoparticles have been successfully synthesized using diverse physiochemical processes, each of which has exerted considerable strain on the surrounding environment. Gold nanoparticles have emerged as a paramount entity among the various metal nanoparticles discussed earlier, primarily owing to their extensive history of therapeutic applications. Gold nanoparticles have been extensively documented and widely reported in the existing body of scholarly literature. Since the year 1996, a substantial corpus of scholarly literature, comprising an estimated count of approximately 87,000 papers, has been disseminated and made available to the academic community and beyond. The investigation of copper, silver, iron, and titanium nanoparticles has been extensively documented in the scientific literature, as highlighted by [26]. These nanoparticles merit individual scrutiny and should be subject to independent research efforts.

In the realm of synthesizing gold nanoparticles, a plethora of diverse methodologies encompassing biological, physical, and chemical approaches have been meticulously devised and refined Fig. 2.

It is an intricate and detailed representation of gold nanoparticles (Au–NPs), providing a multifaceted view of their structure, properties, and applications. The figure begins by depicting the unique physical characteristics of these nanoparticles, such as size, shape, and surface morphology. It illustrates various forms of gold nanoparticles, including spherical, rod-like, and cubic shapes, and explains how their size and shape influence their optical and chemical properties

Fig. 2 Presents a detailed representation of gold nanoparticles (Au–NPs), offering insights into their structure, properties, and applications

3 Biological Synthesis of Nanoparticles

The process of biological synthesis of nanoparticles is widely regarded as a secure, dynamic, and energy-efficient approach to nanoparticle production. The present methodology encompasses a diverse array of biological entities, spanning from prokaryotic organisms to eukaryotic organisms, which are employed for the purpose of synthesizing natural products within living systems. The metabolites present in these sources, including proteins, fatty acids, sugars, enzymes, and phenolic compounds, are known to exert a substantial influence on both the bio-reduction of metallic ions to nanoparticles (NPs) and the subsequent stability of these NPs. The biologically synthesized gold nanoparticles (AuNPs) exhibit a higher degree of stability in comparison to their counterparts generated through alternative methodologies. Gold nanoparticles (AuNPs) have demonstrated high efficiency in their production through chemical synthesis methods. However, a significant concern associated with this approach is the potential generation of hazardous by-products, which pose risks to both human health and the environment. The investigation of alternative pathways for the synthesis of nanoproducts that ensure safety has become a subject of great interest among various biological systems, including plants, bacteria, yeasts, and fungi. These systems are currently being extensively studied as potential sources for the production of gold nanoparticles (AuNPs) [27] Fig. 3.

It is a detailed illustrative guide that comprehensively showcases the diverse forms of gold nanoparticles (AuNPs), emphasizing their distinct shapes, sizes, and structural variations. This figure meticulously categorizes and displays a variety of AuNPs, including spherical nanoparticles, nanorods, nanoshells, nanocages, and nanostars, among others. Each form is visually represented, highlighting the unique geometric

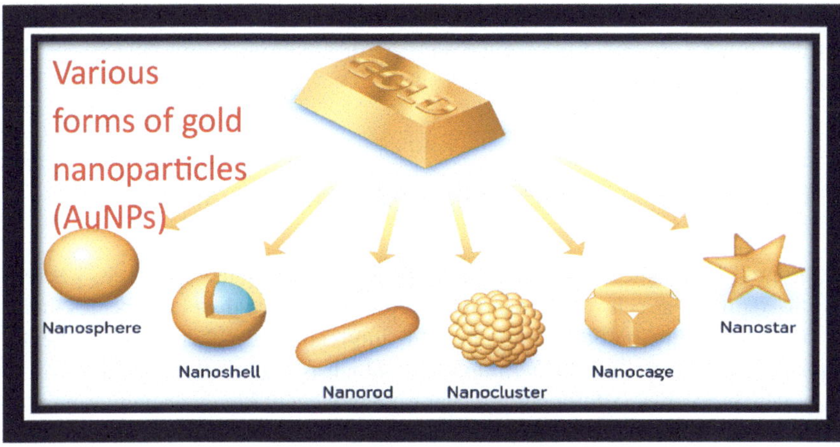

Fig. 3 Serves as a detailed guide showcasing the diverse forms of gold nanoparticles (AuNPs), emphasizing their unique shapes, sizes, and structural variations

structures and surface characteristics that define them. The diagram provides insights into how these different forms are synthesized, noting the specific methods and conditions that lead to each type of nanoparticle. Additionally, the figure elaborates on the specific properties and applications associated with each form of AuNP. For instance, it explains how spherical nanoparticles are commonly used in drug delivery and bioimaging, while nanorods are favored for their enhanced optical properties, making them useful in photothermal therapies. Nanoshells and nanocages are depicted as ideal for targeted cancer therapy due to their ability to absorb and convert light into heat, effectively destroying tumor cells. Nanostars, with their sharp edges and increased surface area, are shown to enhance sensing and catalytic applications. Figure 3 also touches upon the significance of size, surface charge, and coating in determining the behavior and interaction of AuNPs with biological systems. It discusses how these factors influence the stability, biocompatibility, and targeting capabilities of the nanoparticles. By providing a detailed and visual overview of the various forms of AuNPs, Fig. 3 serves as an invaluable resource for researchers, clinicians, and anyone interested in the vast potential of gold nanoparticles in scientific and medical applications

4 Green Synthesis of Plant Extract -AuNPs

The natural world abounds with a plethora of botanical specimens that possess the distinct advantage of being economically viable, highly replicable, ecologically sound, and characterized by a precise purification process when juxtaposed with alternative environmentally conscious biological approaches. The utilization of plant extracts as both reducing agents and stabilizers in the synthesis of gold nanoparticles has garnered significant attention in recent years [28]. This approach, often referred to as the "green pathway," has gained popularity due to its numerous advantages and potential applications.

5 The General Approach to Plant Preparation-AuNPs

The proposed methodology entails the utilization of a targeted approach, wherein specific segments are chosen based on the categorization of the plant species under investigation. In this particular instance, it is imperative to gather the fundamental segment of the *Euphorbia fischeriana* plant, which is commonly employed for its antioxidant properties. Additionally, the pulp component of *Punicagranatum* exhibits antimicrobial characteristics, while the leaf extract derived from *A. noeanum* demonstrates antibacterial attributes. Furthermore, the utilization of papaya juice has been identified as a means of detecting L-Lys, as supported by the works of Qiao and Qi [29–31]. Following the administration of chemical therapy, cancer cells exhibit diminished sensitivity as a consequence of repeated drug exposure. However, it has been observed that the utilization of nanoparticles can significantly enhance intracellular drug accumulation owing to their remarkable ability to selectively target and deliver drugs with greater precision [32]. The utilization of agents and stabilizers is a prominent aspect within various industries. These substances play a crucial role in enhancing the performance and stability of a wide range of products. Agents refer to the biomolecules derived from plant extracts encompass a diverse array of compounds, including polyphenols, flavonoids, polysaccharides, alkaloids, amino acids, vitamins, ketones, phenols, and proteins [33] Fig. 4.

Presents an elaborate and instructive diagram that illustrates the process of synthesizing gold nanoparticles (AuNPs) through plant-based, or "green," methods. This figure systematically outlines the steps involved in this eco-friendly approach, starting from the selection of plant materials to the final production of AuNPs. The diagram begins by showcasing various types of plants—including leaves, fruits, and extracts—that are commonly used as natural reducing agents in the synthesis process. It highlights the procedure of preparing plant extracts, detailing the extraction of phytochemicals that are responsible for the reduction of gold ions to nanoparticles. The core part of the figure focuses on the reaction process, visually representing how gold ions, when mixed with plant extracts, undergo a reduction reaction, leading to the formation of AuNPs. The diagram emphasizes the factors that influence this process, such as pH, temperature, and concentration of the plant extract, and how these conditions affect the size, shape, and yield of the nanoparticles. Additionally, Fig. 3 underscores the advantages of using plant-based methods for AuNP synthesis, such as the avoidance of harmful chemicals, cost-effectiveness, and environmental sustainability. The figure also compares plant-based methods with traditional chemical synthesis, pointing out the differences in terms of safety, energy consumption, and waste production. By providing a clear and comprehensive visual guide to the green synthesis of AuNPs, Fig. 3 serves as an educational tool for researchers and

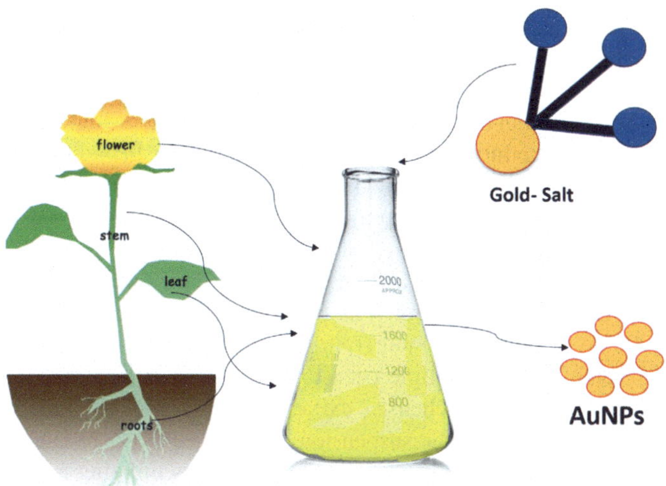

Fig. 4 Presents an informative diagram illustrating the synthesis of gold nanoparticles (AuNPs) using ecofriendly plant-based or "green" methods

students in the field of nanotechnology and green chemistry, offering insights into this innovative and environmentally friendly method of nanoparticle production

In the context of biosynthesis, it is imperative to select a plant species that possesses one or more of the aforementioned chemical compounds, as these compounds have the ability to effectively reduce metal ions to their elemental metal form. The initial step involves the reduction of Au3 + to Au0, a process that has been extensively studied and documented in the literature. Following this reduction, a subsequent procedure is employed to ensure the stability and integrity of the resulting gold nanoparticles (AuNPs). This involves the deliberate application of a protective layer on the outer surface of the gold nanoparticles, effectively impeding their aggregation and promoting their long-term stability.

6 Green Synthesis of AuNPs by Cultivating Plant Cells

The present discourse expounds upon a particular technique employed in the pursuit of cultivating plant cells in an indefinite manner. The hypocotyl regions of seedlings at the age of one week were subjected to germination on a water agar medium. Subsequently, these germinated seedlings were inoculated on MS medium 38, with a pH value of 5.7. The MS medium was supplemented with 30% sucrose, 0.8% agar, 0.2 mg/L indole-3-acetic acid (IAA), and 0.2 mg/L benzyl adenine (BA). Following a period of approximately two weeks, the callus mass was subsequently relocated to a liquid medium of identical composition. The formation of callus clumps was observed within conical flasks, wherein a growth medium was present. These flasks were subjected to agitation in an orbital shaker operating at a speed of 100 revolutions per minute. The temperature was maintained at 27 °C throughout the duration of 48 h, allowing for the development of a cell suspension. The findings of [34] have been substantiated through a meticulous microscopic analysis, which unequivocally establishes the solitary cellular nature of the examined culture. In the context of AuNP synthesis, the cell culture was enriched with KAuCl4 at a concentration ranging from 10 to 200 parts per million (ppm). Subsequently, the culture was subjected to incubation on a shaker for a duration of 24 h while being kept in a lightless environment. The culture (1 ml) was transferred into the Eppendorf tubes using a pipette. Subsequently, centrifugation was performed for a duration of 2 min at a speed of 14,000 revolutions per minute (rpm). Following centrifugation, the medium was carefully removed. On multiple occasions, the utilization of nanopore water was employed to achieve comprehensive extraction of residual quantities of medium and unreacted gold. The cellular suspension was subjected to resuspension in a volume of 100 L of water, followed by subsequent lysis utilizing a sonicator for a duration of 2 min. The discovery of gold nanoparticle formation was achieved through a meticulous examination of the reaction dynamics exhibited by the colour medium, wherein a notable transition from a yellow hue to a distinct red shade was observed and analysed. Subsequently, the resultant colloidal solution comprising gold nanoparticles underwent a series of characterization procedures, as documented in the works of Qiao and Qi and [34].

7 Conditions of Reaction

The determination of the morphology, encompassing the shape and size, of gold nanoparticles is contingent upon various reaction conditions. These conditions include pH levels, temperature, incubation duration, concentration of plant extract, and concentration of metal salt. The morphological characteristics of gold nanoparticles exhibit a notable dependence on the pH of the surrounding medium. Specifically, at a pH of 2, these nanoparticles assume a larger, rod-shaped configuration. In contrast, at pH levels ranging from 3 to 4, a smaller, rod-shaped morphology is observed. Upon reaching a pH of 8, the gold nanoparticles adopt a spherical, oval, or polyhedral shape. Furthermore, at a pH of 9, a predominantly spherical structure is observed. As the pH increases to 10, the gold nanoparticles once again assume a rod-shaped configuration. Finally, at a pH of 11, the nanoparticles exhibit a distinctive nano-wire morphology. The final dimensions of nanoparticles are also subject to the influence of pH, wherein higher pH levels (specifically at pH 3) are associated with a reduction in nanoparticle size. Gold nanoparticles exhibit a diverse range of morphologies, which are contingent upon the temperature conditions within the range of 25–90 °C. These morphologies encompass triangular, pentagonal, hexagonal, and spherical shapes. The acceleration of nanoparticle development is observed in response to elevated temperature and increased salt concentration, wherein the upper limit for temperature is determined to be 50 °C, while the optimal salt concentration is found to be 0.7 mM. The alteration in colour is primarily attributed to a disparity in shape, as indicated by the presence of an absorbance peak within the visible spectrum at wavelengths of 533 and 529 nm. The citation provided, namely [34], indicates that the user is referencing a scholarly publication authored by Bhaskaran and colleagues in the year 2019.

8 Characterizations of the Synthesized Gold Nanoparticles:

The monitoring of the signature of AuNPs was conducted through the utilization of UV–visible spectroscopy, a highly potent analytical technique commonly employed for the comprehensive characterization of colloidal particles.

The investigation involved the utilization of leaf extract to facilitate the synthesis of pure AuNPs. The progress of this synthesis was assessed through the observation of UV–Vis analysis within the wavelength range of 300–800 nm. The UV–Vis spectra were acquired using a Shimadzu UV–Vis spectrophotometer, manufactured in Kyoto, Japan. The instrument operated at a resolution of 1 nm. The determination of the structure and size of the synthesized gold nanoparticles (AuNPs) was accomplished through the utilization of X-ray diffraction spectroscopy. The nanoparticles, which had been subjected to air drying, were subsequently applied as a coating to an X-ray diffraction (XRD) grid. The purpose of this procedure was to investigate the potential formation of gold nanoparticles (AuNPs) using a Philips X-ray diffractometer. The X-ray diffraction (XRD) pattern was meticulously obtained by employing CuKα radiation, which has a wavelength (λ) of 1.54060 A°. The utilization of nickel monochromatic facilitated the acquisition of precise data within the angular range of 2π, spanning from 10 to 80°. The determination of the average crystallite size of the synthesized AuNPs was accomplished through the utilization of Scherer's formula, which is expressed as $D = (0.9\lambda)/\beta Cos\theta$. The present investigation involved a comprehensive examination of the morphological attributes exhibited by AuNPs. This was achieved through the utilization of cutting-edge scientific instrumentation, namely the field emission scanning electron microscope (FESEM) and the high-resolution transmission electron microscope (HRTEM), specifically the JEOL-Model 6390 apparatus.

9 Types of Nanoparticles for Plant Disease Management

Nanoparticles, which typically exhibit dimensions ranging from 10 to 100 nm (nm), possess remarkable chemical, physical, and biological attributes that set them apart from their molecular and bulk counterparts. The utilization of nanoparticles as a standalone approach exhibits promising prospects for their direct application onto plant seeds, foliage, or roots, thereby offering a protective shield against a wide range of pests and pathogens, including insects, bacteria, fungi, and viruses. Extensive investigations have been conducted on the antimicrobial attributes of metal nanoparticles, including but not limited to silver, copper, zinc oxide, and titanium dioxide. These nanoparticles have garnered significant attention in the scientific community due to their demonstrated efficacy against bacteria and fungi. Moreover, their potential as antiviral agents has also been recognized. This section provides a concise synopsis and a contemporary status update on extant literature reviews concerning individual nanoparticles. In recent times, there has been a surge in the utilization of silver nanoparticles, primarily attributed to the emergence of "green synthesis" methodologies for their production, which involve the utilization of various biological entities such as plants, bacteria, fungi, or yeast. The utilization of silver nanoparticles has been demonstrated to exhibit notable antifungal properties, effectively inhibiting the growth and development of various fungal species such as *Alternaria alternata, Sclerotinia sclerotiorum, Macrophomina phaseolina, Rhizoctonia solani, Botrytis cinerea,* and *Curvularia lunata.* This inhibition was assessed through the implementation of a well diffusion assay, a widely employed technique for evaluating the antimicrobial activity of substances. The application of silver nanoparticles onto the leaves of bean plants resulted in the complete suppression of the sun-hemp rosette virus, as evidenced by the absence of any observable viral activity. In a study conducted by Elbeshehy et al., it was demonstrated that faba bean plants, upon being exposed to bean yellow mosaic virus, exhibited significantly enhanced outcomes when treated with silver nanoparticles. Notably, the application of these nanoparticles 24 h after infection yielded superior results compared to pre-infection spray application or simultaneous inoculation. The utilization of silver nanoparticles (AgNPs) in the field of plant disease management has garnered considerable attention due to their remarkable efficacy against both fungal and bacterial pathogens. However, it is imperative to acknowledge the presence of notable challenges that impede their widespread implementation. These challenges primarily revolve around the production process of AgNPs, their potential toxicity, and their interaction with soil ecosystems. Metal nanoparticles are widely employed in various applications due to their unique properties and versatile nature. Among the commonly utilized metal nanoparticles are copper, titanium dioxide, and gold. These nanoparticles have garnered significant attention in the scientific community and have been extensively studied for their potential applications in diverse fields. The utilization of nanoparticles, such as copper and titanium dioxide, in the field of fertilizer application has gained increasing prominence. However, it is noteworthy that limited research has been conducted thus far to explore the potential of these nanoparticles in the domain of plant disease management.

The highlighted research pertains to the investigation of the antimicrobial properties exhibited by titanium dioxide, silver, and copper nanoparticles, as well as the potential insecticidal effects of aluminium nanoparticles when used as dust. Moreover, it has been observed that the incorporation of titanium dioxide nanoparticles into fertilizers has yielded significant advancements in terms of bacterial protection and viral inactivation. The present study elucidates the introduction of poly-dispersed gold nanoparticles into the system through a mechanical abrasive technique. This novel approach was observed to induce the melting and subsequent dissolution of Barley yellow mosaic virus particles, thereby imparting a notable resistance to the host plant. Chitosan, a widely recognized nanoparticle, exhibits a range of advantageous biological characteristics, including biodegradability, biocompatibility, non-allergenicity, and antimicrobial activity. Furthermore, it demonstrates minimal toxicity toward both animals and humans. The induction of viral resistance in plant tissues through the utilization of chitosan nanoparticles has been observed, wherein these nanoparticles serve as a protective shield against infections caused by the mosaic virus affecting various crops such as alfalfa, snuff, peanut, potato, and cucumber. The antimicrobial properties of chitosan nanoparticles have been observed in various instances, including their ability to regulate the growth of Fusarium crown, root rot in tomato, Botrytis bunch rot in grapes, and *Phyricularia grisea* in rice. However, it is worth noting that their efficacy against bacterial pathogens appears to be comparatively limited. Malerba and Cerana have succinctly encapsulated the various potential mechanisms that underlie the antimicrobial properties of chitosan. These mechanisms encompass a range of intricate processes, including agglutination, disruption of the cell membrane, inhibition of H + –ATPase activity, suppression of toxin production and microbial growth, hindrance of messenger RNA and protein synthesis, and impeding the flow of essential nutrients. The antiviral properties of beans have been documented in their ability to combat Bean mild mosaic virus. Similarly, in the case of tobacco, beans have demonstrated efficacy against both Tobacco mosaic virus and Tobacco necrosis virus [35].

10 Plant Pathogens and Plant Diseases

Forest trees exhibit extended rotation periods and hold substantial importance in both environmental and commercial contexts. According to scholarly estimates, it has been posited that forests play a significant role in meeting the needs of the human population, accounting for approximately 16% of these requirements. Furthermore, it has been suggested that forests also contribute to fulfilling the needs of cattle, encompassing approximately 18% of their overall necessities [11]. Both natural forests and plantations are susceptible to damage from a wide range of diseases caused by viral, bacterial, fungal, and other parasitic organisms [36] throughout various regions of the world. Pathogens, as pivotal constituents within ecosystems, assume a paramount position in shaping the intricate tapestry of tree diversity and spatial distribution within the forest ecosystem. The occurrence of disease and pest infestations in trees represents a significant source of stress within the field of forestry. Forest ecosystems are susceptible to a multitude of pathogens, invasive pests, and nutrient deficiencies, thereby exposing them to various challenges. The presence of tree pests and diseases poses a significant challenge to the overall well-being and economic prosperity of forest ecosystems. Numerous bio-pathogens, including but not limited to fungi, bacteria, viruses, oomycetes, phytoplasma, and plant parasitic nematodes, have been identified as causative agents of diseases in forest trees. The etiology of the majority of tree diseases can be attributed to fungal pathogens. Plant diseases like foliar and wood rot are mostly caused by fungi belonging to the taxonomic classes *Ascomycetes* and *Basidiomycetes*. A comprehensive array of symptoms associated with fungal infections in plants encompasses leaf spots, rusts, scorch, blotch, anthracnose, and needle blights. The presence of discernible fruiting bodies, which are characteristic of pathogenic species, may serve as an indicator of the visibility of these fungal pathogens. The prominent instances of fungal pathogens affecting trees encompass *Fusarium, Ganoderma, Rhizctonia, Melampsora, Erysiphe*, and *Armillaria*, among others [11]. Anthracnose, a prevalent affliction observed primarily in deciduous trees, manifests as the formation of darkened and depressed lesions on various plant parts such as flowers, leaves, stems, and fruits. Powdery mildew, a fungal pathogen, induces the formation of a conspicuous white coating on the surface of foliar tissues. This deleterious condition subsequently leads to the manifestation of various detrimental effects, including leaf distortion, yellowing, and ultimately, the demise of the affected leaves. The etiology of this phenomenon can be attributed to a fungal agent, predominantly observed to colonize plant and arboreal substrates within environments characterized by diminished exposure to direct sunlight and elevated levels of atmospheric moisture. Rust, a fungal affliction, manifests as conspicuous yellow spots on the upper leaf surface of various hardwood species.

Bamboo, as a forest species, holds significant importance owing to its remarkable attributes, such as its rapid and cyclical growth, exceptional strength properties, and diverse range of applications. Bamboos are subject to various pathogens, with fungi being the predominant group responsible for culm diseases such as emerging culm rot, culm stain, and culm spot.

It is an extensive compilation that chronicles the most significant historical instances of forest tree diseases that have had a global impact. This table systematically lists and provides detailed accounts of various major tree disease outbreaks, their causative pathogens, the geographic regions they affected, and the time period during which they occurred. Each entry in the table includes the common name of the disease, the scientific name of the pathogen (such as fungi, bacteria, or viruses), and the species of trees that were most affected. The table also delves into the ecological and economic consequences of these outbreaks, highlighting the extent of forest destruction, loss of biodiversity, and the impact on timber and related industries. For instance, it details events like the Chestnut Blight in North America, caused by the fungus *Cryphonectria parasitica*, and the Dutch Elm Disease, caused by the Ophiostoma novo-ulmi fungus, showcasing their devastating effects on chestnut and elm trees, respectively. Additionally, Table 1 discusses the response measures and management strategies that were implemented to combat these diseases, including quarantine measures, breeding of resistant tree varieties, and the use of chemical and biological control methods. The table serves as an important educational and historical resource, offering insights into the patterns and impacts of major forest tree diseases over time. It emphasizes the importance of vigilant monitoring and effective disease management in preserving forest health and sustainability

Additionally, foliar diseases including spot, rust, and blight, as well as rhizome diseases like rhizome rot and sheath diseases such as sheath rot and sheath spot,

Table 1 It is a comprehensive compilation chronicling significant historical instances of forest tree diseases that have impacted global ecosystems

S. no.	Microorganism that causes disease	Term for a disease	Host	Geographical region and chronological sequence
1	*Cronarticumribicola* (Fungus)	White Pine Blister Rust	Pinus parviflora	Baltic, Russia, 1854 Pinus
2	*Erwinia amylovora* (Bacterium)	Fire Blight of Pome	Pea, apple, and Rosaceous sp	New York, 1780
3	*Phytophthora ramorum* (Fungus)	Sudden Oak Death	Oak sp.	Oregon, California, and Europe
4	*Bursaphelenchusxylophilus* (Pine Wood Nematode) spread by *Monochamus spp.* Beetle (Pine Sawyer Beetle)	Pine Wilt	Pine spp.	North America, East Asia, 1940s
5	Phytoplasma	Sandal Spike Disease	Santalum album	Southern India, 1903
6	(i) *Fusarium solani* (Fungus) (ii) *Ganoderma lucidum* (Fungus) (iii) *Phellinus gilvus* (Fungus) (iv) *Meloidogyne javanica* (Nematode)	Shisham Mortality (i) Fusarium Wilt (ii) Ganoderma Root Rot (iii) Phellinus Root and Butt Rot (iv) Root Knot Nematode	Dalbergia sissoo	(i) North India (ii) North-Central India (iii) North-Central India (iv) Dehradun, India
7	*Cryptococcus fagisuga* and *Nectria* Fungus (Insect Fungal Complex)	Beech Bark Disease	Fagus grandifolia	Northern America, 1920–1930s
8	*Ophiostoma-novo-ulmi* (Fungus)	Dutch Elm Disease	Ulmus sp	Northwest Europe, 1910s
9	*Cryphonectriaparasitica* (Fungus)	Chestnut Blight	Castanea sativa	USA, Canada, and Asia, 1904
10	*Phytophthora cinnamomi* (Fungus)	Jarrah Dieback	Eucalyptus marginata	Sumatra, Indonesia, 1922

also affect bamboo [37]. The prevailing afflictions observed in arboreal organisms encompass shot hole, foliar blight, and wilt, predominantly instigated by the pathogenic agents *Erwinia, Xanthomonas,* and *Ralstonia.* According to scientific research, viruses are inanimate objects that exhibit a non-living nature due to their small size and innate reliance on host organisms for survival and replication. Even though these mesobiotic parasites are very small (sub-microscopic), they can only reproduce inside host cells. This means that they need to perform an intracellular multiplication process. The transmission of these entities occurs via direct contact or by means of physical and biological vectors. Numerous viral afflictions, including but not limited to bamboo mosaic potex virus, elm mottle virus, oak mosaic virus, and prunus necrotic ringspot virus, have been identified as the etiological agents responsible for the manifestation of disease symptoms in forest trees. The proliferation of the majority of arboreal species through seed propagation poses challenges owing to limited seed availability. For this reason, using clonal cuttings to make plantlets and preparing seedlings by checking for seed viability are the most important parts of large-scale tree-planting projects. The extensive propagation of seedlings and clonal plantlets within monoculture systems, whether in the nursery field or within mist chambers characterized by high moisture levels and abundant nutrient supply, has been observed to facilitate the proliferation of pathogens. The occurrence of various root diseases in nursery plants is a common phenomenon. These diseases encompass pre-emergence blight/damping-off, post-emergence damping-off, vascular wilt disease, root rot disease, set rot of cuttings, and web blight. The prevalent foliar afflictions encompass a range of maladies, namely leaf spots, leaf blights, leaf scorch, leaf rusts, and powdery mildews. Hence, the application of nano sensors for disease biosensing during the early stages of plant growth holds significant potential within the domain of forestry. The regions of Asia and Europe have experienced significant deforestation, with an estimated loss of approximately 5 million hectares of forested land. This alarming trend can be attributed to the widespread occurrence of forest tree diseases originating from various global sources. The specific outbreaks of these diseases are enumerated in Table 1.

11 Gold Nanoparticle-Based Biosensors

Gold nanoparticles (AuNPs) have emerged as a crucial constituent in a wide range of biological and chemical sensing apparatuses. Nanoparticles' notable characteristics, such as a higher surface area-to-volume ratio, customizable optical properties, and simple synthesis techniques, have facilitated their use in various diagnostic platforms for the detection of plant, animal, or human diseases, biomarkers, and other chemical analytes of significance [38]. Gold nanoparticles (AuNPs) can be synthesized using various techniques, including chemical and photochemical reduction, seed-mediated growth, and green synthesis methods. These methods offer distinct advantages and have been extensively studied in the field of nanotechnology. Chemical reduction involves the use of reducing agents, such as sodium borohydride or citrate, to reduce gold ions into nanoparticles. On the other hand, photochemical reduction utilizes light energy to drive the reduction reaction, often employing a photosensitizer or a photocatalyst. Seed-mediated growth involves the use of pre-formed gold nanoparticles as seeds, which act as nucleation sites.

The Turkevich-Frens method, which entails the reduction of gold salts using trisodium citrate, and the Brust-Schiffrin method, which involves the reduction of gold salts using external thiol ligands, are widely recognized as the predominant reaction methods employed in the synthesis of gold nanoparticles (AuNPs) [31, 39].

11.1 Electrochemical Biosensors

Electrochemical detection methods have been widely recognized for their remarkable attributes, including heightened sensitivity, cost-effectiveness, expeditiousness, and the potential for miniaturization. One of the three main parts of an electrochemical sensor that makes it work is the working electrode (WE), the counter electrode (CE), and the reference electrode (RE). The utilization of glass electrodes, specifically the glassy carbon electrode, is prevalent in various electrochemical applications. These electrodes serve as the working electrode, responsible for facilitating the desired electrochemical reactions. In contrast, platinum wire and Ag/AgCl (saturated KCl) are commonly employed as the counter electrode (CE) and reference electrode (RE), respectively. The CE aids in completing the electrical circuit, while the RE provides a stable reference potential against which the electrochemical measurements can be calibrated. The interaction between the target analytes and the sensor leads to alterations in the current, potential, and impedance parameters. These changes serve as the basis for generating a measurable signal, which can be detected using various approaches. This includes electrochemiluminescence (ECL), electrochemical impedance spectroscopy (EIS), and different types of voltammetry, such as differential pulse voltammetry (DPV), cyclic voltammetry (CV), square wave voltammetry (SWV), and anodic stripping voltammetry (ASV). Moreover, it is imperative to acknowledge that the surface area of the working electrode (WE) holds significant relevance in the determination of the sensor's sensitivity. The utilization of gold nanoparticles (AuNPs) in electrochemical sensors presents a significant advantage due to several factors. For starters, AuNPs have a large surface area to volume ratio, which makes it easier for them to interact with analytes and sense things. This property allows for increased sensitivity and selectivity in detecting target molecules. Furthermore, AuNPs offer ease of modification, enabling the incorporation of various functional groups or biomolecules onto their surfaces. This versatility allows for the customization of AuNPs-based sensors to specifically target and detect different analytes of interest. By tailoring the surface properties of AuNPs, researchers can optimize the sensor's performance and enhance its overall efficiency. Additionally, AuNPs exhibit excellent signal-amplification capabilities. Due to their unique optical and electronic properties, AuNPs can significantly enhance the signal output of electrochemical sensors. This amplification effect leads to improved detection limits and increased accuracy in sensing applications [40]. In conclusion, the application of AuNPs in electrochemical sensors offers a great advantage due to Researchers have found that using gold nanoparticles (AuNPs) can greatly improve the conductivity of sensing surfaces. This is because AuNPs can change the surface and speed up chemical reactions. Putting recognition probes like antibodies, aptamers, and peptides on the surface of gold nanoparticles (AuNPs) has been shown to make them more specific to the analytes of interest.

In addition, it is noteworthy to mention that the utilization of gold nanoparticles (AuNPs) in this context serves as an electrochemical indicator, operating on the basis of a redox reaction involving the conversion of Au^0 to Au^{3+}. In the field of electrochemical biosensors, there are three main ways to detect the signals that AuNPs (gold nanoparticles) emit. Firstly, the direct detection of the oxidation signal of AuNPs without any prior treatment has been employed [41]. Secondly, the electro-oxidation of AuNPs to gold ions is achieved by utilizing hydrochloric acid (HCl) [23], 83]. Lastly, the treatment of AuNPs in HBr/Br2 solutions has also been explored as a means of signal detection [42]. A group of researchers successfully demonstrated the binding event between the single-stranded DNA binding protein (SSB) of *Escherichia coli* and complementary single-strand oligonucleotides conjugated to gold nanoparticles (AuNPs). It was the AuNPs-conjugated SSB that was used as the label for hybridization, and then measurements were taken to see how the AuNPs' oxidation signal changed. The researchers successfully determined the limit of detection (LOD) for the sensor to be 2.17 pM of the target DNA, as reported in reference [43].

11.2 Enzyme Biosensors

Enzymes exhibit distinctive catalytic activity and specificity toward their respective target substrates, thereby rendering them highly valuable in the realm of biomedical applications. The subject of enzyme immobilization has garnered significant interest in the scientific community owing to its potential for enabling repeated or prolonged utilization of enzymes for specific applications, as well as enhancing their stability [44]. Notwithstanding, the exorbitant costs associated with production and the utilization of intricate separation methodologies occasionally serve as deterrents to their widespread adoption. Numerous methodologies for enzyme immobilization have been extensively documented in the existing body of scholarly literature. Predominantly, these techniques encompass covalent bonding, adsorption, cross-linking, entrapment, and self-assembled monolayers. The phenomenon of covalent bonding is characterized by the mutual sharing of an electron pair between an enzyme and the platform upon which the molecule is immobilized. This intricate interaction results in a robust and versatile attachment, exhibiting both high strength and multipotent capabilities. The covalent interaction is typically executed through a biphasic process. Initially, the substrate, consisting of Au nanoparticles, undergoes activation via the introduction of linker molecules. These linker molecules, such as carbodiimide or thiol, serve to facilitate the subsequent covalent bonding event. Furthermore, it is important to consider the complementary end of the linker molecule in relation to the enzymes employed for the purpose of detection.

11.3 Immunosensors

Immunological assays, commonly referred to as immunoassays, represent a subset of analytical techniques that have been meticulously developed and have emerged as pivotal tools for quantifying various analytes. The aforementioned methodologies have garnered significant utilization within the realm of biomedical diagnostics, identification of foodborne impurities, and analysis of the environment. In recent times, there has been a notable surge in interest surrounding the advancement of novel assays that possess commendable attributes such as heightened sensitivity, specificity, and cost-effectiveness. The utilization of biological recognition probes, such as antibodies, peptides, and aptamers, in conjunction with nanoparticles has facilitated the advancement of exceptionally sensitive methodologies. The biomolecules under consideration exhibit the remarkable ability to preserve their conformational orientation while also engaging in interactions with specific target antigens. The quantification of analyte concentrations can be achieved through the utilization of signal detection mechanisms facilitated by nanoparticles. Amongst a plethora of metallic nanoparticles, it is noteworthy to mention that gold nanoparticles (AuNPs) have emerged as the most frequently utilized labels in various applications, including immunosensors and DNA sensors [45, 36].

11.4 DNA Sensors

The utilization of DNA probes that exhibit complementarity to a distinct DNA sequence of a pathogen, coupled with the incorporation of gold nanoparticles (AuNPs) as labels, has garnered significant attention and adoption within the realm of biosensing for plant pathogens. The utilization of DNA sequences as detection probes in biosensors confers a notable advantage in terms of achieving high specificity. DNA biosensors, as utilized in the field of analytical science, serve as valuable tools for the purpose of detecting specific DNA sequences associated with diverse pathogens found in clinical, food, and environmental samples. Electrochemical DNA biosensors possess a multitude of advantageous attributes, primarily stemming from their cost-effectiveness, heightened sensitivity, and expeditious signal-generation capabilities. The utilization of gold nanoparticles (AuNPs) in the context of DNA biosensors assumes a crucial role in facilitating the immobilization of DNA molecules onto the electrode surface. This immobilization process is instrumental in enabling the detection of DNA hybridization events, wherein the AuNPs serve as a label [11]. The utilization of gold nanoparticles (AuNPs) offers a notable advantage in terms of providing an enhanced surface area for the conjugation of the mercaptohexyl group at the 5'-phosphate of the DNA molecule. This particular conjugation strategy serves to augment the detection of nucleic acids, thereby improving the overall sensitivity and accuracy of nucleic acid detection methodologies. Previous research has extensively investigated the surface density of the colloidal Au-modified electrode when combined with ssDNA, revealing a remarkable value of 1.0 × 1014 molecules/cm^2. This value was found to be tenfold greater than that observed on a bare gold electrode, as reported in reference [11].

12 Pathogen Biosensing

Biosensors play a crucial role in the expeditious and punctual identification of various plant quarantine pathogens, thereby mitigating the potential introduction of foreign pathogens into novel ecosystems. Gold nanoparticles (AuNPs) have gained significant attention in the scientific community due to their extensive utilization as a means to label antibodies that are specific to target pathogens. Moreover, these AuNPs have been extensively explored and developed as diagnostic devices, showcasing their potential in the field of medical diagnostics.

12.1 Bacterial Pathogen Detection

The identification and detection of bacterial pathogens can be achieved through two distinct approaches: capturing the entire bacterium or selectively targeting the DNA sequence unique to the specific bacterium in question. The process of capturing the DNA probe typically exhibits a high degree of sensitivity and specificity toward the targeted disease. The utilization of colloidal gold nanoparticles (AuNPs) was employed as a means to mark single-stranded DNA (ssDNA) probes that exhibit specificity toward *Acidovorax avenae* subspecies *citrulli,* the causative agent of bacterial fruit blotch. Consequently, a DNA sensor based on a strip format was fabricated, enabling swift and on-site detection of the aforementioned pathogen. In this study, the internal transcribed spacer (ITS) region was employed as the basis for the development of a pathogen-specific probe. The dipstick method employed in this study demonstrated a commendable level of sensitivity, enabling the qualitative detection of target DNA at a concentration as low as 4 nanomolars (nM). By examining the optical density and concentration of the target DNA, it was also possible to observe the potential for semi-quantitative detection. The limit of detection (LOD) was determined to be 0.48 nanomolars (nM) according to the reference [46]. In a further illustration of the identification of Pseudomonas syringe pathovars, which give rise to extensive bacterial afflictions in agricultural crops, DNA probes labelled with gold nanoparticles (AuNPs) were employed in a colorimetric assay for the detection of pathogenic DNA molecules. The design of specific primers was undertaken based on the conserved N-terminal region of the hrcV gene. Subsequently, probes were meticulously designed, incorporating thiol-capping at either the 50 or 30 ends of the probes. The utilization of colorimetric detection resulted in a discernible alteration in colour, transitioning from a red hue, indicative of the non-hybridized state of the DNA probe labelled with gold nanoparticles (AuNPs), to a purple shade, signifying the successful hybridization of the probe with the target DNA. This transformative colour change facilitated the identification of pathogenic DNA within the examined sample [47]. The detection of the soil bacterium *Ralstonia solanacearum*, known for its ability to induce wilt disease in potatoes, was carried out using a comparable approach. The present study describes a novel approach for the direct detection of unamplified DNA from *Ralstonia solanacearum (R. solanacearum)* using a colorimetric assay based on gold nanoparticles (AuNPs) conjugated with specific probes. The assay relies on the principle that the hybridization of a specific AuNP-bound probe with the target DNA hinders the aggregation of AuNPs under acidic conditions. Consequently, the solution retains its characteristic red colour, serving as an indicator of the presence of the specific R. solanacearum DNA. In the absence of positive samples, wherein hybridization does not occur, the aggregation of AuNPs takes place, leading to a discernible alteration in coloration, specifically transitioning to a shade of purple, thereby enabling detection. The study revealed that the nano-biosensor exhibited remarkable attributes in terms of speed, sensitivity, and specificity in detecting R. solanacearum in soil samples, obviating the need for DNA amplification procedures.

In the present study, a novel electrochemical biosensor was devised employing a sandwich immunoassay methodology. This approach entailed the utilization of a capture antibody, a detection antibody, gold nanoparticles (AuNPs), and the enzyme horseradish peroxidase (HRP). The primary objective of this biosensor was to facilitate the detection of the quarantine bacterial pathogen *Pantoea stewartii* subspecies *stewartii*-NCPPB 449 (PSS), which is notably accountable for the onset of Stewart's vascular wilt in maize. The utilization of gold nanoparticles (AuNPs) in this study capitalizes on their remarkable properties of high conductivity and surface area. These properties are leveraged to enhance the electrochemical signal, thereby amplifying the detection sensitivity. Specifically, the AuNPs are employed to conjugate the horseradish peroxidase (HRP)-labelled anti-poly (styrene sulfonate) (PSS) detection antibody. This conjugated complex is subsequently employed for the purpose of detecting the presence of bacterial cells that are bound to the capture antibody. In the current investigation, the response to the addition of HRP substrate and the evaluation of its subsequent catalytic activity were both assessed. Notably, a discernible correlation was observed between the concentration of bacteria and the observed linear relationship. The biosensing assay that was developed exhibited a limit of detection of 7.8 × 103 cfu/mL, surpassing the conventional ELISA method. Additionally, it demonstrated a lack of specificity toward other chosen pathogens [48]. In a separate investigation, the utilization of colloidal gold nanoparticles was employed as a means to mark monoclonal detection antibodies that were specifically developed to target *Pantoea* subspecies *stewartii* (NCPPB 449). This labelling technique was implemented within a strip-based immunoassay, which relied on the principles of a sandwich enzyme-linked immunosorbent assay (ELISA). The study involved the examination of various dilutions of pathogens in order to assess the test's sensitivity. The limit of detection was determined to be 1 × 105 colony-forming units per milliliter (cfu/mL) in both standard and spiked samples. As previous research [49] has shown, there were no instances of cross-reactivity with other pathogens.

Surface-enhanced Raman scattering (SERS) is an invaluable technique that affords the opportunity for highly sensitive and specific detection of molecules derived from biological or chemical sources. This method has gained a lot of popularity as a way to find and study pathogenic microorganisms that cause diseases in plants, animals, people, and even foodborne illnesses. By employing a synergistic approach that combines surface-enhanced Raman scattering (SERS) methodology with an isothermal DNA amplification technique, the present study aims to investigate and elucidate the underlying mechanisms and potential applications of this novel hybrid methodology. The utilization of SERS, a highly sensitive spectroscopic technique that enhances the Raman scattering signal, in conjunction with an isothermal DNA amplification technique that allows for rapid and efficient amplification of target DNA sequences at a constant temperature, holds great promise in various fields of research and technology. This integrated approach has the potential to revolutionize their study, Lau et al. successfully devised a multiplex point-of-care system for the simultaneous detection of *Botrytis cinerea, Pseudomonas syringae*, and the fungal pathogen *Fusarium oxysporum* in both tomatoes and *Arabidopsis thaliana* [50].

The global impact of these pathogens on various crop species necessitates the urgent and expeditious identification and monitoring of their presence, as this plays a crucial role in mitigating the detrimental effects on crop yield. The utilization of gold nanoparticles (AuNPs) in the context of pathogen detection has been explored, wherein these nanoparticles were effectively labelled with a Raman reporter that specifically targets the pathogen of interest. Additionally, a DNA capture probe was employed to further enhance the functionality of these AuNPs, allowing them to serve as surface-enhanced Raman scattering (SERS) nanotags. The utilization of recombinase polymerase amplification (RPA) technique was employed in order to facilitate the amplification of pathogen-specific amplicons. These amplicons were characterized by a 10 nucleotide 50 overhang and a biotin label positioned at the opposite end. The amplification process was conducted using pathogen genomic DNA that had been isolated from the afflicted plant. The subsequent step involved the hybridization process of RPA (recombinase polymerase amplification) products with Surface-Enhanced Raman Scattering (SERS) nanotags. This hybridization was achieved by utilizing the 50 overhang present in RPA amplicons and its corresponding complementary DNA probe bound to gold nanoparticles (AuNPs). By capturing the biotin-RPA-SERS products with streptavidin-tagged magnetic beads, it was then possible to detect these hybridized products. The captured products were then subjected to laser excitation within a portable Raman spectrometer, enabling the analysis of the resulting Raman scattering signals. The experimental results revealed the presence of discernible peaks corresponding to particular pathogens through the utilization of surface-enhanced Raman scattering (SERS) nanotags and the hybridization of recombinase polymerase amplification (RPA) amplicons. Specifically, the presence of *Botrytis cinerea, Pseudomonas syringae,* and *Fusarium oxysporum f.sp. conglutinans* was successfully detected. The system that was devised exhibited a commendable level of sensitivity, enabling the detection of B. cinerea DNA even at a minimal quantity of two copies. Moreover, its versatility was demonstrated through successful multiplexing in both the tomato and Arabidopsis thaliana systems. Notably, the system's efficacy was validated in external laboratory environments, further affirming its practicality and applicability.

In a subsequent investigation, an analogous approach involving target DNA amplification using RPA (recombinase polymerase amplification) was integrated with an electrochemical sensing platform based on AuNPs (gold nanoparticles). This amalgamation facilitated the accurate and selective identification of *Pseudomonas syringae*, a prevalent bacterium known for its pathogenic effects on plants. In a study conducted by Lau et al., the researchers discovered that after amplification of the target DNA region using RPA (recombinase polymerase amplification), it was subjected to hybridization with capture DNA probes that were labelled onto AuNPs (gold nanoparticles).

Subsequently, the resulting conjugate was separated by exploiting the binding affinity between biotin tags present in RPA amplicons and streptavidin-tagged magnetic beads. The entire complex underwent a heat treatment process in order to denature double-stranded DNA (dsDNA) and facilitate the release of gold nanoparticles (AuNPs). This release of AuNPs was crucial for inducing the electrochemical reduction of gold ions (Au^{+3}) to elemental gold (Au^0) on a screen-printed carbon electrode. The reduction process was quantitatively assessed using a technique called differential pulse voltammetry (DPV). As the quantity of AuNPs increases, a corresponding increase in the concentration of target DNA within the sample is observed, thereby facilitating the detection of pathogen presence. The assay was conducted over a period of 60 min. Notably, the amplification of target DNA via recombinase polymerase amplification (RPA) exhibited heightened sensitivity, surpassing that of end-point polymerase chain reaction (PCR) and gel electrophoresis by a factor of 100. Similarly, the electrochemical assay demonstrated enhanced sensitivity, surpassing end-point PCR and gel electrophoresis by a factor of 100. In the conducted study, a novel AuNP-electrochemical biosensor was successfully developed and evaluated. The biosensor exhibited remarkable sensitivity, surpassing existing methods by a factor of 10,000.

This heightened sensitivity enabled the biosensor to detect the presence of pathogens in the initial stages of infection, as demonstrated through experimentation on *P. syringae*-infected A. thaliana specimens [40]. In an endeavour to enhance the efficiency of detecting the potato wilt pathogen R. solanacearum using a biosensor-based technique called lateral flow immunoassay (LFIA), efforts were made to decrease the limit-of-detection value. The AuNPs (gold nanoparticles) underwent a process known as gold enhancement to achieve this [23]. Table 2 showcases the comprehensive analysis of plant pathogens and the diverse array of sensing techniques employed for their detection, along with their corresponding limits of detection (LOD).

Table 2 Presents an in-depth overview of the various sensing methodologies and nanomaterials used for detecting plant pathogens

S. no.	Species	Plant pathogens/Diseases	Method of sensing	Nanomaterial employed	LOD	References
1	Potato	Ralstonia solanacearum	Lateral flow immunoassay	Enlarging AuNPs	3×10^4 cells/mL	[51]
2	Banana plants	Pseudocerocospora fijiensis Black Sigatoka (Leave Streak Disease)	Surface plasmon resonance based immunosensor	Cell wall protein HF1 of P. Fijiensis immobilized onto gold chip	11.7 μg/mL	[45]
3	Wheat	Karnal Bunt Disease	Surface Plasmone Resonance (SPR)	AuNPs	NA	[33]
4	Solanaceae plant	Bacterial Spot Disease by Xanthomonas axonopodis	Fluorescence-linked immunosorbent assay	Fluorescence silica nanoparticles	NA	[52]
5	Tomato	Tomato Yellow Leaf Curl Virus (TYLCV)	Localized surface plasmon resonance	AuNps with colorimetric nano-biosensing	5 ng	[25]
6	Lime	Witches' Broom Disease (Candidatus Phytoplasma aurantifolia)	Fluorescence resonance energy transfer (FRET)	Quantum dot (QD)-based nano-biosensor	5 ca. P. aurantifolia/μL	[53]
7	Fruits	Acidovorax avenae subsp. citrulli	Dipstick method	Colloidal gold nanoparticles	0.48 nM of DNA	[46]
8	Potatoes and Tomatoes	Late Blight in Potatoes and Tomatoes (caused by Phytophthora Infestans)	PCR with AuNPs based lateral flow biosensor	AuNPs	0.1 pg/mL range	[54]
9	Rose	Powdery Mildew	Relative fluorescence units	Colloidal nanosilver (1.5 nm diameter)	4.2 μM Ag ions	[55]
10	Papaya	Cucumber Mosaic Virus (CMV) and Papaya Ring Spot Virus (PRSV)	Amperometry detection	Nanowire based biosensor	0.1 mA/mL	[56]

12.2 Detection of Fungal Pathogens

Pathogenic fungal species have been identified as significant contributors to the substantial decline in crop quality and yields, thereby posing a considerable threat to the economic stability of the global agricultural sector. According to available reports, it has been observed that a considerable number of approximately 8000 species belonging to the fungal kingdom and oomycetes group have been identified as causative agents of diseases in various plant species and agricultural crops [57]. Fungal pathogens have the ability to induce diseases in plants throughout various stages of their growth, either independently or in conjunction with other types of phytopathogens, within the context of natural environmental settings [58]. The diseases most frequently attributed to pathogenic fungi encompass a range of afflictions, including anthracnose, blight, canker, damping off, dieback, gall, leaf spot, powdery mildew, rust, root rot, scab, and wilt [59]. Various conventional methods have been devised for the identification and detection of fungal species. These include polymerase chain reaction (PCR) techniques such as nested PCR, multiplex PCR, quantitative PCR, and magnetic-capture hybridization PCR. Additionally, isothermal amplification methods such as loop-mediated isothermal amplification (LAMP) have also been developed for this purpose. These techniques offer valuable tools for accurately detecting and characterizing fungal species. Nevertheless, it is important to note that the implementation of these techniques necessitates the involvement of personnel possessing a high level of expertise. Additionally, the protocols associated with these methodologies can be quite laborious, requiring meticulous attention to detail. Consequently, the time required to obtain the desired results may be extended. In recent times, the emergence of nanotechnology-based biosensors has presented a promising avenue for the identification and monitoring of detrimental fungal pathogens. *Phytophthora infestans*, a highly destructive fungal pathogen, serves as a primary causative agent of late blight, a severe disease affecting both potatoes and tomatoes. This pathogen poses a significant threat to global agriculture, necessitating urgent attention and effective management strategies. Hence, the prompt and timely identification of *P. infestans* assumes paramount importance as a pivotal measure to curtail the dissemination of the disease.

The integration of universal primer-mediated asymmetric PCR with gold nanoparticles (AuNPs)-based lateral flow assay (LFA) was successfully achieved by Zhan et al. for the purpose of visually detecting *P. infestans* [54]. In this study, the researchers employed the technique of asymmetric polymerase chain reaction (PCR) to amplify a substantial quantity of single-stranded DNA (ssDNA). Subsequently, the researchers conducted sandwich hybridization in a lateral flow assay (LFA) setup. In the context of the experiment, it was observed that when target DNA was present, a sandwich-type hybridization reaction occurred between the AuNP probe, target DNA, and capture probe located on the test line of the lateral flow assay (LFA). This reaction resulted in the accumulation of AuNPs, leading to the formation of a distinct red visible line. Through evaluation of the red line's signal intensity, the lateral flow assay (LFA) quantification was possible. The limit of detection (LOD) was determined to be 0.1 picograms per microliter (pg/μL).

Pseudocercospora fijiensis, a hemi-biotrophic fungus, is primarily responsible for the harmful disease known as black sigatoka. This pathogen poses a significant threat to banana plantations on a global scale. The identification and monitoring of this fungal pathogen are of utmost significance in order to mitigate potential losses and effectively curtail its transmission to adjacent agricultural crops. In order to address the challenge posed by this particular pathogen, [45] devised a highly specialized surface plasmon resonance (SPR) immunosensor. The sensor was created through the process of covalent immobilization, where a polyclonal antibody (specifically anti-HF) was attached to an Au-coated chip. This antibody was produced against the HF1 cell wall protein of *P. fijiensis*. Utilizing a mixed self-assembled monolayer (SAM) of alkanethiols, as described in a prior study [45], the attachment was successful. The limit of detection (LOD) for the surface plasmon resonance (SPR) immunosensor was determined to be 11.7 μg/mL. The linear response range for the cell wall antigen was observed to be from 39.1 to 122 μg/mL. The findings of the study revealed the absence of any matrix effects during the analysis of real leaf banana extracts.

In-depth presentation of the array of sensing methodologies and nanomaterials currently in use for the detection of plant pathogens. This comprehensive table categorizes and describes various innovative techniques and the types of nanomaterials employed in each method. It outlines different sensing approaches like optical, electrochemical, and biosensing methods, detailing how they function and their suitability for specific types of pathogens, including fungi, bacteria, and viruses affecting plants. For each sensing methodology, the table lists nanomaterials such as gold nanoparticles, carbon nanotubes, quantum dots, and graphene, and elaborates on their role in enhancing the sensitivity and specificity of pathogen detection. The table highlights the advantages of each combination of sensing technique and nanomaterial, such as improved detection limits, rapid response times, and the ability to detect pathogens in various environmental conditions. Additionally, Table 2 provides case studies or examples of successful applications of these technologies in real-world scenarios, demonstrating their effectiveness in early detection and prevention of plant diseases. This information is crucial for researchers, agronomists, and agricultural professionals, as it offers insights into the latest advancements in plant pathology diagnostics. The table underscores the significance of integrating nanotechnology in agriculture for efficient and accurate disease management, contributing to the overall health and productivity of crops

In their recent study, [60] devised a novel approach for the detection of *Leptosphaeria maculans*, a highly destructive phytopathogen affecting oilseed rape. The researchers employed an innovative technique involving the utilization of AuNPs-enhanced dynamic microcantilever (MCL) in conjunction with isothermal recombinase polymerase amplification (RPA). This combination of technologies holds promise for the accurate and sensitive detection of the aforementioned pathogen, thereby aiding in the mitigation of its detrimental impact on oilseed rape crops [60]. The findings of this study were published in a scientific journal, providing valuable insights for researchers and practitioners in the field. The production of a phytotoxin, specifically sirodesmin PL, by *L. maculans* has been the subject of extensive investigation. Furthermore, it has been observed that this organism is capable of persisting for extended durations under normal atmospheric conditions. The obtained detection results indicate that the sensitivity of the RPA-MCL assay surpasses that of the previously investigated fluorescence RPA assay. The limit of detection (LOD) achieved by the RPA-MCL assay is one copy of *L. maculans* DNA. In addition, the expeditious and highly responsive identification of fungal spores has garnered significant attention owing to the possible detrimental effects on both agricultural practices and public well-being. In their study, [61] devised a novel methodology to identify *Aspergillus niger* spores. The approach involved utilizing specific peptides as a recognition probe and employing AuNPs as the detection label [61]. The peptides facilitate a prompt interaction with the *A. niger* spore, leading to a discernible alteration in the intensity of colour observed in the supernatant subsequent to the settling of the spores. The colorimetric assay demonstrated a remarkable level of sensitivity, capable of detecting as few as 50 spores in less than 10 min.

12.3 Detection of Viruses

In contrast to alternative nanomaterials, AuNPs exhibit distinct advantages as a tool for virus detection, as elucidated in the preceding sections. The ease of synthesis, surface modifications, stability, biocompatibility, and high absorption coefficient are key factors that contribute to the wide range of applications of these materials in detection platforms. The vivid crimson hue is readily observable and establishes enduring bio-conjugates with various biological entities, including DNA, antibodies, and proteins. This property facilitates the utilization of gold nanoparticles (AuNPs) for the detection of viruses with exceptional sensitivity and specificity. The employed methodologies are predicated upon the utilization of colorimetric, fluorescence, or electrochemical techniques, among others, to detect and quantify the target analytes. Presently, there exist multiple methodologies for the identification and detection of plant viruses utilizing gold nanoparticles (AuNPs). The employed methodologies are predicated upon colorimetric, fluorescence, or electrochemical principles, among others.

The Cucumber Green Mottle Mosaic Virus (CGMMV) is responsible for inducing pronounced mosaic symptomatology in watermelon and cucumber plants. This viral pathogen has the ability to spread through various means, including infected cucumber seeds, leaves, and soil. Given the notable transmissibility that viruses exhibit, it is essential to emphasize the significance of early detection in successfully managing and reducing their subsequent dissemination within crop fields. In their study, [62] devised a straightforward and highly responsive label-free colorimetric detection technique for cucumber green mottle mosaic virus (CGMMV). The method involved utilizing unmodified gold nanoparticles (AuNPs) as a colorimetric probe [62]. The fundamental principle underlying the assay is predicated upon the interaction between the target products of CGMMV in RT-PCR and the probes that are specific to the species. This interaction subsequently leads to a discernible alteration in coloration subsequent to the induction of salt (NaCl). In the typical experimental setup, the species-specific probes are observed to bind to the surface of gold nanoparticles (AuNPs), resulting in an enhanced ability of the nanoparticles to resist aggregation induced by sodium chloride (NaCl).

The method that was devised for this study did not require the use of costly instrumentation and demonstrated the ability to visually detect the presence of 30 picograms per microliter of Cucumber Green Mottle Mosaic Virus (CGMMV) RNA without the aid of any additional equipment. The assay demonstrated a remarkable specificity of 100%, indicating its ability to accurately identify the target analyte without any false positives. Moreover, the assay exhibited excellent reproducibility, suggesting consistent and reliable results across multiple experimental replicates. Localized surface plasmon resonance (LSPR) is a notable property exhibited by gold nanoparticles (AuNPs) and is currently being utilized in the advancement of colorimetric bio-sensing techniques. The geometric properties of the gold nanoparticles (AuNPs), such as shape and size, are what primarily influence the Localized Surface Plasmon Resonance (LSPR) phenomenon. Additionally, the properties of the surrounding medium also play a significant role in modulating the LSPR response of AuNPs. In their study, Razmi et al. (year) utilized the localized surface plasmon resonance (LSPR) phenomenon exhibited by unmodified gold nanoparticles (AuNPs) for the purpose of detecting the genome of the tomato yellow leaf curl virus (TYLCV) in plants that were infected with the virus [25]. A DNA probe that specifically targeted the coat protein of the tomato yellow leaf curl virus (TYLCV) was meticulously designed and subsequently employed for hybridization. This probe was carefully designed to possess complementary nucleotide sequences to the coat protein of TYLCV. The hybridization process involved the utilization of the designed DNA probe, which was then combined with the total DNA extracted from the infected sample. The hybridization process was executed, followed by cooling the sample to ambient temperature. The subsequent introduction of AuNPs resulted in a noticeable alteration in coloration. The observed alteration in color resulting from the presence of AuNPs indicated the potential occurrence of viral infection, as visually discerned. This finding was further corroborated through the utilization of UV–Vis spectroscopy. A comparable visual colorimetric technique has been devised for the detection of Begomovirus in chili and tomato plants. This method demonstrates a sensitivity of 500 ag/μL of begomo viral DNA [61]. The results of the comparative screening conducted on chili plants for begomoviral infection using both PCR and the AuNP assay revealed that the AuNP assay exhibited a higher efficacy (77.7%) compared to the more frequently employed PCR methods (49.4%). Advanced detection methods have been employed in scientific investigations through the utilization of attenuated total reflection (ATR)-based evanescent wave absorption monitoring of localized surface plasmon resonance (LSPR) of gold nanoparticles (AuNPs) [63]. The investigation of the binding kinetics of gold nanoparticles (AuNPs) has been conducted using an amine-functionalized surface refractive index sensor. This sensor was specifically designed to monitor the absorption peak of localized surface plasmon resonance (LSPR) during the binding process. The employed methodology was utilized in order to detect the presence of single-stranded DNA specific to the chili leaf curl virus. The limit of detection (LOD) achieved for the target viral DNA was determined to be 1.0 μg/mL.

The successful identification of the banana bunchy top virus (BBTV) was accomplished using an enhanced dot immunobinding assay (DIBA) based on gold nanoparticles (AuNPs). This novel approach offers a faster and more streamlined alternative to the traditional enzyme-linked immunosorbent assay (ELISA) method [64]. In this study, the gold nanoparticles (AuNPs) were successfully linked to the primary antibody. The limit of detection (LOD) of the detection indicator binding assay (DIBA) was determined to be at a dilution of 10^{-2}. Grapevine leafroll-associated virus 3 (GLRaV-3) is a highly impactful pathogen that has been identified as a major contributor to the substantial reduction in grape yield. In order to achieve comprehensive management of this viral infection, it is imperative to employ on-site analytical techniques that exhibit a heightened level of sensitivity. The utilization of immunochromatographic assays (ICA) or lateral flow assays (LFA) in the realm of pathogen and biomarker detection has garnered significant attention and widespread adoption. The ICA (Immunochromatographic Assay) method utilizes gold nanoparticles (AuNPs) and offers a straightforward means of interpreting results, distinguishing between the presence or absence of viral particles. The ICA exhibits several notable advantages, including a brief analysis duration ranging from 10 to 15 min, a simplified sample preparation process, and straightforward result interpretation. Byzova and colleagues [49] devised an independent component analysis (ICA) technique to expedite the identification of Grapevine Leafroll-associated Virus 3 (GLRaV-3). This approach was built upon the sandwich immunoassay format, as detailed in their study [49]. The researcher compared the three AuNP preparations, with average diameters of 51.0 7.9, 28.3 3.3, and 18.5 3.3 nm, respectively. The results demonstrated that the AuNPs with the largest average diameter of 51.0 ± 7.9 nm exhibited the highest sensitivity for detecting GLRaV-3, even at maximal dilutions. The assay demonstrated a sensitivity of 100% and a specificity of 92% when compared to the gold-standard ELISA method. Additionally, when compared to the PCR method, the assay showed a sensitivity of 93% and a specificity of 92% (Fig. 3B). The utilization of AuNPs in the LFA methodology occasionally presents challenges in terms of sensitivity and limit of detection. In the pursuit of heightened sensitivity, numerous signal enhancement techniques have been devised and implemented in the realm of plant virus detection. Panferov et al. developed the silver enhancement method for the detection of potato leafroll virus (PLRV). The silver enhancement process relies on the reduction of silver ions occurring on the surface of gold nanoparticles (AuNPs), resulting in an intensified coloration of the AuNPs. The successful accomplishment of this task involved the utilization of a composite comprising silver lactate and hydroquinone, followed by the introduction of a buffer solution containing chloride ions to halt the reduction of silver. The findings of the study indicate that the silver-enhanced lateral flow immunoassay (LFIA) exhibited a significantly higher sensitivity, approximately 15 times greater, with a limit of detection (LOD) of 0.2 ng/mL and a testing time of 15 min. In comparison, the conventional LFIA demonstrated a LOD of 3 ng/mL and a testing time of 10 min. Moreover, the improved lateral flow immunoassay (LFIA) exhibits the ability to detect the presence of potato leafroll virus (PLRV) in leaf extracts obtained from infected potato plants, even at higher dilutions compared to the enzyme-linked immunosorbent assay (ELISA). In a separate investigation, [65] conducted a study employing magnetic nanoparticles (MNPs) and AuNPs to achieve dual enhancement of the lateral flow assay (LFA) [65]. The enrichment of the target analyte was accomplished by utilizing specific antibodies that were conjugated to magnetic nanoparticles.

This conjugation allowed for the selective capture and concentration of the target analyte. To further enhance the signal, the magnetic nanoparticles were aggregated through the use of gold nanoparticles. This aggregation process facilitated an amplification of the signal, thereby improving the sensitivity and detection capabilities of the assay. The employed methodology aimed at detecting the presence of potato virus X (PVX) through a specific strategy. The achieved sensitivity of 0.25 ng/mL was notably higher compared to the non-enhanced lateral flow assay (LOD = 8 ng/mL), demonstrating a 32-fold increase in sensitivity. In the realm of immunoassays, there exists a diverse array of techniques. One such technique involves the utilization of electrochemical immunosensors that incorporate gold nanoparticles (AuNPs). These immunosensors have the distinct advantage of enabling the creation of label-free assays, which possess several desirable characteristics, including reduced analysis time and enhanced simplicity, when compared to assays that rely on labelling strategies [66, 67]. In their study, [68] created a label-free impedimetric biosensor that was successfully used to detect tristeza, a disease caused by the citrus tristeza virus (CTV). The sensing platform utilized in this study was constructed using a screen-printed carbon electrode (SPCE) that was modified with gold nanoparticles (AuNPs). In order to improve the conductivity of the electrode, thiolated ssDNA was attached to the surface of AuNPs. We looked into how well the probe DNA hybridized with the target DNA by using EIS measurements in the Fe (CN6)4 − /Fe (CN6)3 − redox system. The sensor exhibited the capability to identify the CTV nucleic acids within a linear range spanning from 0.1 to 10 μM, even when other non-specific DNAs were present [68]. Watermelon mosaic virus (WMV) is a significant pathogen that affects plants belonging to the Cucurbitaceae and Leguminosae families. This viral infection leads to the development of mosaic-like patterns on the leaves, accompanied by a decrease in leaf area and fruit production. Consequently, the overall yield of affected crops is severely compromised. In their study, [69] proposed a novel approach involving nicking and polymerization techniques to achieve ultrasensitive electrochemical detection of WMV (Watermelon mosaic virus) [69]. The detection platform utilized in this study is founded upon the exonuclease and polymerase activity of T4 DNA polymerase. Additionally, it employs the Mg^{2+}-dependent DNA zyme-assisted cascade amplification strategy as well as the hemin/quadruplex DNA zyme-assisted cascade amplification strategy. A nicking endonuclease identified and cleaved the hybridized DNA that contained the target WMV sequence, referred to as HP1, along with P1. The DNA segments were cut in half from 30 to 50 bases away, and the process stopped at the 30-terminal G locus by using the exonuclease activity of T4 DNA polymerase. The synthesis of the Mg^{2+}-dependent DNA zyme was carried out by employing T4 DNA polymerase subsequent to the introduction of dNTPs. These dNTPs were able to form a hybridization complex with the substrate sequence that had been immobilized on an Au electrode. This hybridization event then triggered the initiation of the cleavage round. Following the subsequent release of the caged G-quadruplex sequence, it underwent a conformational change and formed a hemin/G-quadruplex-based DNA zyme. This molecular transformation led to the production of electrochemical signals. The assay demonstrated a linear dynamic range of detection spanning from 50 fM to 1 nM, with a limit of detection (LOD) of 50 fM.

13 Prospects and Possibilities in Forestry

Forests are known to harbour complex ecological systems that encompass not only the diverse populations of trees, herbs, and wildlife but also provide sustenance for all forms of life, along with their essential support mechanisms. The productivity of forests is a critical factor in assessing their socio-economic significance, as they offer a wide range of products and services to humanity. Forest productivity refers to the dynamic process by which forests generate biomass, store carbon, and maintain overall environmental well-being. The occurrence of a disturbance within a forest ecosystem has direct implications for environmental oscillations, leading to consequential actions that disrupt the overall strength and composition of the forest [10]. The initial measure in addressing the matter of disease in natural forests, including trees outside forests (TOF), involves the accurate identification of the pathogens and their corresponding physiological characteristics. The utilization of conventional diagnostic methods in plant pathology poses several challenges, including time consumption, limited accuracy, and high costs when implemented on a larger scale. Consequently, there is a pressing need for the development and adoption of emerging low-cost methods that can enhance the precision and efficiency of plant pathogen diagnosis. Nanoparticle-based biosensors have emerged as a significant tool for enhancing the overall quality of life and its various components. These biosensors are being extensively utilized in a wide range of applications, including clinical, ecological, and quality-measured domains, on a global scale. The utilization of various nanomaterials, including metal-based materials, carbon allotropes, polymers, and composites, in the form of nanoparticles, nanotubes, nanorods, and nanowires, has significantly expanded the possibilities for improving the speed and reproducibility of plant disease detection. The utilization of nano-based biosensors has demonstrated its significance in the diagnosis of various clinical, conservational, and food quality attributes. Nevertheless, the existing literature pertaining to plant biosensors remains relatively limited in comparison. The utilization of this approach has been observed in agricultural crops and plantations across various regions globally. However, its application in forest areas is characterized by significant diversity in natural forests. To effectively implement this approach in forest areas, it is crucial to identify hotspots within protected areas. Furthermore, an integrated approach, combined with institutional management, should be adopted to ensure its efficient utilization. Commercially available nanoparticle-based diagnostic sensors have been developed by companies such as pocket-diagnostic and Loewe. The utilization of nanoparticles in this compact diagnostic kit involves their integration with the extracted genetic material from plant leaves. This amalgamation facilitates the generation of prompt results, indicating the presence or absence of specific diseases within a time frame of 3–10 min. Moreover, significant advancements have been made in the field of smart sensor technology, particularly in the development of wearable sensors. These sensors have the capability to monitor and analyse both the environmental conditions and the physiological status of plants. The utilization of these technologies offers numerous advantages, including the ability to obtain real-time updates on the health conditions of plants within both indoor agricultural facilities and natural forest environments. The wearable sensors possess a compact and lightweight design, exhibiting both stretchability and biocompatibility. In their study, Lee et al. (year) conducted research on the development and integration of a nanomaterial-based (specifically, single-walled carbon nanotubes, or SWCNTs) graphitic electrode into plant cells. The objective of their investigation was to explore the potential of this electrode for real-time monitoring of toxic gases. Skin-like, flexible, and wearable sensors have the potential to be utilized for disease detection or the early identification of pathogen attacks in various environments, including agricultural fields and forested areas. In addition to the aforementioned progress, it is worth noting that further advancements have been made in the field of lab-on-drone technology. This innovative approach integrates both sensing and robotic technologies, resulting in a synergistic combination that holds great potential for various applications. The proposed methodology facilitates the process of collecting samples and transferring them to the lab-on-a-chip (LOC) microfluidic devices. Subsequently, the obtained results can be effectively interpreted by leveraging the capabilities of a smartphone. The utilization of sensor signals and the advanced image processing capabilities of mobile phones hold significant potential for assisting farmers and forest researchers in combating plant diseases. The necessity of conducting research lies in the quantification of the influence that changes have on forests, as well as the insects and pathogens associated with them. This research is crucial in order to accurately predict potential disturbance events and the associated risks that forests may face. The implementation of this approach is expected to contribute significantly to the promotion of forest sustainability and the enhancement of long-term environmental resilience within forest ecosystems.

14 Applications of Nanoparticles into Plants

The utilization of AuNPs has been suggested as a viable option in the context of agricultural crops and the germination of seeds from endangered plant species [70, 71]. The investigation focused on examining the impact of gold nanoparticles (AuNPs) that were synthesized using an extract derived from Terminalia arjuna fruit. Specifically, the study aimed to understand the influence of these AuNPs on the germination process of Gloriosa superba seeds as well as the subsequent growth of the plant's leaves. The application of gold nanoparticles (AuNPs) has been observed to yield positive outcomes in various aspects of plant development, including seed germination, node elongation, and vegetative growth. The results of the study indicate a significant correlation between the concentration of gold nanoparticles (AuNPs) in the environment and both the germination of Gloriosa superba seeds and their subsequent vegetative growth. Furthermore, [72] made a noteworthy observation regarding the application of AuNPs at concentrations of 10 and 25 mg/L on *Brassica juncea* plants. This application has demonstrated an augmentation in the chlorophyll content of the plants. In light of the evidence presented, it can be inferred that the utilization of AuNPs yields beneficial effects on the growth and development of plants.

15 Conclusion and Future Perspectives

In a nutshell, various researchers have effectively accomplished the environmentally friendly synthesis of gold nanoparticles (ranging from 3 to 100 nm) through the utilization of a wide range of plant extracts. The utilization of various plant parts has been observed, with leaves being identified as the most frequently employed component. The majority of the studies conducted primarily reported the presence of a zone of inhibition, indicating a gap in the comprehensive assessment of antimicrobial activity. However, a limited number of studies did include the determination of minimum inhibitory concentration (MIC) and minimum bactericidal concentration (MBC) values. In order to fully optimize the potential benefits of antimicrobial studies, it is imperative to complement them with comprehensive toxicity studies. At this stage, it is premature to make definitive statements regarding the toxicity of nanoparticles (NPs), as the majority of existing studies are still in their preliminary phases. The stability of nanoparticles (NPs) is a crucial aspect that demands significant attention and consideration in various research studies. The optimization of the synthesis process for nanoparticles (NPs) is imperative due to the significant impact that various factors, such as size, shape, and reaction variables, have on the biological activity of NPs. Nanotechnology-based formulations have emerged as a potential strategy for combating drug-resistant pathogens and mitigating the persistent challenge of resistance proliferation. The potential to manipulate the characteristics of gold nanoparticles (Au NPs) through size and shape modifications presents a captivating avenue for exploring novel applications in the foreseeable future. The utilization of docking programs in the investigation of the mechanisms of natural products (NPs) and Phyto active components holds great potential for expanding the realm of intelligent antimicrobials, which can be effectively employed to counteract drug resistance. Moreover, it is worth noting that theoretical frameworks in drug design can be effectively employed to propose the potential utilization of green synthesized nanoparticles (NPs) as innovative pharmaceutical agents. The field of nanobiology has witnessed significant advancements in the biosynthesis of nanoparticles (NPs). This has opened up numerous possibilities for exploring novel and environmentally friendly preparatory methods.

References

1. Bachheti RK, Bachheti A, Husen A (2024) Metal and metal-oxide based nanomaterials (Synthesis, Agricultural, Biomedical and Environmental Interventions). Springer Nature Singapore Pte Ltd., Singapore
2. Chawla S, Singh S, Husen A (2023) Smart nanomaterials targeting pathological hypoxia. Springer Nature Singapore Pte Ltd., Singapore. ISBN: 978-981-99-1717-4. https://doi.org/10.1007/978-981-99-1718-1
3. Husen A (2019) Natural product-based fabrication of zinc oxide nanoparticles and their application. In: Husen A, Iqbal M (eds) N*anomaterials* and plant potential. Springer International Publishing AG, Cham, pp 193–291. https://doi.org/10.1007/978-3-030-05569-1_7
4. Husen A (2020) Introduction and techniques in nanomaterials formulation, In: Husen A, Jawaid M (eds) Nanomaterials for agriculture and forestry applications. Elsevier Inc., Cambridge, MA, USA, pp 1–14. https://doi.org/10.1016/B978-0-12-817852-2.00001-9
5. Husen A (2022) Engineered nanomaterials for sustainable agricultural production, soil improvement and stress management. Elsevier Inc., Cambridge, MA 02139, USA. https://doi.org/10.1016/C2021-0-00054-7
6. Husen A (2023a) Secondary metabolites based green synthesis of nanomaterials and their applications. Springer Nature Singapore Pte Ltd., Singapore. https://doi.org/10.1007/978-981-99-0927-8
7. Husen A (2023b) Nanomaterials from agricultural and horticultural products. Springer Nature Singapore Pte Ltd., Singapore. https://link.springer.com/book/9789819934348
8. Husen A (2023c) Nanomaterials and Nanocomposites exposures to plants (Response, Interaction, Phytotoxicity and Defense Mechanisms). Springer Nature Singapore Pte Ltd., Singapore. https://doi.org/10.1007/978-981-99-2419-6
9. Husen A, Bachheti RK, Bachheti A (2023) Current trends in green nano-emulsions (Food, Agriculture and Biomedical Sectors). Springer Nature Singapore Pte Ltd., Singapore
10. Taghiyari HR, Morrell JF, Husen A (2023) Emerging nanomaterials (Opportunities and Challenges in Forestry Sectors). Springer Nature Switzerland AG, Gewerbestrasse, Cham, Switzerland. https://doi.org/10.1007/978-3-031-17378-3
11. Pingarrón JM, Yáñez-Sedeño P, González-Cortés A (2008) Gold nanoparticle-based electrochemical biosensors. Electrochim Acta 53:5848–5866
12. Aldewachi H, Chalati T, Woodroofe MN, Bricklebank N, Sharrack B, Gardiner P (2017) Gold nanoparticle-based colorimetric biosensors. Nanoscale 10:18–33
13. Guo X (2012) Surface plasmon resonance based biosensor technique: a review. J Biophotonics 5:483–501
14. Holzinger M, Le Goff A, Cosnier S (2014) Nanomaterials for biosensing applications: a review. Front Chem 2:63
15. Elmer W, White JC (2018) The future of nanotechnology in plant pathology. Annu Rev Phytopathol 56:111–133
16. Fang Y, Ramasamy RP (2015) Current and prospective methods for plant disease detection. Biosensors 5:537–561
17. Albrecht A, Kandji ST (2003) Carbon sequestration in tropical agroforestry systems. Agric Ecosyst Environ 99:15–27
18. Bonan GB (2008) Forests and climate change: forcings, feedbacks, and the climate benefits of forests. Science 320:1444–1449
19. Zomer RJ, Neufeldt H, Xu J, Ahrends A, Bossio D, Trabucco A, Van Noordwijk M, Wang M (2016) Global tree cover and biomass carbon on agricultural land: the contribution of agroforestry to global and national carbon budgets. Sci Rep 6:29987
20. Boyd IL, Freer-Smith PH, Gilligan CA, Godfray HCJ (2013) The consequence of tree pests and diseases for ecosystem services. Science 342:1235773
21. Berger S, Sinha AK, Roitsch T (2007) Plant physiology meets phytopathology: plant primary metabolism and plant–pathogen interactions. J Exp Bot 58:4019–4026

22. Boa E (2003) An illustrated guide to the state of health of trees: recognition and interpretation of symptoms and damage. FAO, Rome, Italy
23. Lau HY, Wu H, Wee EJH, Trau M, Wang Y, Botella JR (2017) Specific and sensitive isothermal electrochemical biosensor for plant pathogen DNA detection with colloidal gold nanoparticles as probes. Sci Rep 7:38896
24. Mandal AK, Ghorai S, Husen A (2023) Functionalized smart nanomaterials for point-of-care testing. Springer Nature Singapore Pte Ltd., Singapore
25. Razmi A, Golestanipour A, Nikkhah M, Bagheri A, Shamsbakhsh M, Malekzadeh-Shafaroudi S (2019) Localized surface plasmon resonance biosensing of tomato yellow leaf curl virus. J Virol Methods 267:1–7
26. De Souza CD, Nogueira BR, Rostelato MEC (2019) Compounds: review of the methodologies used in the synthesis gold nanoparticles by chemical reduction. 798:714–740
27. Madhusudhan A, Purohit SD, Prasad R, Husen A (2024) Functional smart nanomaterials and their theranostics approaches. Springer Nature Singapore Pte Ltd., Singapore
28. Fu L-H, Yang J, Zhu J-F, Ma M-G (2017) Synthesis of gold nanoparticles and their applications in drug delivery. In: Metal nanoparticles in pharma. Springer, Berlin, p 155–191
29. Khatua EP, Rajamani P, Patel A, Kumar J, Naik A et al (2020) Phytosynthesis, characterization and fungicidal potential of emerging gold nanoparticles using Pongamia pinnata leave extract: a novel approach in nanoparticle synthesis. 31(1):125–131
30. Siddiqi KS Husen A (2017) Recent advances in plant-mediated engineered gold nanoparticles and their application in biological system. AJJoTEiM Biol 40:10–23
31. Yuan Y, Wang Y, Wang H, Hou S (2019) Gold nanoparticles decorated on single layer graphene applied for electrochemical ultrasensitive glucose biosensor. J Electroanal Chem 855:113495
32. Beik J, Khateri M, Khosravi Z, Kamrava SK, Kooranifar S, Ghaznavi H et al (2019) Gold nanoparticles in combinatorial cancer therapy strategies. 387:299–324
33. Singh S, Singh M, Agrawal VV, Kumar A (2010) An attempt to develop surface plasmon resonance based immunosensor for Karnal bunt (Tilletia indica) diagnosis based on the experience of nano-gold based lateral flow immuno-dipstick test. Thin Solid Films 519:1156–1159
34. Bhaskaran S, Sharma N, Tiwari P, Singh SR, Sahi SVJ (2019) Fabrication of innocuous gold nanoparticles using plant cells in culture. 9(1):1–9
35. Kumar A, Omprakash MK, Kumar M, Rai CP (2022) Future of nanotechnology in plant disease management. Vigyan Varta 3(3):110–113
36. Omidfar K, Zarei H, Gholizadeh F, Larijani B (2012) A high-sensitivity electrochemical immunosensor based on mobile crystalline material-41-polyvinyl alcohol nanocomposite and colloidal gold nanoparticles. Anal Biochem 421:649–656
37. Kainthola C, Harsh NSK, Ginwal HS, Tripathi A, Saha R, Pandey A (2022) New report of fusarium flocciferum causing rot of emerging culm in dendrocalamus strictus. Indian For 147:1220–1225
38. Biju V (2014) Chemical modifications and bioconjugate reactions of nanomaterials for sensing, imaging, drug delivery and therapy. Chem Soc Rev 43:744–764
39. Turkevich BJ, Stevenson PC, Hillier J (1951) A Study of the Nucleation and Growth Processes in the Synthesis of Colloidal Gold. Anal Chem 47:475
40. Ebrahimi M, Norouzi P, Safarnejad MR, Tabaei O, Haji-Hashemi H (2019) Fabrication of a label-free electrochemical immunosensor for direct detection of Candidatus Phytoplasma Aurantifolia. J Electroanal Chem 851:113451
41. Ozsoz M, Erdem A, Kerman K, Ozkan D, Tugrul B, Topcuoglu N, Ekren H, Taylan M (2003) Electrochemical genosensor based on colloidal gold nanoparticles for the detection of Factor V Leiden mutation using disposable pencil graphite electrodes. Anal Chem 75:2181–2187
42. Rochelet-Dequaire M, Limoges B, Brossier P (2006) Subfemtomolar electrochemical detection of target DNA by catalytic enlargement of the hybridized gold nanoparticle labels. Analyst 131:923–929
43. Kerman K, Morita Y, Takamura Y, Ozsoz M, Tamiya E (2004) Modification of Escherichia coli single-stranded DNA binding protein with gold nanoparticles for electrochemical detection of DNA hybridization. Anal Chim Acta 510:169–174

44. Liu G, Lin Y (2007) Nanomaterial labels in electrochemical immunosensors and immunoassays. Talanta 74:308–317
45. Luna-Moreno D, Sánchez-Álvarez A, Islas-Flores I, Canto-Canche B, Carrillo-Pech M, Villarreal-Chiu JF, RodríguezDelgado M (2019) Early detection of the fungal banana black sigatoka pathogen pseudocercospora fijiensis by an SPR immunosensor method. Sensors 19:465
46. Zhao W, Lu J, Ma W, Xu C, Kuang H, Zhu S (2011) Rapid on-site detection of Acidovorax avenae subsp. citrulli by gold-labeled DNA strip sensor. Biosens Bioelectron 26:4241–4244
47. Vaseghi A, Safaie N, Bakhshinejad B, Mohsenifar A, Sadeghizadeh M (2013) Detection of Pseudomonas syringae pathovars by thiol-linked DNA–Gold nanoparticle probes. Sens Actuators B Chem 181:644–651
48. Zhao Y, Liu L, Kong D, Kuang H, Wang L, Xu C (2014) Dual amplified electrochemical immunosensor for highly sensitive detection of pantoea stewartii sbusp. stewartii. ACS Appl Mater Interfaces 6:21178–21183
49. Byzova N, Vinogradova S, Porotikova E, Terekhova U, Zherdev A, Dzantiev B (2018) Lateral flow immunoassay for rapid detection of grapevine leafroll-associated virus. Biosensors 8:111
50. Lau HY, Wang Y, Wee EJH, Botella JR, Trau M (2016) Field demonstration of a multiplexed point-of-care diagnostic platform for plant pathogens. Anal Chem 88:8074–8081
51. Razo SC, Panferova NA, Panferov VG, Safenkova IV, Drenova NV, Varitsev YA, Zherdev AV, Pakina EN, Dzantiev BB (2019) Enlargement of gold nanoparticles for sensitive immunochromatographic diagnostics of potato brown rot. Sensors 19:153 [CrossRef]
52. Yao KS, Li SJ, Tzeng KC, Cheng TC, Chang CY, Chiu CY, Liao CY, Hsu JJ, Lin ZP (2009) Fluorescence silica nanoprobe as a biomarker for rapid detection of plant pathogens. Adv Mater Res 79–82:513–516
53. Rad F, Mohsenifar A, Tabatabaei M, Safarnejad MR, Shahryari F, Safarpour H, Foroutan A, Mardi M, Davoudi D, Fotokian M (2012) Detection of candidatus phytoplasma aurantifolia with a quantum dots fret-based biosensor. J Plant Pathol 94(3):525–534. http://www.jstor.org/stable/45156279
54. Zhan F, Wang T, Iradukunda L, Zhan J (2018) A gold nanoparticle-based lateral flow biosensor for sensitive visual detection of the potato late blight pathogen, Phytophthora infestans. Anal Chim Acta 1036:153–161
55. Choi O, Deng KK, Kim NJ, Ross L, Surampalli RY, Hu Z (2008) The inhibitory effects of silver nanoparticles, silver ions, and silver chloride colloids on microbial growth. Water Res 42:3066–3074
56. Ariffin SAB, Adam T, Hashim U, Sfaridah SF, Zamri I, Uda MNA (2014) Plant diseases detection using nanowire as biosensor transducer. Adv Mater Res 832:113–117
57. Fisher MC, Gurr SJ, Cuomo CA, Blehert DS, Jin H, Stukenbrock EH, Stajich JE, Kahmann R, Boone C, Denning DW et al (2020) Threats posed by the fungal kingdom to humans, wildlife, and agriculture. MBio 11:e00449-e520
58. Hariharan G, Prasannath K (2021) Recent advances in molecular diagnostics of fungal plant pathogens: a mini review. Front Cell Infect Microbiol 10:829
59. Iqbal Z, Khan MA, Sharif M, Shah JH, ur Rehman MH, Javed K (2018) An automated detection and classification of citrus plant diseases using image processing techniques: A review. Comput Electron Agric 153:12–32
60. Lei R, Wu P, Li L, Huang Q, Wang J, Zhang D, Li M, Chen N, Wang X (2021) Ultrasensitive isothermal detection of a plant pathogen by using a gold nanoparticle-enhanced microcantilever sensor. Sens Actuators B Chem 338:129874
61. Lee JI, Jang SC, Chung J, Choi WK, Hong C, Ahn GR, Kim SH, Lee BY, Chung WJ (2021) Colorimetric allergenic fungal spore detection using peptide-modified gold nanoparticles. Sens Actuators B Chem 327:128894
62. Wang L, Liu Z, Xia X, Yang C, Huang J, Wan S (2017) Colorimetric detection of Cucumber green mottle mosaic virus using unmodified gold nanoparticles as colorimetric probes. J Virol Methods 243:113–119

63. Das S, Agarwal DK, Mandal B, Rao VR, Kundu T (2021) Detection of the chilli leaf curl virus using an attenuated total reflection-mediated localized surface-plasmon-resonance-based optical platform. ACS Omega 6:17413–17423
64. Malhotra BD, Ali MA (2018) Nanomaterials in biosensors: fundamentals and applications. Nanomater Biosens 37:1–74
65. Razo SC, Panferov VG, Safenkova IV, Varitsev YA, Zherdev AV, Dzantiev BB (2018) Double-enhanced lateral flow immunoassay for potato virus X based on a combination of magnetic and gold nanoparticles. Anal Chim Acta 1007:50–60
66. Bhardwaj J, Devarakonda S, Kumar S, Jang J (2017) Development of a paper-based electrochemical immunosensor using an antibody-single walled carbon nanotubes bio-conjugate modified electrode for label-free detection of foodborne pathogens. Sens Actuators B Chem 253:115–123
67. Wang L, Shan J, Feng F, Ma Z (2016) Novel redox species polyaniline derivative-Au/Pt as sensing platform for label-free electrochemical immunoassay of carbohydrate antigen 199. Anal Chim Acta 911:108–113
68. Khater M, de la Escosura-Muñiz A, Quesada-González D, Merkoçi A (2019) Electrochemical detection of plant virus using gold nanoparticle-modified electrodes. Anal Chim Acta 1046:123–131
69. Wang Y, Li B, Liu J, Zhou H (2019) T4 DNA polymerase-assisted upgrade of a nicking/polymerization amplification strategy for ultrasensitive electrochemical detection of Watermelon mosaic virus. Anal Bioanal Chem 411:2915–2924
70. Alabdallah NMA, Hasan MM (2021) Plant-based green synthesis of silver nanoparticles and its effective role in abiotic stress tolerance in crop plants. Saudi J Biol Sci. https://doi.org/10.1016/j.sjbs.2021.05.081
71. Gopinath K, Gowri S, Karthika V, Arumugam A (2014) J Nanostruct Chem 4:115
72. Arora S, Sharma P, Kumar S, Nayan R, Khanna PK, Zaidi MGH (2012) Gold nanoparticle induced enhancement in growth and seed yield of Brassica juncea. Plant Growth Regul 66(3):303–310
73. Husen A, Iqbal M (2019) Nanomaterials and plant potential. Springer International Publishing AG, Cham, Switzerland. https://doi.org/10.1007/978-3-030-05569-1
74. Husen A, Siddiqi KS (2023) Advances in smart nanomaterials and their applications. Elsevier Inc., Cambridge, MA, USA. https://doi.org/10.1016/C2021-0-02202-1
75. Husen A, Jawaid M (2020) Nanomaterials for agriculture and forestry applications. Elsevier Inc., Cambridge, MA, USA. https://doi.org/10.1016/C2018-0-02349-X
76. Husen A, Siddiqi KS (2014) Phytosynthesis of nanoparticles: concept, controversy and application. Nanoscale Res Lett 9(229):1–24. https://doi.org/10.1186/1556-276X-9-229
77. Husen A, Jawaid M (eds) Nanomaterials for agriculture and forestry applications. Elsevier, Amsterdam, The Netherlands. ISBN 9780128178522
78. Jin-Chul K, Madhusudhan A, Husen A (2021) Smart nanomaterials in biomedical applications. Springer Nature Switzerland AG, Cham, Switzerland https://doi.org/10.1007/978-3-030-84262-8
79. Khaledian S, Nikkhah M, Shams-bakhsh M, Hoseinzadeh S (2017) A sensitive biosensor based on gold nanoparticles to detect Ralstonia solanacearum in soil. J Gen Plant Pathol 83:231–239
80. Lavanya R, Arun V (2021) Detection of Begomovirus in chilli and tomato plants using functionalized gold nanoparticles. Sci Rep 11:14203
81. Lipińska W, Grochowska K, Siuzdak K (2021) Enzyme immobilization on gold nanoparticles for electrochemical glucose biosensors. Nanomaterials 11:1156
82. Teimouri M, Khosravi-Nejad F, Attar F, Saboury AA, Kostova I, Benelli G, Falahati M et al (2018) Gold nanoparticles fabrication by plant extracts: synthesis, characterization, degradation of 4-nitrophenol from industrial wastewater, and insecticidal activity–a review. 184:740–753
83. Majumder S, Johari S (2018) Development of a gold-nano particle based novel dot immunobinding assay for rapid and sensitive detection of Banana bunchy top virus. J Virol Methods 255:23–28

84. Manion PD (1991) Tree disease concepts. Prentice Hall, Englewood Cliffs, NJ, USA, ISBN 9780139294235
85. Parthasarathy S, Thiribhuvanamala G, Muthulakshmi P, Angappan K (2021) Diseases of forest trees and their management. CRC Press, Boca Raton, FL, USA. ISBN 9781003173861
86. Qiao J, Qi LJT (2020) Recent progress in plant-gold nanoparticles fabrication methods and bio-applications, p 121396
87. Shahriari M, Hemmati S, Zangeneh A, Zangeneh MM (2019) Biosynthesis of gold nanoparticles using Allium noeanum Reut. ex Regel leaves aqueous extract; characterization and analysis of their cytotoxicity, antioxidant, and antibacterial properties. 33(11):e5189
88. Sloan S, Sayer JA (2015) Forest resources assessment of 2015 shows positive global trends but forest loss and degradation persist in poor tropical countries. For Ecol Manag 352:134–145
89. Wang Y, Alocilja EC (2015) Gold nanoparticle-labeled biosensor for rapid and sensitive detection of bacterial pathogens. J Biol Eng 9:16
90. Yu T, Xu C, Qiao J, Zhang R, Qi LJCCL (2019) Green synthesis of gold nanoclusters using papaya juice for detection of L-lysine. 30(3):660–663

Comparative Evaluation of the Phytotoxicity Between Chemically and Green Synthesized Au–NP's

Murugan Karuvelan, Salai S. Sumukhi, N. Selvakumar, Ramachandran Chelliah, Ghazala Sultan, Momna Rubab, and Deog Hwan Oh

Abstract The major growing demand and promising adaptability in a wide range of applications, there has been an upsurge in extensive research on metal nanoparticles. Among these because of its versatility and unique physicochemical properties, gold nanoparticles have garnered significant heeds in all its applications. Due to the increasing usage of Gold nanoparticles in agriculture, there has been an increase in concerns of phytotoxicity. The two major synthesis methods of nanoparticles include the chemical synthesis and green synthesis method. The former is a conventional method that has been applied in engineering gold nanoparticles and the products thus synthesized have precise control over its shape and size but its potential environmental repercussions are a matter of concern. The latter on the other hand offers a more sustainable and ecofriendly approach. In order to comparatively evaluate the phytotoxic effects of AuNP's synthesized by chemical and green methods, certain plant physiological parameters such as growth potential, seed germination, biomass, etc., are to be taken into account. The information provided with the above-mentioned

M. Karuvelan · S. S. Sumukhi
Department of Microbiology, School of Life Sciences, Vels Institute of Science and Technology and Advanced Studies, Chennai, Tamil Nadu, India

N. Selvakumar
Centre for Ocean Research, Sathyabama Research Park, Sathyabama Institute of Science and Technology, Rajiv Gandhi Salai, Jeppiar Nagar, Chennai 119, India

R. Chelliah (✉) · D. H. Oh
Department of Food Science and Biotechnology, College of Agriculture and Life Sciences, Kangwon National University, Chuncheon 24341, South Korea
e-mail: ramachandran865@gmail.com

R. Chelliah
Saveetha School of Engineering, (SIMATS) University, Sriperumbudur 600124, India

G. Sultan
Department of Computer Science, Faculty of Science, Aligarh Muslim University, Aligarh 202002, Uttar Pradesh, India

M. Rubab
Department of Food Engineering and Biotechnology, University of Engineering and Technology, New Campus, Lahore 54000, India

A. Husen (ed.), *Plant Response to Gold Nanoparticles*, Smart Nanomaterials Technology, https://doi.org/10.1007/978-981-97-7174-5_16

assessments along with certain characterization techniques like TEM, SEM, UV–Vis spectrophotometry, etc., provides a much clearer idea about the mechanism of interaction between nanoparticle and plant cell. The findings reveal significant differences in the phytotoxic effects of chemical and green synthesized nanoparticles. While there is an increased phytotoxic effect exhibited by chemically synthesized AuNPs, the green synthesized particles showed milder consequences. This chapter focuses on the comparison of phytotoxic effects of chemical and green synthesis methods and accentuates the need to consider the methods of synthesis and evaluating potential environmental impacts along with the other important aspects of engineering nanoparticles.

Keywords Nanoparticles · Phytotoxicity · Gold · Chemical synthesis · Green synthesis · Environment

1 Introduction

Nanoparticles are a wide class of minute particles that have at least one dimension less than 100 nm. These materials gained importance when researchers found that the physicochemical properties of a substance can be affected by its size. Due to their size and high surface area, they possess unique physical and chemical properties and because of this reason, they possess a wide range of applications in various areas like energy-based research, medicine, catalysis, agriculture, imaging, and environmental applicationsThe rapid advancement of technology and growing interest in nanomaterials have led to a paradigm shift in various scientific fields. Gold nanoparticles (Au–NPs) have emerged as exceptional candidates due to their unique properties and versatility in applications such as electronics, catalysis, medicine, and environmental remediation. In recent years, there has been a growing interest in gold nanoparticles due to their unique properties and potential applications. Researchers have been focusing on the synthesis, characterization, and applications of these nanoparticles using natural ingredients such as plant extracts and marine organisms. Plant extracts containing phytoconstituents have been found to be effective in capping and reducing agents, making them ideal for the synthesis of gold nanoparticles [1]. These nanoparticles have shown promising results in various fields, particularly in anti-aging products, where they have demonstrated better efficacy compared to purely natural ingredients. GNPs possess excellent biocompatibility and resistance to oxidation in the body, making them well-suited for use in biomedical applications. They can be easily modified with chemotherapeutic agents and have the ability to improve the delivery and targeting of drugs. Additionally, GNPs have the ability to absorb infrared radiation and exhibit optical properties, making them valuable in various biomedical applications, particularly in the treatment of cancer. In the case of lung cancer, GNPs play a theranostic role by providing nanoscale formulations that can penetrate tumor cells and release drugs specifically at the site of the tumor. They can be utilized as carriers for chemotherapeutic agents, enhancing the effectiveness of lung cancer treatment. The unique characteristics of GNPs, such as their adjustable nano size and interaction with visible light, make them a promising tool for biomedical applications in lung cancer. Gold nanoparticles have a wide range of uses in cosmetics. For example, one study discovered that gold nanoparticles modified with sulfadiazine could effectively measure mercury ions in cosmetics. Another study examined the effects of gold nanoparticles as additives in creams and found that they improved collagen quality, skin moisturization, and the function of the epidermis and dermis. Researchers also explored the potential of gold nanoparticles derived from snail slime as multifunctional ingredients in cosmetics, with antioxidant and photostability properties suitable for sunscreen products. Additionally, gold nanoparticles obtained from a mixture of gold and plant extract rich in flavonoids were suggested for use in cosmetics [2].

Gold nanoparticles (Au–NPs) have gained considerable attention in the field of photovoltaics, particularly in the advancement of next-generation solar cells. Their unique characteristics make them valuable components for improving the efficiency and performance of solar energy conversion devices. Au–NPs can greatly enhance the light absorption of solar cells. This enhancement is achieved through the localized surface plasmon resonance (LSPR) effect, where the oscillation of conduction electrons in the Au-NPs interacts with incoming light. This interaction leads to increased light scattering and absorption within the active layer of the solar cel [3]. Various sources have been used to synthesize gold nanoparticles (AuNPs) with antioxidant activity. For example, Au/CeO_2 nanoparticles have shown enhanced antioxidant activity due to their core shell and porous structure, which allows for effective exposure of catalytic sites and higher content of Ce (III) and oxygen vacancy. Hesperidin gold nanoparticles have also demonstrated significant antioxidant activity by scavenging free radicals of DPPH, ABTS, and hydrogen peroxide. Green synthesized AuNPs using Glaucium flavum leaf extract have shown effective antioxidant activity with low cytotoxicity and acceptable biocompatibility. Additionally, AuNPs synthesized from Sargassum longifolium have exhibited significant antioxidant activity by scavenging DPPH and H_2O_2 radicals. Lastly, gold nanoparticles synthesized from Physalis minima have shown antioxidant properties by utilizing flavonoids, phenols, and polyphenols as reducing agents [4]. Numerous studies have investigated the use of gold nanoparticles as catalysts in different applications. These studies have focused on understanding the physical and chemical properties of gold nanoparticles in ores and catalysts, such as their shape and composition. Researchers have also explored a fluorous biphasic approach for recycling noble metal catalysts, including gold nanoparticles, which has shown promising results in catalytic oxidation reactions. Additionally, conductive polymers supported with gold nanoparticles have demonstrated efficient catalytic performance in the reduction of 4-nitrophenol, suggesting their potential use in treating industrial wastewater. Furthermore, gold nanoparticles decorated on zinc oxide nanorods have been found to enhance the degradation of organic pollutants under sunlight, making them a promising option for environmental decontamination. These studies highlight the wide range of applications and catalytic abilities of gold nanoparticles in various fields [5]. Gold nanoparticles have been extensively researched for their applications in security and anti-counterfeiting measures. One method involves using gold nanoclusters with strong red emission, which are synthesized quickly and in an environmentally friendly manner using microwave irradiation. Another approach is the development of anti-counterfeiting inks that utilize gold nanoparticles functionalized with aryl diazonium salts. These inks provide distinct Raman fingerprints for security labels. Gold nanoparticles can also be used as authentication tags on various types of paper, offering unique optical signatures and long-lasting printed labels. Additionally, gold nanoparticle self-assemblies have been utilized as "stealth nanobeacons" for authentication purposes. These self-assemblies exhibit distinguishable surface-enhanced Raman scattering (SERS) signals, enabling rapid and unclonable authentication of items. These studies highlight the potential of gold nanoparticles in enhancing security and anti-counterfeiting measures [6]. Gold nanoparticles (AuNPs) have been extensively studied for their potential applications in the food industry, particularly in the areas of food safety and quality control. AuNPs possess unique properties, such as aggregation, fluorescence quenching, and broad absorption at the surface plasmon band, which make them effective nanosensors for detecting contaminants, pesticides, heavy metals, and adulterants in food. They can be utilized as colorimetric sensors, where a change in color indicates the presence of a specific contaminant or allergen. Additionally, AuNPs can be modified to directly or indirectly sense specific targets through various mechanisms, such as nucleic acid hybridization, antigen–antibody recognition, and enzyme inhibition. These characteristics make AuNPs a promising tool for rapid, cost-effective, and on-site detection of food contaminants and allergens, thereby ensuring food safety in the food and beverage industry [7]. Gold nanoparticles (AuNPs) have demonstrated potential in the field of gene delivery. When combined with pulsed electric fields, particularly using microsecond protocols, they can enhance the efficacy of gene delivery. The effectiveness of AuNPs in gene delivery is heavily influenced by their surface charge and size. AuNPs possess unique physicochemical characteristics, low cytotoxicity, and optical properties that make them suitable for targeted drug or gene delivery and detection applications. Peptide-modified AuNPs have been developed for gene delivery, where peptides act as modifiers to control the morphology and size of AuNPs, as well as stabilizers to prevent aggregation and precipitation. Through electrostatic interactions with DNA, AuNPs can form condensed complexes, facilitating efficient gene delivery. In summary, AuNPs offer a potential solution for enhancing the delivery of RNA and nucleic acids in the treatment of genetic disorders [8]. Gold nanoparticles have found widespread use in various fields, including art conservation. Their unique optical properties and ability to be modified to carry drugs and release them in specific tissues make them valuable for targeted drug delivery in cancer treatment.

Furthermore, the application of nanotechnologies and nanostructured materials in conservation treatments, such as cleaning and consolidation, has shown promising results. The use of gold nanoparticles in art conservation presents challenges and opportunities, as conservators continually adapt or develop new solutions and formulations for specific conservation issues. Therefore, the development of new nanomaterials and preservation methods for artwork is crucial, and material scientists are being called upon to contribute to this global effort [9]. Gold nanoparticles have demonstrated promise in the field of environmental remediation. They have been utilized in the creation of composite materials that can effectively adsorb contaminants like Cu2+ from water solutions. Researchers have proposed the use of ultrasound-assisted synthesis to create a montmorillonite-Au (MMT–Au) composite, which has shown a higher capacity for adsorbing Cu2+ compared to MMT alone. Additionally, scientists have explored green synthesis methods using plant extracts to produce metal/metal oxide nanoparticles (MONPs) for environmental remediation. Plant materials contain phytochemicals that greatly influence the bioactivity of MONPs. Gold nanoparticles, along with other types of nanoparticles, have been extensively used for environmental clean-up, particularly in soil and water remediation. They offer a promising solution for removing and mitigating toxic pollutants from the environment. Overall, gold nanoparticles have demonstrated potential in environmental remediation, both in composite materials and in their pure form. The use of green methods for their synthesis holds promise for sustainable remediation strategies [10]. Gold nanoparticles (AuNPs) have demonstrated significant potential in the field of photothermal therapy (PTT) for the treatment of cancer. They possess remarkable optical properties and can be utilized for molecular detection, biological imaging, and targeting cancer cells. The photothermal characteristics of AuNPs enable them to absorb light and convert it into heat, resulting in the destruction of tumor cells. Various types of AuNPs, including nanorods, nanocages, nanoshells, and nanospheres, have been synthesized and explored for their effectiveness in PTT. Furthermore, researchers have investigated the combination of chemotherapy with PTT to enhance the therapeutic outcomes. Strategies such as ROS-responsive nanosystems and light-induced AuNP aggregation have been developed to improve the efficacy of PTT. These advancements in AuNP-mediated PTT offer promising prospects for more efficient and targeted cancer treatment [11].

Electrochemical sensors based on gold nanoparticles were also developed to quantify ferulic acid in cosmetic products. These studies demonstrate the various ways in which gold nanoparticles can be utilized in the cosmetics industry. In electronics, GNPs have been utilized in sensors, conductive films, and electronic devices due to their unique optical and electrical properties. GNPs also have catalytic applications, where they can act as catalysts in various chemical reactions, including oxidation, reduction, and hydrogenation. Additionally, GNPs have been investigated for their potential in environmental remediation, such as the degradation of organic dyes and the removal of heavy metals from contaminated water. Marine organisms have also been explored for the biosynthesis of gold nanoparticles, as they possess antibacterial, antioxidant, and anti-inflammatory properties that can be beneficial in medical applications. Furthermore, the conjugation of phenolic compounds onto gold nanoparticles has been studied, leading to enhanced bioactivities such as antimicrobial and anticancer effects. Overall, these studies highlight the potential of gold nanoparticles synthesized from natural sources for a wide range of applications, including drug delivery systems. However, as these nanoparticles enter ecosystems, it is crucial to examine their potential consequences for the environment and its inhabitants. Nanoparticles interact with biological systems differently than their larger counterparts, and their impact on plant life has become a topic of increasing interest. Plants are essential for maintaining ecological balance and supporting human livelihoods, so any disruption to their growth and health can have far-reaching effects.

This review aims to comprehensively investigate the phytotoxicity of gold nanoparticles, specifically comparing two synthesis methods: chemical synthesis and green synthesis. Chemical synthesis, which involves harsh chemicals and high temperatures, is the traditional approach, while green synthesis is an environmentally conscious method that uses plant extracts, microbes, or other eco-friendly sources to synthesize nanoparticles. By comparing these two methods, we hope to uncover any differences in the phytotoxic effects of Au-NPs on plants. This comparative evaluation addresses environmental concerns by determining whether green synthesis methods, which minimize chemical pollutants and energy consumption, result in Au-NPs with reduced phytotoxicity compared to chemically synthesized nanoparticles. Additionally, this research contributes to the field of green nanotechnology, exploring sustainable approaches to nanomaterial synthesis that align with global environmental objectives. Through careful experimentation, data analysis, and consideration of ecological factors, this study aims to understand the complex interaction between Au-NPs and plant biology. It will provide insights into phytotoxicity mechanisms and potential ecological implications. Ultimately, the findings from this investigation can guide future nanomaterial design, risk assessment, and environmentally responsible nanoparticle production, promoting a harmonious coexistence between nanotechnology and the natural world (Table 1).

2 Synthesis Methods

Gold nanoparticles (Au–NPs) have gained significant attention due to their unique properties and find applications in various fields such as electronics, drug delivery, imaging, agriculture, and environmental remediation [12–16]. The synthesis of Au–NPs is a critical aspect of their effective utilization and has evolved over the years. It has evolved significantly, and various methods are available, including chemical, green, electrochemical, and photochemical approaches. Each method offers unique advantages in terms of control over size, shape, and environmental impact, enabling the effective utilization of gold nanoparticles in diverse applications. Chemical synthesis methods include the Turkevich method, which involves the reduction of chloroauric acid by a strong reducing agent in the presence of heat. This method offers control over nanoparticle size and produces well-dispersed spherical nanoparticles. The Brust-Schiffrin method is a two-phase synthesis technique that utilizes a water-immiscible organic solvent and a reducing agent to produce monodisperse Au-NPs. It allows for excellent control over size and shape. Seed-mediated growth is a versatile approach that involves the initial formation of small Au–NP "seeds"

Table 1 Application and explanation of gold nanoparticles

S. no.	Applications	Explanation
1	Biomedical	Gold nanoparticles are widely used in biomedical industries in many aspects like; Drug delivery, cancer therapy, diagnostic techniques, and gene delivery
2	Catalysis	Gold nanoparticles are used as catalysts in a variety of chemical reactions, including oxidation, reduction, and hydrogenation processes. Their catalytic activity is highly dependent on size and shape
3	Electronics	Au-NPs are utilized in the development of nanoscale electronic devices and sensors due to their excellent electrical conductivity and size-dependent electronic properties
4	Remediation	Gold nanoparticles can be used for the removal of pollutants from water and air. They can catalyze the degradation of organic contaminants and adsorb heavy metals
5	Solar cells	Au-NPs are employed in solar cells to enhance light absorption and increase energy conversion efficiency
6	Imaging	Gold nanoparticles exhibit unique optical properties, such as localized surface plasmon resonance (LSPR), which make them suitable for applications in sensors, imaging, and as contrast agents in microscopy
7	Food	Au-NPs are used for food packaging and as sensors to detect contaminants and spoilage in food products
8	Art conservation	Gold nanoparticles are used in the restoration and preservation of artworks and historical artifacts, helping to stabilize and protect delicate surfaces
9	Anticounterfeit	Au-NPs can be incorporated into security inks and labels to provide authentication and prevent counterfeiting
10	Photothermal therapy	Gold nanoparticles can be used in the treatment of other diseases and infections through photothermal therapy, where they are heated by laser irradiation to kill pathogens

followed by their growth into larger nanoparticles. This method offers precise control over size, shape, and surface properties. Green synthesis methods involve the use of biological entities such as plant extracts or microorganisms to reduce and stabilize Au ions into nanoparticles. Plant-mediated synthesis utilizes plant extracts rich in natural reducing agents, offering eco-friendliness and potential for large-scale production. Microbial synthesis utilizes microorganisms with enzymes that play a crucial role in nanoparticle formation, providing a sustainable and cost-effective alternative. Electrochemical synthesis involves the application of an electrical potential to a solution containing Au ions, leading to the reduction and deposition of Au atoms on an electrode surface. This method offers precise control over nanoparticle size and shape, making it suitable for tailored applications. Photochemical synthesis utilizes ultraviolet or visible light irradiation to reduce Au ions to nanoparticles. It allows for rapid

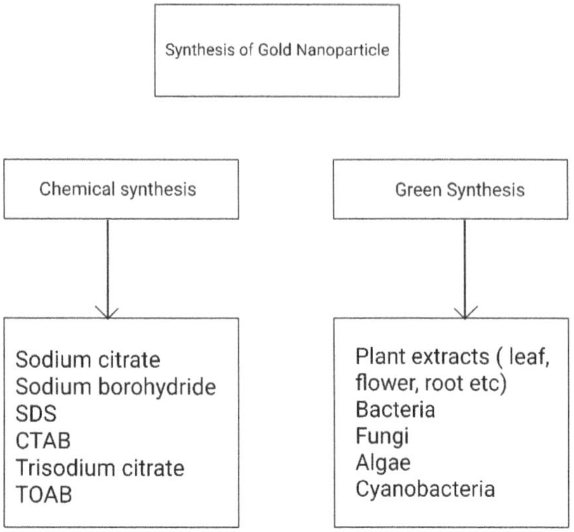

Fig. 1 Chemical and gold synthesis

synthesis and the possibility of producing anisotropic shapes through precise light control (Fig. 1).

3 Chemical Synthesis of Gold Nanoparticles (GN)

Gold nanoparticles (GNPs) can be synthesized using various chemical methods, including the Turkevich synthesis, reduction with $NaBH_4$ (with or without citrate), seeding growth, ascorbic acid-based synthesis, green synthesis, and the Brust-Schiffrin method [17]. These methods offer different approaches to synthesizing, modifying the surface, and characterizing GNPs. By manipulating synthesis parameters such as reducing agents, capping agents, and reaction conditions, the size and shape of GNPs can be controlled. Additionally, the polarity of solvents can be utilized in the chemical reduction method to tailor the size of GNPs, with higher polarity resulting in smaller and spherical nanoparticles, and lower polarity resulting in larger nanoparticles with varying shapes. To determine the size, shape, and surface coverage of GNPs, various characterization techniques such as UV–Vis, dynamic light scattering (DLS), and transmission electron microscopy (TEM) are employed. These chemical synthesis methods play a crucial role in developing GNPs with specific properties for applications in medicine, biology, and other industries. The chemical synthesis method utilizes chemicals which commonly include citrates, hydroxides, sodium borohydride, SDS, CTAB, etc., which are not only expensive but also toxic to the environment and ecosystem. To curb these disadvantages, the green synthesis of gold nanoparticles has been employed predominantly these days.

4 Green Synthesis of Gold Nanoparticles (GN)

Gold nanoparticles can be synthesized using green methods, which involve using natural sources like citrus fruits, plant extracts, and flower extracts as reducing and capping agents. These methods are considered eco-friendly and have gained attention due to their low cost, availability, and non-toxic nature. For instance, citrus fruit juice can be used as a reducing agent without the need for chemical reagents or protective agents. Flower extracts can also serve as both reducing and capping agents, resulting in stable gold nanoparticles. Additionally, chitosan, a biocompatible polymer, has been utilized as a reducing and capping agent for gold nanoparticles, showing potential applications in drug delivery systems. Overall, green synthesis methods provide a simple, fast, and environmentally friendly approach to synthesizing gold nanoparticles. The synthesis process for preparing gold nanoparticles using natural citrus fruit juice as the main ingredient involves several steps. First, the citrus fruits are cleaned, dried, and juiced. The juice is then centrifuged to obtain a supernate. This supernate is diluted with deionized water to create a juice solution. In a separate container, a chloroauric acid solution is magnetically stirred for 2 min under room temperature conditions. After that, the juice solution is rapidly added to the chloroauric acid solution, and stirring continues for 20 min. The completion of the reaction is indicated by the appearance of a wine red color. The resulting gold nanoparticle solution is then preserved in a refrigerator at 4 °C. This green synthesis method is simple, fast, low-cost, and has minimal pollution to the environment [18]. Previous research describes a green method for synthesizing gold nanoparticles using Peltophorum pterocarpum flower extracts, essential oil menthapiperita, Lawsonia inermis leaf extract, pomegranate peel extract, Terminalia arjuna fruit extracts,3-dihydroxyfumaric acid (DHFA), etc. [19]. Microorganisms are also used in the green synthesis of gold nanoparticles.

In recent years, there has been a growing interest in using microbial synthesis methods to produce gold nanoparticles. Fungi, bacteria, and yeast have all been studied for their ability to synthesize these nanoparticles. Fungi-mediated synthesis of gold nanoparticles has several advantages, including high efficiency, low energy consumption, and simplicity of operation. Similarly, plant extracts, bacteria, and yeast have been used to synthesize silver nanoparticles, with yeast showing promise in controlling the size and shape of the particles. Additionally, gamma ray radiation has been used to synthesize colloidal silver nanoparticles, which have antimicrobial properties. The use of microbial sources for synthesizing gold nanoparticles is environmentally sustainable and green, as it utilizes non-toxic chemicals from natural sources. Overall, microbial synthesis methods offer a promising approach for producing gold nanoparticles with various applications in medicine and other fields.

5 Characterization Techniques

Understanding the physical and chemical characteristics of silver nanoparticles is crucial for determining how they might be used to control plant diseases. Silver nanoparticles have been created and characterized using a variety of techniques, such as reduction processes and sono-electrochemical deposition. Transmission electron microscopy (TEM), UV–Vis spectroscopy, Fourier transform infrared spectroscopy (FTIR), atomic force microscopy (AFM), X-ray diffractometry (XRD), and scanning probe microscopy (SPM) are among the characterization methods used. To explore the surface topography and morphology of the nanoparticles, AFM and SPM methods are used. The nanoparticles' structural structure is revealed via XRD analysis. Understanding the characteristics and behavior of silver nanoparticles is critical for their applications in a variety of disciplines. This is where these characterization techniques come into play. For studying the formation of AuNPs Information on the size, structure, stabilization, and aggregation of NPs would be available via UV–vis.

Additionally, TEM study would enable visualization while XRD analysis would shed light on the nanoparticles' structural structure. Insights into the crystalline structure of the silver nanoparticles would also be provided by XRD analysis. The size distribution and stability of the nanoparticles are also determined using dynamic light scattering analysis. On the other hand, AFM provides details regarding the NPs' size, form, physical structure, and morphology. The chemical makeup, surface functionalization, and stability of silver nanoparticles, including emission, absorption, photoconductivity, the distinctive bonding involved, stability of surface charge, etc., are studied using methods like FITR, XPSM, and Zeta potential.

6 Phytotoxicity Assessment

6.1 Seed Germination Assay

This test can assess the rate of germination of seeds which in turn reflects on the root and shoot growth, ultimately providing information about phytotoxicity. The effects of gold nanoparticles on seed germination have been extensively studied, with varying results. Some studies have shown positive effects, while others have shown negative effects. The size, shape, and concentration of the nanoparticles, as well as the specific plant species and growth media, can all influence the outcome. Gold nanoparticles synthesized using a seed-mediated growth approach have been found to have a positive impact on seed germination. However, exposure to gold nanoparticles can also be toxic and inhibit root elongation. It is important to note that most studies have been conducted in controlled environments, and further research is needed to fully understand the impact of gold nanoparticles on seed germination in agricultural soil.

6.2 Growth Inhibition Assay

The phytotoxicity of gold nanoparticles (AuNPs) has been examined in various plant species. In a study Allium cepa (A. cepa) was used to assess the effects of different sizes of citrate-capped AuNPs, which resulted in chromosomal aberrations and oxidative stress another study compared the phytotoxicity of neem-coated and citrate-coated silver nanoparticles (Ag NPs) and found that citrate-coated Ag NPs had a significant impact on seedling growth, while neem-coated Ag NPs showed mild toxicity likely due to nutrient supplements from neem.

6.3 Genetic Analysis

Arabidopsis thaliana seedlings grown on nutrient agar plates containing 100 mg/L gold experienced a 75% decrease in root lengths. Both roots and shoots contained oxidized gold, while gold nanoparticles were only found in the root tissues. A microarray-based study discovered that genes involved in plant stress response, such as glutathione transferases, cytochromes P450, glucosyl transferases, and peroxidases, were up-regulated when exposed to gold. On the other hand, a small number of genes encoding proteins responsible for transporting copper, cadmium, iron, and nickel ions, as well as aquaporins, were significantly down-regulated. Aquaporins, which regulate the movement of water molecules in and out of cells, were found to be down-regulated by gold treatment. This inhibition of water uptake and movement within the plant could explain the observed toxicity symptoms in gold-treated Arabidopsis seedlings. Similar to silver and mercury, gold has been shown to hinder aquaporin function. Gold nanoparticles were only observed in plants exposed to ionic gold in solution, suggesting that plants primarily take up gold in its ionic form. The down-regulation of specific metal transporters and aquaporins may contribute to the reduced uptake of gold by plants [20] (Fig. 2).

7 Mechanism of Action of Gold Nanoparticles (GN)

Gold nanoparticles (AuNPs) have been discovered to possess antimicrobial properties through various mechanisms. The oxidation state of gold, its ability to generate reactive oxygen species (ROS), and its interaction with bacterial components all contribute to causing damage to bacterial walls and DNA. Moreover, the small size of AuNPs enables them to attach to the surface of bacterial pathogens, disrupt the cell membrane, and produce ROS, ultimately leading to the destruction or impairment of bacterial cells. Gold nanoparticles also exhibit catalytic activity, facilitating the oxidation of organic radicals and the formation of acetone and acetanilide hydroxylation products. In terms of antifungal activity, AuNPs synthesized using *Allium*

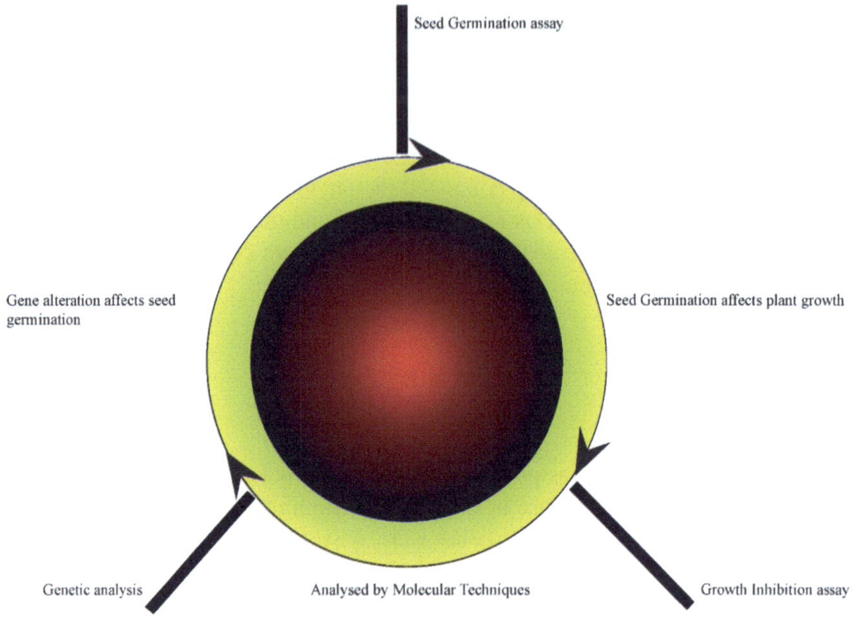

Fig. 2 Phytotoxic assessment

sativum extract have been observed to generate ROS, alter fungal cell morphology, and harm the cell membrane, resulting in the death of microbial cells. The generation of ROS, interaction with proteins, and nonoxidative mechanisms are proposed as the primary mechanisms underlying the antimicrobial and anticancer effects of metal nanoparticles, including gold nanoparticles. By electrostatic contact with the cell membrane, stimulation of the production of reactive oxygen species (ROS), and subsequent DNA damage, gold nanoparticles can cause damage to the bacterial cell wall. In Escherichia coli, extensive genomic and proteomic studies have revealed that Au (III) is more hazardous than Au (I), having an impact on oxidative stress, bacterial energy metabolism, biosynthetic activities, and cell transport. Gold nanoparticles have the ability to alter the mitochondrial membrane potential and increase the production of ROS in eukaryotic cells, but they have little effect on the viability, morphology, or focal adhesions of these cells. Ionic gold nanoparticles that resemble antibiotics can display potent antibacterial activity against bacteria while being safe for human cells under specific circumstances [21]. The lysis or destruction of bacterial cells by SGNPs involves a number of mechanisms, including membrane rupture, oxidative stress, and damage to DNA and proteins [22]. Gold nanoparticles have been discovered to act as catalysts for the oxidation of organic radicals. For example, they can catalyze the oxidation of 2-propanol to acetone, resulting in a significant increase in acetone formation by approximately 340%. However, in the case of acetanilide hydroxylation during water radiolysis, the presence of both oxygen and gold nanoparticles unexpectedly leads to a decrease in hydroxylation product formation. This is surprising because oxygen typically stimulates hydroxylation reactions in the absence of gold nanoparticles. It is proposed that in the presence of both oxygen and nanoparticles, oxygen first attaches to acetanilide OH-adducts, and then the nanoparticles catalyze the oxidation of peroxyl-type radicals, which does not result in the formation of hydroxylation products [23].

8 Interaction of Gold Nanoparticles and Plant Tissues

Gold nanoparticles (AuNPs) have the ability to interact with plant cells and tissues by either depositing on the surface of roots or being taken up by plant root cells. The internalization and subcellular localization of AuNPs within plant root cells are influenced by the surface charge of the nanoparticles. The surface coatings of AuNPs, such as branched polyethyleneimine (b–PEI), can also affect their interactions with plant cells and tissues. Different surface charges of AuNPs result in distinct accumulation patterns on the root surface and within plant cells. Furthermore, the surface charge of AuNPs plays a role in their impact on plant physiology and biochemistry, including the concentrations of hydrogen peroxide and enzymatic antioxidant activities. Therefore, when evaluating the potential toxicity and effects of AuNPs on food crops and the environment, it is important to consider their unique properties, particularly their surface charge [24].

9 Environmental Impacts and Potential Transformations

The synthesis of nanoparticles often involves the use of various chemicals, some of which can be harmful to plants. It is important to consider the specific chemicals used, their concentrations, and the duration and route of exposure (e.g., through leaves, soil, or roots). Different plants may also have different sensitivities to these chemicals. Numerous studies have been conducted to evaluate the toxicity of nanoparticles and their precursors on plants. Some commonly used chemicals in nanoparticle synthesis, such as metal salts (e.g., silver nitrate, copper sulfate) or reducing agents (e.g., sodium borohydride, sodium citrate), can negatively impact plant growth and development. The potential toxic effects of chemicals used in the chemical synthesis of nanoparticles on plants depend on various factors. These include the nature and concentration of the chemicals, the specific plant species, and the environmental conditions. If the chemicals are not properly removed or neutralized during the synthesis process, they can be released when the nanoparticles come into contact with plant tissues. Residual chemicals, such as reducing agents or stabilizing agents, can be toxic to plants. The concentration of these chemicals is crucial, as even low concentrations can have harmful effects on plants, especially with prolonged exposure. The chemical properties of the residual chemicals also play a role, as they can disrupt cellular processes, affect nutrient uptake, or cause oxidative stress in plants.

Different plant species may have varying levels of sensitivity to specific chemicals, and environmental factors like soil type, pH, temperature, and moisture levels can further influence the extent of phytotoxicity. One major phytotoxic effect is the accumulation of ions and nanoparticles in the plant system. The effects of AuNPs on biochemical parameters in plants, such as hydrogen peroxide concentrations in roots and enzymatic antioxidant activities, were found to be significantly influenced by the surface charge of the nanoparticles. The deposition of AuNPs on the surface of plant roots was observed regardless of the type of AuNPs used. However, the internalization of AuNPs into plant root cells and their subcellular localization were affected by the surface charge which is usually dasic in chemically synthesised nanoparticles. Overall, this study highlights the importance of considering the physical and chemical properties of AuNPs in their interactions with agricultural crops and emphasizes the need to assess their food safety concerns based on their unique properties. However, the effects of AuNPs on the concentrations of H_2O_2 and CAT activity in plant root tissues varied significantly. Additionally, AuNPs with different surface charges exhibited distinct accumulation patterns on the surface of plant roots and different internalization and subcellular localizations. These findings highlight the importance of surface properties in determining the impact of AuNPs on food crops and emphasize the need to consider the unique properties of engineered nanoparticles in assessing their environmental fate and impact. The hydrodynamic size of bPEI–AuNPs showed substantial aggregation immediately after dilution in water, and after five days, the hydrodynamic size of all AuNPs increased except for bPEI–AuNPs, which still had the largest hydrodynamic size [24]. Green synthesized gold nanoparticles, on the other hand, are completely eco-friendly and the best suitable and sustainable option for the synthesis of nanoparticles with no compromise on the quality, quantity or function of the nanoparticles.

10 Advantages of Green Synthesis Methods

The synthesis of nanoparticles using green methods has several advantages. Firstly, it is a more environmentally friendly approach compared to traditional physical and chemical methods. This is because it reduces the use of toxic chemicals and minimizes environmental toxicity. Secondly, green synthesis is cost-effective and economically viable. It utilizes living organisms such as plants, bacteria, and algae as reducing and stabilizing agents, which are readily available and renewable resources. Additionally, green synthesis provides a safer alternative for nanoparticle production. It reduces the risk of exposure to hazardous materials and ensures the biocompatibility of the synthesized nanoparticles. Moreover, green synthesized nanoparticles have shown promising applications in various fields such as electronics, sensors, catalysis, optics, antimicrobial agents, and energy processing. Overall, the green synthesis of nanoparticles offers a sustainable and efficient approach for their synthesis, with reduced environmental impact and enhanced biocompatibility [25].

11 Conclusion

This study's exploration into the phytotoxic effects of gold nanoparticles (Au–NPs), synthesized both chemically and through eco-friendlier green methods, sheds vital light on their environmental impact and the importance of responsible usage. The research delves into how different synthesis techniques influence the properties of Au-NPs and, consequently, their interactions with plant life. It is observed that chemically synthesized Au–NPs typically possess attributes that heighten their phytotoxicity in comparison to those produced via green synthesis. The latter employs natural reducing agents, resulting in nanoparticles that are more benign to the environment. The study also finds that the physical characteristics of Au–NPs, such as size and shape, play a significant role in their phytotoxicity. Particularly, smaller and anisotropic nanoparticles have shown increased phytotoxic effects. Additionally, the chemical synthesis process often involves surface modifications and stabilizers, which may intensify the adverse interactions between Au-NPs and plant tissues.

Understanding the phytotoxicity of Au–NPs is essential for assessing their overall impact on various ecosystems, especially concerning applications in agriculture and environmental remediation. The findings highlight that green synthesis approaches are more aligned with sustainable practices, primarily due to their lower toxicity potential. Looking ahead, the study emphasizes the need for further research that examines the effects of Au–NP synthesis methods across various plant species and environmental conditions, focusing on long-term ecological impacts. This future research is crucial for developing comprehensive mitigation strategies and establishing guidelines for the safe and responsible utilization of Au–NPs in environmental contexts. Such measures are imperative to ensure that the integration of these nanoparticles into our ecosystems is done in a way that minimizes harm and maximizes benefits.

References

1. Koliyote S, Shaji J (2023) A recent review on synthesis, characterization and activities of gold nanoparticles using plant extracts. Indian J Pharm Educ Res 57
2. Anne-Laure M, Sophie G (2021) Gold nanoparticles comprising a plant extract and their cosmetic use. Nanobiotechnology 10(2)(2018):e1480. https://doi.org/10.1002/wnan.1480
3. Gupta V, Srivastava A, Jain R, Sharma VK, Kumar L (2022) Optimization and development of ITO-free plasmonic gold nanoparticles assisted inverted organic solar cells. Opt Mater. https://doi.org/10.21203/rs.3.rs-1249295/v1
4. Rajeshkumar S, Parameswari RP, Jayapriya J, Tharani M, Ali H, Aljarba NH, Alkahtani S, Alarifi S (2022) Apoptotic and antioxidant activity of gold nanoparticles synthesized using marine brown seaweed: an in vitro study. BioMed Res Int. https://doi.org/10.1155/2022/5746761
5. Sanakulov KS, Vorobyov AL, Kozyrev E, Lianzi Z (2022) Physical and chemical properties of gold nanoparticles in ores and catalysts. Устойчивое развитие горных территорий. https://doi.org/10.21177/1998-4502-2022-14-4-676-684

6. Fukuoka T, Yamaguchi A, Hara R, Matsumoto T, Utsumi Y, Mori Y (2015) Application of gold nanoparticle self-assemblies to unclonable anti-counterfeiting technology. https://doi.org/10.1109/ICEP-IAAC.2015.7111051
7. Rastogi S, Kumari V, Sharma V, Ahmad FJ (2021) Gold nanoparticle-based sensors in food safety applications. Food Anal Methods. https://doi.org/10.1007/S12161-021-02131-Z
8. Xiaomin X, Yang L, Yujie Y, Jin-ming W, Meiwen C, Li S (2022) One-pot synthesis of functional peptide-modified gold nanoparticles for gene delivery. Colloids Surf, A. https://doi.org/10.1016/j.colsurfa.2022.128491
9. Serafini I, Ciccola A (2019) Nanotechnologies and nanomaterials: an overview for cultural heritage. https://doi.org/10.1016/B978-0-12-813910-3.00014-8
10. Jorepalli S, Somala AR, Annavaram V, Koduru JR (2022) Green synthesis of metal nanoparticles for environmental remediation. https://doi.org/10.1016/b978-0-12-824547-7.00011-4
11. Park Y, Young, Mee, Jung., Sing, Yian, Chew., Hyuk, Sang, Yoo (2023) Reactive oxygen species-responsive clicked assembly of gold nanoparticles to enhance photothermal therapy. J Mater Chem B. https://doi.org/10.1039/d3tb00500c
12. Husen A (2017) Gold nanoparticles from plant system: synthesis, characterization and their application. In: Ghorbanpourn M, Manika K, Varma A (eds) Nanoscience and plant–soil systems vol 48. Springer International Publishing AG, Cham, pp 455–479
13. Husen A, Rahman QI, Iqbal M, Yassin MO (2019) Plant-mediated fabrication of gold nanoparticles and their applications. In: Husen A, Iqbal M (eds) N*anomaterials* and plant potential. Springer International Publishing AG, Cham, pp 71–110 https://doi.org/10.1007/978-3-030-05569-1_3
14. Madhusudhan A, Bandi R, Sharma G, Dadigala R, Husen A, Kim JC (2022) Current trends in engineered gold nanoparticles for cancer therapy. In: Kim JC, Alle M, Husen A (eds) Smart nanomaterials in biomedical applications. nanotechnology in the life sciences. Springer, Cham, pp 1–40. https://doi.org/10.1007/978-3-030-84262-8_1
15. Siddiqi KS, Husen A (2016) Engineered gold nanoparticles and plant adaptation potential. Nanoscale Res Lett 11(400):1–10
16. Siddiqi KS, Husen A (2017) Recent advances in plant-mediated engineered gold nanoparticles and their application in biological system. J Trace Elem Med Biol 40:10–23
17. Mehdizadeh F, Barzegar-Jalali M, Izadi E, Osouli-Bostanabad K, Mohaghegh S, Shakeri MS, Nazemiyeh H, Omidi Y, Adibkia K (2022) Green and chemical reduction approaches for facile pH-dependent synthesis of gold nanoparticles. Inorganic Nano-Metal Chem. https://doi.org/10.1080/24701556.2022.2078357
18. Yu J, Guan HN, Wu S, Chi D (2015) Green synthesis method of gold nanoparticles
19. Madheswaran B, Saravanan K, Shanmugam S (2016) Green synthesis of gold nanoparticles by using peltophorum pterocarpum flower extracts. Nano Biomed Eng. https://doi.org/10.5101/NBE.V8I4.P213-218
20. Taylor AF et al (2014) Investigating the toxicity, uptake, nanoparticle formation and genetic response of plants to gold. PLOS one 9(4):e93793
21. Paesa M et al (2023) Elucidating the mechanisms of action of antibiotic-like ionic gold and biogenic gold nanoparticles against bacteria. J Colloid Interface Sci 633:786–799
22. Das G, Patra JK (2023) Evaluation of antibacterial mechanism of action, tyrosinase inhibition, and photocatalytic degradation potential of sericin-based gold nanoparticles. Int J Mole Sci 24(11):9477
23. Shcherbakov V, Denisov SA, Mostafavi M (2021) The mechanism of organic radical oxidation catalysed by gold nanoparticles. Phys Chem Chem Phys 23(46):26494–26500
24. Xingmao M, Bryan Q (2016) Effects of surface charge on the fate and phytotoxicity of gold nanoparticles to phaseolus vulgaris
25. AL-Shammari RHH, Abdulkareem AF (2020) Green synthesis of nanoparticles by different microorganisms. Int J Sci Res Arch 6(2)(2022):212–217

Gold Nanoparticles as the Sterilant in Large-Scale Micropropagation of Plants

Abdelatti I. Nowwar, Azamal Husen, and Salem S. Salem

Abstract One of the most challenging circumstances in micropropagation is microbial contamination (fungi, bacteria, etc.), which results in a decline in plant trait and destruction of valuable stocks. As a result, a crucial stage in the effectiveness of plant micropropagation is the disinfection of explant surface and culture medium sterilization. The activities of additives and plant development aids may have decreased due to the typical autoclaving media approach. Typical surface cleaners like mercuric chloride are considered harmful to human health and are not environmentally friendly. Gold nanoparticles (AuNPs) have recently been demonstrated to be an efficient sterling for the culture media and explants, as well as to positively promote plant regeneration and elicit the development of secondary products. Current chapter highlights recent research studies of AuNPs in the large-scale micropropagation of plants as the sterilant.

Keywords Gold nanoparticles · Plants · In vitro propagation · Culture condition · Various explant growth

1 Introduction

Plant micropropagation, an effective technique for producing genetically homogeneous, acquired secondary compounds and plants devoid of illness in large numbers in vitro in large numbers in vitro, is preferred over traditional methods like cuttings

A. I. Nowwar · S. S. Salem (✉)
Botany and Microbiology Department, Faculty of Science, Al-Azhar University, Cairo 11884, Egypt
e-mail: salemsalahsalem@azhar.edu.eg

A. Husen
Wolaita Sodo University, PO Box 138, Wolaita, Ethiopia

Department of Biotechnology, Graphic Era (Deemed to Be University), Dehradun, Uttarakhand 248002, India

A. Husen (ed.), *Plant Response to Gold Nanoparticles*, Smart Nanomaterials Technology, https://doi.org/10.1007/978-981-97-7174-5_17

and seeds because it has a number of benefits [1]. However, the likelihood of microbiological contaminants that inhibit the growth of the plants in the culture medium is one of the remaining drawbacks of micropropagation [2]. Therefore, a crucial step in plant micropropagation is sterilizing the culture material. Disinfected medium, however, may lessen the effectiveness of plant hormones and nutritional elements in culture media [3].

Autoclaves are typically employed in micropropagation laboratories to culture material disinfected because they have a long and trustworthy history. Although efficient, autoclaving is a significant cost for many laboratories of micro-propagation and results in the creation of harmful byproducts of dissolution [4]. Chemical sterilization utilizing substances like sodium hypochlorite (NaClO) [5], H_2O_2 [6], and chlorine dioxide (ClO_2) [7] can take the place of physical sterilization via autoclaving. Sterilization by chemical compounds can be quite successful in disinfecting, however it is typically harmful to the plants and reduces the effectiveness of multiplication.

In recent years, authors from all around the world have become attractive in the connections between nanomaterials and plant growth [8–14]. There have been numerous studies into the metal nanoparticles effects on plants [15–17]. Researchers have demonstrated that the structure and concentrations of the nanoparticles, and the types of plants employed influence both advantageous and detrimental consequences of nanomaterials on plant development [18, 19–23]. Because of the wide range of applications that nanomaterials provide in many fields of study and enterprise, many techniques have been created for the briefing of nanomaterials [24–38].

Nanoparticles, especially gold nanoparticles (AuNPs), when added to the culture media at the proper proportions, serve as plant disinfectants and enhance the process of plant regeneration in many plant micropropagation applications [39]. Other applications, like improving microponic systems and obtaining secondary products from cells, are also effective [40]. Additionally, research is being done on how to use them to advance genetic engineering through gene transformation [41]. This chapter intends to record knowledge regarding the use of AuNPs as the sterilant in plant micropropagation and to highlight current accomplishments.

2 Sterilization Methods Used in Plant Micropropagation

2.1 Autoclaving

Autoclaving or steam sterilization is the most popular way to sterilize with water vapor under pressure. The autoclave can be used to sterilize cotton swabs, gauze, lab water, plastic caps, glassware, filters, pipettes, water, small tools, and nutrient medium. By exposing them for 10–15 min to the extremely hot steam of an autoclave, nearly all microorganisms are rendered inert. All items need to be sterilized for 15–20 min at 121 °C and 15 pressure [42]. Glassware is replaced by high-quality plastic labware that can be repeatedly autoclaved, such as polypyrene, polymethylpentene, polyallomer, Tfzel, ETPE, and Teflon FEP. There is also a wide range of pre-sterilized culture vessels that can be used to pour autoclaved media. It is also possible to sterilize a small amount of a medium using some home pressure cookers, which operate on the same principles as an autoclave [43].

Media components may suffer negative consequences from autoclaving. Long-term heating under pressure could have a negative impact on the quality of the media. For instance, some compounds' characteristics and the pH of media can be changed [44]. Steam sterilization may cause the formation of glucose and fructose as a result of the hydrolysis of sucrose [45]. Gibberellic acid and other heat-sensitive plant growth regulators may lose the majority of their activity. High temperatures cause some vitamins to degrade, salts can precipitate and the characteristics of agar can change when sterilization times are too long [46].

2.2 Explant Surface Disinfection

Prior to in vitro culture, the experimental material must first be surface sterilized to eliminate all surface microorganisms. Surface contaminations can be removed using a variety of chemicals and techniques. This is done by briefly exposing plant cells to one or more sterilizing chemicals, followed by washing with sterile water. The most often used compounds include mercuric chloride, H_2O_2, ethanol, fungicides, and antibiotics. They also include sodium and calcium hypochlorite. Other substances, including chlorine dioxide and isothiocyanates, have been developed as surface cleaners [47, 48].

It will be easier to choose the optimal sterilization agent(s) for the explant if you have full insight of the advantage and disadvantage of the several sterilization agents. Bacteria can be killed very effectively using hypochlorites, e.g,. sodium hypochlorite, which is a potent oxidizing agent. In addition to being a potent sterilizer, ethanol is also harmful to plants. A solution (70%) works to swiftly denature proteins and render lipophilic viruses inactive. Applying ethanol at a 70% concentration for a few minutes or just a few seconds before or after a hypochlorite treatment will boost its efficiency. Mercuric chloride has a wide range of activity and is primarily bacteriostatic. Mercuric substances are thought to impede enzyme function by attaching to sulfhydryl groups [49]. Despite mercuric chloride is an efficient decontaminant, it is extremely toxic and unfriendly to the environment. Since antibiotics have detrimental effects on explant growth and are poisonous to them, they are rarely utilized. Also, antibiotic-resistant bacterial strains may arise as a result of ongoing consumption of antibiotics [50].

There are benefits and drawbacks to the present disinfectants used in laboratories. Gold compounds, particularly gold nanoparticles, have gained popularity as a disinfectant for plant micropropagation due to their antibacterial properties. gold compounds are more efficient, environmentally friendly, and in harmony with a variety of tissue types when used in their ideal quantities (Fig. 1).

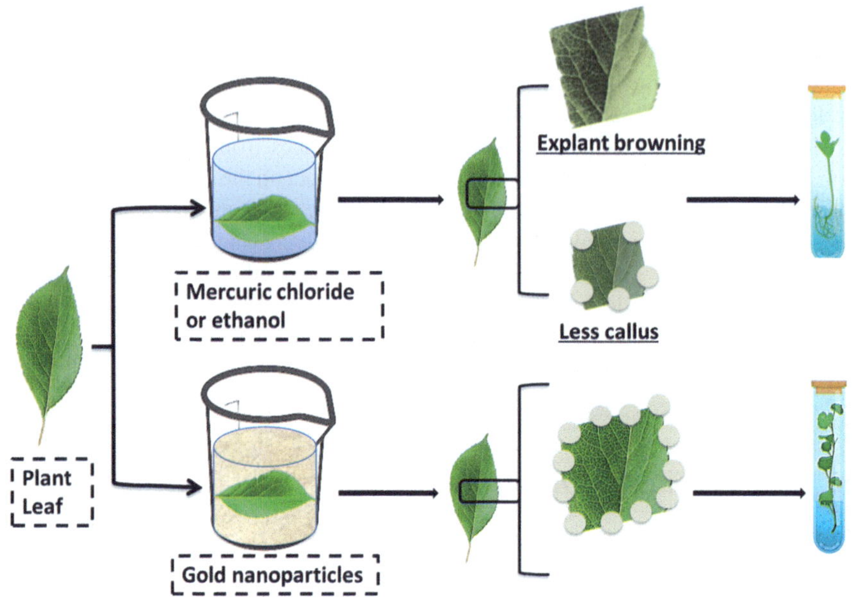

Fig. 1 Gold nanoparticles as explant surface disinfection and media additive

3 Use of AuNPs in Large-Scale Plant Micropropagation

3.1 AuNPs as an Explant Disinfectant

In recent years, gold nanoparticles (AuNPs) are commonly utilized as an agent of antimicrobial and act as an explant antiseptic. AuNPs antibacterial effectiveness is influenced by a variety of elements, including the synthesis process, metallic silver content, particle shape, size distribution, and coating agent [51]. Due to their enormous surface area, which facilitates greater microbial adhesion, AuNPs demonstrate more potent antibacterial activities than gold salts. Bacteria easily interact with AuNPs that are less than 15 nm. When AuNPs are adhered to bacterial cell membranes, gold ions are released into the bacterial cells, increasing bactericidal activity [52].

In micropropagation of plant, AuNPs have been utilized to disinfect surfaces with success. In this regard, [53] studied the efficacy of treatments using AuNP as disinfectant agents of explants of *Eugenia involucrate*. They found that AuNPs significantly decreased the contamination compared to that treated with ethanol and sodium hypochlorite. Furthermore, [54] reported that application of AuNPs into culture media significantly improved the growth, development of narrow-leave, stimulated formation of shoots, enhanced plant weight, increased the number of trichomes and increased formation of oil glands of *Lavandula angustifolia* plant propagated in vitro.

The recent studies that use AuNPs as an explant disinfectant are listed in Table 1. Further, depending on the researchers, the application of AuNPs at a suitable concentration can produce favorable results and show to have less negative effects on the growth of explants.

Table 1 Selected recent studies on the use of AuNPs as a surface sterilant of explants

Species	Treatment and outcomes	References
Scoparia dulcis L	The characteristic feature of the synthesized gold nanoparticles (AuNPs) used was spherical shape with an average diameter of 3.5 nm. The Minimum Inhibitory Concentration of AuNPs against all endophytic bacterial contaminants was found to be 0.25 mg/ml. Tolerance studies showed that AuNPs were bactericidal in nature against all the isolated bacterial endophytes	[55]
Prunella vulgaris	Calli culture of *Prunella vulgaris* were exposed to differential ratios of AuNPs (1:1, 1:2, 1:3, 2:1, and 3:1) along with naphthalene acetic acid (2.0 mg NAA). Two concentrations (1 and 2 mg/6 μl) of all treated calli cultures and wild plants were used against *Escherichia coli, Pseudomonas aeruginosa, Salmonella typhi, Bacillus atrophaeus, Bacillus subtilis, Agrobacterium tumefaciens, Erwinia caratovora and Candida albicans*. The calli exposed to AuNPs and 2.0 mg naphthalene acetic acid displayed the highest activity (25.7 mm) against *Salmonella typhi* than other extracts, which was considered the most susceptible species, while *Agrobacterium tumefaciens and Candida albicans* were the most resistance species	[56]
Rauvolfia serpentina (Linn.) Benth. Ex Kurz	AuNPs were synthesized using the ethyl acetate extract of *Rauvolfia serpentina* callus tissue. AuNPs recorded antibacterial activity against gram positive (*Staphylococcus aureus*) and gram-negative (*Escherichia coli*) bacteria	[57]
Chrysanthemum morifolium (Ramat)	The gold (AuNPs) nanoparticles were purchased from Nanoparticles Innovation NPIN s.c. (Poland). AuNPs (20 nm-in-size) showed encouraging antibacterial and antifungal efficacy, even at lower concentrations (10 and 30 ppm) for a brief period of disinfection. In addition, there was no apparent injury to *Chrysanthemum* plant tissue	[58]

3.2 AuNPs as a Disinfectant in Autoclaved Media

In autoclaved medium, AuNPs can also serve as a disinfectant. Many explants, especially those from woody species, develop bacterial and fungal diseases that are internal to the tissue. Surface sterilization is insufficient to eradicate these microorganisms. AuNPs can be added to culture media to either eliminate or significantly slow the growth of microbes, producing "clean" cultures [59]. Natural honey-derived AuNPs showed strong antibacterial efficacy against pathogenic bacteria, including those that were drug-resistant to multiple antibiotics [60]. Sreelakshmi et al. [60] showed that AuNPs made from the Indian plant *Salicornia brachiata* have strong antibacterial action against a number of harmful bacteria, including *Pseudomonas aeruginosa* and *Escherichia coli*. AuNPs produced from *Mentha piperita* were also discovered to be effective against E. *coli* and *Staphylococcus aureus* [61]. Using extracts from plants known to have antibacterial properties to create AuNPs has the potential to create nanoparticles with strong antibacterial properties.

Furthermore, AuNPs added to culture media at the right concentration not only sterilizes but also promote plant growth [62]. Fazal et al. [63] studied the effects of AuNPs (30 μg l^{-1}) on the progress of callus culture and construction of secondary metabolites of *Prunella* vulgaris L. in vitro. They reported that the application of AuNPs to the [64] medium increased callus proliferation, plant biomass, total protein contents, superoxide dismutase and peroxidase enzymes and the contents phenolic and flavonoid in callus cultures as compared to the controls. The recent studies of enhanced growth response from different explant systems using AuNPs in the culture media are shown in Table 2.

3.3 AuNPs as a Media Sterilizing Agent in Non-autoclaved Media

In the cultivation of plantlets employing microponic systems, the addition of NPs to culture media is very beneficial [4]. Superior grade plantlets made from in-vitro cultures can be produced using a technology called microponics that combines micropropagation and hydroponics. The plant propagation rooting stage is the ideal time to use the microponic technique. Bacterial and fungal infections can be minimized with the inclusion of AuNPs as a media sanitization and in combination with liquid nutritive media free a carbon source, like sucrose (Fig. 2). The duration of plant propagation as a whole is drastically shortened, and production costs are also reduced through combining the acclimation phase with rooting initiation [71]. Even if it works, autoclaving is a significant investment for different tissue culture operations [72]. The regular utilization of the autoclaving process is reduced when AuNPs are directly added to media. Treatment with AuNPs can lessen the detrimental impacts of the traditional autoclaving technique in media preparation because a high temperature treatment of the media is not required, which is beneficial for plant growth

Table 2 Selected recent studies of enhanced growth response from different explant systems using AuNPs in the culture media

Plant species	Treatment and outcomes	References
Lamprocapnos spectabilis 'Valentine'	An approach known as chemical reduction was used to create AuNPs from aqueous solution of hydrate gold (III). In vitro plant development and multiplication were positively impacted by the addition of AuNPs (50, 75, or 100 ppm) to the culture medium. The addition of AuNPs into the culture medium, especially at 75 ppm, provokes the proliferation of shoots and multiplication ratio. Gold nanoparticles stimulate the elongation and branching of roots. Gold nanoparticles affect also the profile of metabolites in bleeding heart and improve the survival and quality of acclimatized plants. The amount of polyphenols (6.14–6.96 mg g^{-1} FW) in all shoots derived from AuNP-treated explants was higher than that of the untreated control (5.46 mg g^{-1} FW)	[65]
Arabidopsis thaliana ecotype Columbia 0 (Col-0)	Sterilized seeds were sown on agar plates containing 1/2 MS plant medium with gold nanoparticles (AuNPs), (in range from 0 to 20 mg/L), to a final concentration of 10 mg/L. Exposure of plants to moderate concentrations of AuNPs resulted in enhanced growth of the plants with longer primary roots, more and longer lateral roots, and increased rosette diameter, and reduced oxidative stress responses elicited by the immune-stimulatory PAMP flg22	[62]
Hordeum vulgare L	Barley (*Hordeum vulgare* L.) seeds were cultivated in AuNPs-containing nutrient medium (at concentrations 0, 1, 3, 5, 8, and 10 mg mL^{-1}) for 2 weeks before the seedlings were transferred into Au-NP-free media and grown for another 3 weeks. The exposure of barley seeds with AuNPs up to a concentration of 1 mg mL^{-1} in the nutrient medium showed a significant increase in seed germination, fresh biomass per plant, and leaf and root lengths	[66]

(continued)

Table 2 (continued)

Plant species	Treatment and outcomes	References
Chrysanthemum grandiflorum / Ramat./Kitam. 'Bydgoszczanka' and *Gerbera jamesonii* H. Bol. 'Suri'	Gold (AuNPs) and silver (AgNPs) nanoparticles at the concentrations of 10, 30, and 50 ppm were prepared by the dilution of stock 100 ppm colloid with sterile distilled water. In comparison to AgNPs, AuNPs are less toxic to in vitro isolated *Chrysanthemum* and *Gerbera* plant explants. Adventitious root regeneration in Chrysanthemum was restricted after exposure to AgNPs at concentrations of 10 and 30 ppm. Chrysanthemums exposed to AuNPs (at 10 and 30 ppm) had roots that were significantly larger in diameter (0.73 and 0.74 mm for 10 and 30 ppm, respectively) than AgNPs treatment. In *gerbera* plant, similarly to *chrysanthemum*, silver nanoparticles reduced the number of regenerated roots (3.30 and 3.20 root per one explant at 10 and 30 ppm, respectively). The highest number of roots was recorded on the explants treated with 10 ppm AuNPs (7.50 root per one explant)	[58]
Gloriosa superba L	AuNPs were green synthesized from the fruit extract of *Terminalia arjuna*. The *Gloriosa superba* seeds were treated with two different concentrations of AuNPs (500 and 1,000 μM). AuNPs exposure at 1,000 μ M concentration has the most significant effect on seed germination rate and vegetative growth of *Gloriosa superba* plant	[67]
Brassica napus L	AuNPs obtained from Nanoparticles Innovation NPIN s.c. (Lodz, Poland). AuNPs at concentrations of 0, 50, and 100 ppm were used in vitro. The level of chlorophylls, carotenoids, anthocyanins, free sugars, H_2O_2 increased due to the application of AuNPs. However, the contents of phenolic compounds do not change	[68]
Eruca sativa	Seeds of *Eruca sativa* were germinated on Murashige and Skoog (MS) medium incorporated with gold nanoparticles suspension (30 μg/ml). AuNPs decreased seed germination, shoot and root lengths, seed vigor index, total antioxidant capacity, total phenolic and flavonoid contents, increased DPPH, SOD, and POD activities, no change in protein content	[69]

(continued)

Table 2 (continued)

Plant species	Treatment and outcomes	References
Musa acuminate	In vitro micropropagated *Musa acuminate* (banana) was achieved upon supplementation with green-synthesized biogenic *Caralluma tuberculata* gold nanoparticles (Ca-AuNPs) and poly (ethylene glycol) methacrylate coated Ca AuNPs nanocomposites (PEGMA-AuNPs) in the Murashige and Skoog (MS) media. The maximum fresh, dry weight, and shoot length were achieved upon 80 ppm PEGMA-Ca-AuNPs supplementation. In the physiological analysis, higher chlorophyll content, abundant soluble sugar, and elevated moisture content also appeared upon 80 ppm PEGMA AuNPs addition. Phytohormone analysis showed over-accumulation of IAA content, while on contrary, decreased MDA and H_2O_2 content, higher production of phenolics, flavonoids, phenylalanine ammonia lyase activity, and antioxidant activity was noticeable upon 80 ppm PEGMA-AuNPs supplementation. Moreover, enhanced antioxidant enzyme activities including CAT, SOD, POD, and APx were detected at 80 ppm PEGMA-Ca-AuNPs supplementation in MS media	[70]

and development. It is possible to apply this strategy to liquid culture systems and open culture setups. Plant shoots were raised in the *Chrysanthemum microponic* system using media enriched with various AuNPs levels that were applied right to the boiled culture medium without the use of an autoclave [4]. Tymoszuk and Miler [58] studied the effects of 10 and 30 ppm AuNPs on the adventitious roots regeneration in chrysanthemum and gerbera,and shoots regeneration in Streptocarpus. They reported that at 10 ppm, AuNPs induced rhizogenesis in gerbera, while at 10 and 30 ppm in chrysanthemum, they promoted the development of roots with the largest diameter. The findings of this study also demonstrate that *Streptocarpus hybridus* Voss micropropagation is more effective when AuNPs (10 ppm) are added.

4 Conclusion

The findings of the current chapter point to useful uses for AuNPs as the disinfectant in large-scale micropropagation of plants. AuNPs has the ability to sterilize culture media without the use of a traditional autoclave. The amount of physical space required for plant growth is used to its fullest when larger culture systems are used. The addition of AuNPs to micropropagation techniques can lower overall

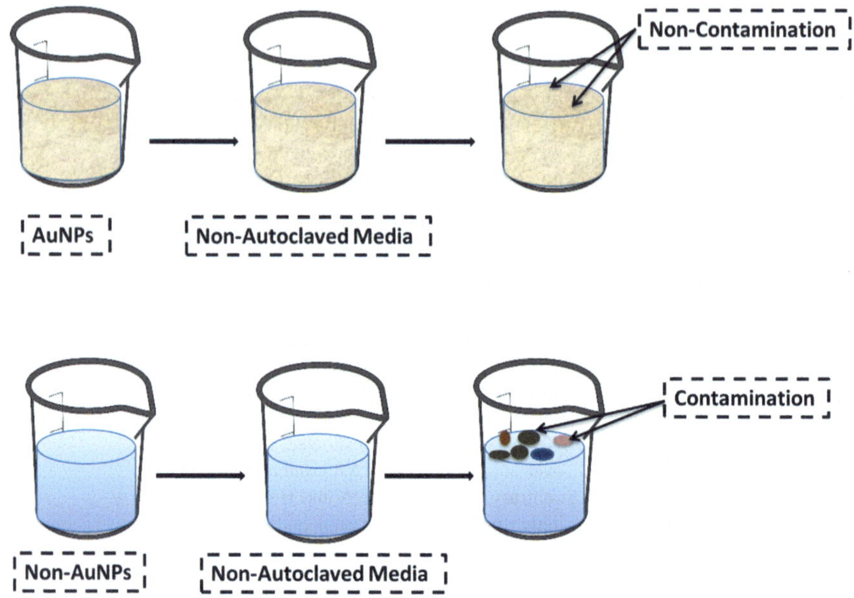

Fig. 2 Gold nanoparticles (AuNPs) as a media disinfectant agent in non-autoclaved media

operating costs while improving plant growth. Finally, the application of nanoparticles in plant science will be further expanded through ongoing study employing various approaches. Plant micropropagation will be successful if the interactions among the applied NPs and endogenous plant components are properly understood theoretically.

References

1. Hala SMS, ELNagar MM, Mohamed MH, Badr LA, Aliwa ME (2023) Micro propagation of Egyptian taro using tissue culture technique. Benha J Appl Sci 8(5):55–65. https://doi.org/10.21608/bjas.2023.202809.1137
2. Leifert C, Cassells AC (2001) Microbial hazards in plant tissue and cell cultures. In Vitro Cellular Dev Biol Plant 37:133–138
3. Saad AIM, Elshahed AM (2012) Plant tissue culture media. In: Recent advances in plant in vitro culture, pp 30–40
4. Eisa EA, Tilly-Mándy A, Honfi P, Shala AY, Gururani MA (2022) Chrysanthemum: a comprehensive review on recent developments on in vitro regeneration. Biology 11(12):1774
5. Pais AK, da Silva AP, de Souza JC, Teixeira SL, Ribeiro JM, Peixoto AR, da Paz CD (2016) Sodium hypochlorite sterilization of culture medium in micropropagation of Gerbera Hybrida Cv. Essandre. African J Biotechnol 15(36):1995–1998
6. Vargas DP, Formoso RS, Dutra LF, Mayer NA, dos Santos J, Ueno B (2016) Chemical sterilization of in vitro culture for peach rootstock. In: Colloquium agrariae, vol 12. Universidade do Oeste Paulista (UNOESTE), pp 1–6

7. Cardoso JC, Sheng Gerald LT, Teixeira da Silva JA (2018) Micropropagation in the twenty-first century. Plant Cell Culture Protocols 17–46
8. Abdelfattah NAH, Yousef, MA, Badawy AA, Salem SS (2023) Influence of biosynthesized magnesium oxide nanoparticles on growth andphysiological aspects of cowpea (*Vigna unguiculata* L.) plant cowpea beetle and cytotoxicity. Biotechnol J. https://doi.org/10.1002/biot.202300301
9. Hashem AH, Abdelaziz AM, Attia MS, Salem SS (2022) Selenium and nano-selenium-mediated biotic stress tolerance in plants. In: Hossain MA, Ahammed GJ, Kolbert Z, El-Ramady H, Islam T, Schiavon M (eds) Selenium and nano-selenium in environmental stress management and crop quality improvement. Springer International Publishing, Cham, pp 209–226. https://doi.org/10.1007/978-3-031-07063-1_11
10. Salem SS, Husen A (2022) Effect of engineered nanomaterials on soil microbiomes and their association with crop growth and production. In: Engineered nanomaterials for sustainable agricultural production, soil improvement and stress management, pp 311–336. https://doi.org/10.1016/B978-0-323-91933-3.00010-6
11. Salem SS, Husen A (2023) Zinc oxide nanoparticles: plant response, interaction, phytotoxicity, and defence. In: Husen A (eds) Nanomaterials and nanocomposites exposures to plants. smart nanomaterials technology. Springer, Singapore. https://doi.org/10.1007/978-981-99-2419-6_12
12. Badawy AA, Abdelfattah NAH, Salem SS, Awad MF, Fouda A (2021) Efficacy assessment of biosynthesized copper oxide nanoparticles(CuO–NPs) on stored grain insects and their impacts on morphological and physiological traits of wheat (*Triticum aestivum L.*) plant. Biology 10(3). https://doi.org/10.3390/biology10030233
13. Husen A (2022) Engineered nanomaterials for sustainable agricultural production, soil improvement and stress management. Elsevier Inc., Cambridge, MA, USA. https://doi.org/10.1016/C2021-0-00054-7
14. Husen A (2023) Nanomaterials and nanocomposites exposures to plants (Response, Interaction, Phytotoxicity and Defense Mechanisms). Springer Nature Singapore Pte Ltd., Singapore. https://doi.org/10.1007/978-981-99-2419-6
15. Rastogi A, Zivcak M, Sytar O, Kalaji HM, He X, Mbarki S, Brestic M (2017) Impact of metal and metal oxide nanoparticles on plant: a critical review. Front Chem 5:78
16. Sanzari I, Leone A, Ambrosone A (2019) Nanotechnology in plant science: to make a long story short. Front Bioeng Biotechnol 7:120. https://doi.org/10.3389/fbioe.2019.00120
17. Salem SS (2023) A mini review on green nanotechnology and its development in biological effects. Arch Microbiol 205:128. https://doi.org/10.1007/s00203-023-03467-2
18. Kim DH, Gopal J, Sivanesan I (2017) Nanomaterials in plant tissue culture: the disclosed and undisclosed. RSC Adv 7(58):36492–36505
19. Ali S, Mehmood A, Khan N (2021) Uptake, translocation, and consequences of nanomaterials on plant growth and stress adaptation. J Nanomater 2021:1–17
20. Badawy AA, Husen A, Salem SS (2024) Use of nanobiotechnology in augmenting soil–plant system interaction for higher plant growth and production. InEssential Minerals in Plant-Soil Systems (pp. 423–443). Elsevier. https://doi.org/10.1016/B978-0-443-16082-0.00006-0
21. Salem SS, Saied E, Shah MP (2024) Advanced (nano) materials. In Development in wastewater treatment research and processes (pp. 93–115). Elsevier. https://doi.org/10.1016/B978-0-323-99278-7.00011-0
22. Attia MS, Salem SS, Elakraa AA, Abdel-Maksoud MA, Malik A, Kiani BH, Malash MN, El-Sayyad GS (2024) Promising antagonistic effect of bimetallic silver-selenium nanoparticles against Ralstonia solanacearum-causing wilt disease in eggplant (Solanum melongena L.). Physiological and Molecular Plant Pathology.1;133:102369. https://doi.org/10.1016/j.pmpp.2024.102369
23. Salem SS, Mekky AE (2024). Biogenic Nanomaterials: Synthesis, Characterization, and Applications. In: Shah, M.P., Bharadvaja, N., Kumar, L. (eds) Biogenic Nanomaterials for Environmental Sustainability: Principles, Practices, and Opportunities. Environmental Science and Engineering. Springer, Cham. https://doi.org/10.1007/978-3-031-45956-6_2

24. Dezfuli AAZ, Abu-Elghait M, Salem SS (2023) Recent insights into nanotechnology in colorectal cancer. Appl Biochem Biotechnol. https://doi.org/10.1007/s12010-023-04696-3
25. Elfadel RG, Refat HM, Abd El-Wahab H, Salem SS, Owda ME, Abdel Reheim MAM (2023) Preparation of new surface coating based onmodified oil-based polymers blended with ZnO and CuZnO NPs for steel protection. Sci Rep 13:7268. https://doi.org/10.1038/s41598-023-34085-z
26. El-Khawaga AM, Elsayed MA, Gobara M, Soliman AA, Hashem AH, Zaher AA, Mohsen M, Salem SS (2023) Green synthesized ZnO nanoparticles by *Saccharomyces cerevisiae* and their antibacterial activity and photocatalytic degradation. Biomass Conv Bioref. https://doi.org/10.1007/s13399-023-04827-0
27. Said A, Abu-Elghait M, Atta HM, Salem SS (2023) Antibacterial activity of green synthesized silver nanoparticles using Lawsonia inermis against common pathogens from urinary tract infection. Appl Biochem Biotechnol. https://doi.org/10.1007/s12010-023-04482-1
28. Salem SS (2022) Baker's yeast-mediated silver nanoparticles: characterisation and antimicrobial biogenic tool for suppressing pathogenic microbes. BioNanoScience 12(4):1220–1229. https://doi.org/10.1007/s12668-022-01026-5
29. Salem SS (2022) Bio-fabrication of selenium nanoparticles using baker's yeast extract and its antimicrobial efficacy on food borne pathogens. Appl Biochem Biotechnol 194(5):1898–1910. https://doi.org/10.1007/s12010-022-03809-8
30. Salem SS (2023) Application of Nano-materials. In: Raja R, Hemaiswarya S, Narayanan M, Kandasamy S, Jayappriyan K (eds) Haematococcus. Springer, Singapore. https://doi.org/10.1007/978-981-99-2901-6_10
31. Salem SS, Fouda A (2021) Green synthesis of metallic nanoparticles and their prospective biotechnological applications: an overview. Biol Trace Elem Res 199(1):344–370. https://doi.org/10.1007/s12011-020-02138-3
32. Salem SS, Hammad EN, Mohamed AA, El-Dougdoug W (2023) A comprehensive review of nanomaterials: types, synthesis, characterization, and applications. Biointerface Res Appl Chem 13(1). https://doi.org/10.33263/BRIAC131.041
33. Soliman MKY, Abu-Elghait M, Salem SS, Azab MS (2022) Multifunctional properties of silver and gold nanoparticles synthesis by *Fusarium pseudonygamai*. Biomass Convers Biorefin. https://doi.org/10.1007/s13399-022-03507-9
34. Soliman MKY, Salem SS, Abu-Elghait M, Azab MS (2023) Biosynthesis of silver and gold nanoparticles and their efficacy towards antibacterial, antibiofilm, cytotoxicity, and antioxidant activities. Appl Biochem Biotechnol 195(2):1158–1183. https://doi.org/10.1007/s12010-022-04199-7
35. Sharaf MH, Nagiub AM, Salem SS, Kalaba MH, El Fakharany EM, Abd El-Wahab H (2022) A new strategy to integrate silver nanowires with waterborne coating to improve their antimicrobial and antiviral properties. Pigm Resin Technol. https://doi.org/10.1108/PRT-12-2021-0146
36. Soliman MK, Hashem AH, Al-Askar AA, AbdElgayed G, Salem SS (2024) Green synthesis of silver nanoparticles from Bauhinia variegata and their biological applications. Green Processing and Synthesis. Sep 12;13(1):20240099. https://doi.org/10.1515/gps-2024-0099
37. Almuhayawi MS, Alruhaili MH, Soliman MK, Tarabulsi MK, Ashy RA, Saddiq AA, Selim S, Alruwaili Y, Salem SS (2024) Investigating the in vitro antibacterial, antibiofilm, antioxidant, anticancer and antiviral activities of zinc oxide nanoparticles biofabricated from Cassia javanica. PloS one. Oct 1;19(10):e0310927. https://doi.org/10.1371/journal.pone.0310927
38. Abd-Elkhalek HF, Badawy AA, Al-Askar AA, Abd Elgawad H, Hashem AH, Salem SS (2024) Biosynthesis and characterization of selenium and silver nanoparticles using Trichoderma viride filtrate and their impact on Culex pipiens. Green Processing and Synthesis. Jun 11;13(1):20240025. https://doi.org/10.1515/gps-2024-0025
39. Aboyewa JA, Sibuyi NRS, Meyer M, Oguntibeju OO (2021) Green synthesis of metallic nanoparticles using some selected medicinal plants from Southern Africa and their biological applications. Plants 10(9):1929

40. Rivero-Montejo SD, Vargas-Hernandez M, Torres-Pacheco I (2021) Nanoparticles as novel elicitors to improve bioactive compounds in plants. Agriculture 11(2):134
41. Wang JW, Grandio EG, Newkirk GM, Demirer GS, Butrus S, Giraldo JP, Landry MP (2019) Nanoparticle-mediated genetic engineering of plants. Mol Plant 12(8):1037–1040
42. Burger DW (1988) Guidelines for autoclaving liquid media used in plant tissue culture. HortScience 23(6):1066–1068
43. Swenson VA, Stacy AD, Gaylor MO, Ushijima B, Philmus B, Cozy LM, Videau NM, Videau P (2018) Assessment and verification of commercially available pressure cookers for laboratory sterilization. PLoS ONE 13(12)
44. Fodor A, Vellai T, Hess C, Makrai L, Dublecz K, Pál L, Molnár A, Klein MG, Tarasco E, Józsa S (2023) XENOFOOD—an autoclaved feed supplement containing autoclavable antimicrobial peptides—exerts anticoccidial GI activity, and causes bursa enlargement, but has no detectable harmful effects in broiler cockerels despite in vitro detectable cytotoxicity on LH. Pathogens 12(3):458
45. Büter B, Pescitelli SM, Berger K, Schmid JE, Stamp P (1993) Autoclaved and filter sterilized liquid media in maize anther culture: significance of activated charcoal. Plant Cell Rep 13:79–82
46. Sathyanarayana BN, Varghese DB (2007) Plant tissue culture: practices and new experimental protocols. IK International Pvt Ltd.
47. Lazo-Javalera MF, Troncoso-Rojas R, Tiznado-Hernández ME, Martínez-Tellez MA, Vargas-Arispuro I, Islas-Osuna MA, Rivera-Domínguez M (2016) Surface disinfection procedure and in vitro regeneration of grapevine (Vitis Vinifera L.) axillary buds. Springerplus 5:1–9
48. Duan Y, Zhang H, Sun M, Zhao F, Xue T, Xue J (2019) Use of chlorine dioxide to sterilize medium for tissue culture of potato. Sci Rep 9(1):10232
49. Cappelletti S, Piacentino D, Fineschi V, Frati P, D'Errico S, Aromatario M (2019) Mercuric chloride poisoning: symptoms, analysis, therapies, and autoptic findings. a review of the literature. Crit Rev Toxicol 49(4):329–341
50. Andersson DI (2003) Persistence of antibiotic resistant bacteria. Curr Opin Microbiol 6(5):452–456
51. Rai A, Prabhune A, Perry CC (2010) Antibiotic mediated synthesis of gold nanoparticles with potent antimicrobial activity and their application in antimicrobial coatings. J Mater Chem 20(32):6789. https://doi.org/10.1039/c0jm00817f
52. Zhang Y, Shareena TP, Dasari HD, Hongtao Y (2015) Antimicrobial activity of gold nanoparticles and ionic gold. J Environ Sci Health C 33(3):286–327
53. Gallon FI, da Silva PR, da Silva DC, de Castro FB, Stefenon VM (2018) Explants Sterilization through metal nanoparticles for in vitro mass propagation of Eugenia Involucrata. Plant Cell Culture Micropropagation 14(2):45–55. ISSN 1808-9909
54. Jadczak P, Kulpa D, Bihun M, Przewodowski W (2019) Positive effect of AgNPs and AuNPs in in vitro cultures of Lavandula Angustifolia Mill. Plant Cell, Tissue Organ Culture (PCTOC) 139:191–197
55. Rakhimol KR, Ashitha A, Thomas S, Kalarikkal N, Jayachandran K (2023) Casein stabilized metal and metal oxide nanoparticle to eradicate the endophytic bacterial contamination from in vitro culturing of Scoparia Dulcis L. Plant cell, tissue and organ culture (PCTOC), pp 1–9. https://doi.org/10.1007/s11240-023-02541-3.
56. Ahmad N, Muhammad J, Khan K, Ali W, Fazal H, Ali M, Rahman LU, Khan H, Uddin MN, Abbasi BH, Hano C (2022) Silver and gold nanoparticles induced differential antimicrobial potential in calli cultures of Prunella Vulgaris. BMC Chem 16(1):1–11. https://doi.org/10.1186/s13065-022-00816-y
57. Kishore RP, Vikram V, Sundaram LR (2020) Evaluation of antioxidant and antibacterial potential and α-amylase expression in the leaf callus tissue of Rauvolfia Serpentina (Linn.) Benth. Ex Kurz. Int J Pharm Pharmacol 4:144. https://doi.org/10.31531/2581-3080.1000144 2 R. Serpentina
58. Tymoszuk A, Miler N (2019) Silver and gold nanoparticles impact on in vitro adventitious organogenesis in Chrysanthemum, Gerbera and Cape Primrose. Sci Hortic 257. https://doi.org/10.1016/j.scienta.2019.108766

59. Bahrulolum H, Nooraei S, Javanshir N, Tarrahimofrad H, Mirbagheri VS, Easton AJ, Ahmadian G (2021) Green synthesis of metal nanoparticles using microorganisms and their application in the agrifood sector. J Nanobiotechnol 19(1):1–26
60. Sreelakshmi CH, Datta KKR, Yadav JS, Reddy BV (2011) Honey derivatized Au and Ag nanoparticles and evaluation of its antimicrobial activity. J Nanosci Nanotechnol 11(8):6995–7000
61. MubarakAli D, Thajuddin N, Jeganathan K, Gunasekaran M (2011) Plant extract mediated synthesis of silver and gold nanoparticles and its antibacterial activity against clinically isolated pathogens. Colloids Surf, B 85(2):360–365
62. Ferrari E, Barbero F, Busquets-Fité M, Franz-Wachtel M, Köhler H-R, Puntes V, Kemmerling B (2021) Growth-promoting gold nanoparticles decrease stress responses in arabidopsis seedlings. Nanomaterials 11(12):3161
63. Fazal H, Abbasi BH, Ahmad N, Ali M (2016) Elicitation of medicinally important antioxidant secondary metabolites with silver and gold nanoparticles in callus cultures of Prunella Vulgaris L. Appl Biochem Biotechnol 180:1076–1092
64. Murashige T, Skoog F (1962) A revised medium for rapid growth and bio assays with tobacco tissue cultures. Physiol Plant 15(3):473–497
65. Kulus D, Tymoszuk A, Jedrzejczyk I, Winiecki J (2022) Gold nanoparticles and electromagnetic irradiation in tissue culture systems of bleeding heart: biochemical, physiological, and (Cyto) genetic effects. Plant Cell, Tissue Organ Culture (PCTOC) 149(3):715–734
66. Feichtmeier NS, Walther P, Leopold K (2015) Uptake, effects, and regeneration of barley plants exposed to gold nanoparticles. Environ Sci Pollut Res 22(11):8549–8558. https://doi.org/10.1007/s11356-014-4015-0
67. Gopinath K, Gowri S, Karthika V, Arumugam A (2014) Green synthesis of gold nanoparticles from fruit extract of terminalia arjuna, for the enhanced seed germination activity of gloriosa superba. J Nanostruct Chem 4(3). https://doi.org/10.1007/s40097-014-0115-0
68. Tomaszewska-Sowa M, Lisiecki K, Pańka D (2022) Response of rapeseed (Brassica Napus L.) to silver and gold nanoparticles as a function of concentration and length of exposure. Agronomy 12(11):2885
69. Zaka M, Abbasi BH, Rahman L-u, Shah A, Zia M (2016) Synthesis and characterisation of metal nanoparticles and their effects on seed germination and seedling growth in commercially important eruca sativa. IET Nanobiotechnol 10(3):134–140. https://doi.org/10.1049/iet-nbt.2015.0039
70. Anwar N, Wahid J, Uddin J, Khan A, Shah M, Shah SA, Subhan F, Khan MA, Ali K, Rauf M, Arif M (2021) Phytosynthesis of Poly (Ethylene Glycol) Methacrylate-Hybridized gold nanoparticles from C. Tuberculata: their structural characterization and potential for in vitro growth in banana. In Vitro Cellular Dev Biol Plant 57(2):248–260. https://doi.org/10.1007/s11627-020-10150-4
71. Hahn EunJoo, Bae JongHyang, Lee YongBeom (2000) Growth and photosynthetic characteristics of chrysanthemum plantlets as affected by PH and EC of the nutrient solution in microponic culture. J Korean Soc Hortic Sci 41(1):12–15
72. Chen C (2016) Cost analysis of plant micropropagation of phalaenopsis. Plant Cell. Tissue Organ Culture (PCTOC) 126:167–175

Ethical Issues Surrounding the Utilization of Gold Nanoparticles in Plant and Soil Systems

Bhupal Bhattacharya, Rittick Mondal, Amit Kumar Mandal, and Debasis Biswas

Abstract As the incorporation of nanotechnology into agricultural practices gathers traction, the application of gold nanoparticles (AuNPs) in plant and soil systems emerges as a frontier with enormous promise. In an attempt to clarify the nuances related to AuNPs' introduction into the delicate balance of plant-soil interactions, this study paper explores the ethical issues surrounding their use in agriculture. The ethical debate centered around three major domains: the environmental impact, human health and safety, and socioeconomic elements. The environmental implications of AuNPs presence in soils encompass an assessment of the potential ecological repercussions on biodiversity, non-target organisms, and the overall ecological well-being. The research also examines the intricate network of human health and safety issues, such as the creation of strong safety measures, consumption of crops treated with nanoparticles, and possible hazards related to exposure to nanoparticles during production process. The socio-economic component also explores questions of affordability, accessibility, and the fair sharing of the advantages and hazards of implementing AuNPs. In spite of the fact that this technology has the potential to be revolutionary, this research supports a reasonable approach that places a strong emphasis on ethical and responsible behaviour.

Keywords Ethical issues · Gold nanoparticles · Environmental impact · Sustainable agriculture · Plant-soil interactions

---

B. Bhattacharya
Department of Law, Raiganj University, Raiganj 733134, West Bengal, India

R. Mondal · A. K. Mandal (✉)
Department of Sericulture, Raiganj University, Raiganj 733134, West Bengal, India
e-mail: amitmandal08@gmail.com

D. Biswas
Department of Economics, Raiganj University, Raiganj 733134, West Bengal, India

A. Husen (ed.), *Plant Response to Gold Nanoparticles*, Smart Nanomaterials Technology, https://doi.org/10.1007/978-981-97-7174-5_18

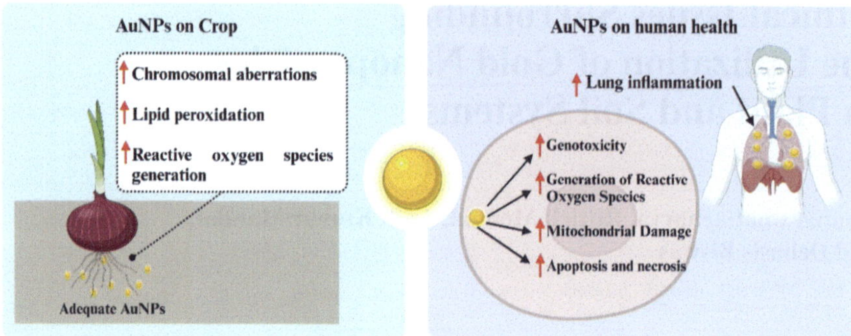

Fig. 1 Effect of gold nanoparticles (AuNPs) on plant and human system

1 Introduction

The use of nanotechnology in agricultural practices has opened up new avenues for innovation, with gold nanoparticles (AuNPs) appearing as a particularly promising tool for improving plant and soil systems [1]. An examination of the ethical implications associated with the utilization of AuNPs is crucial as scientists exploit their distinctive characteristics to tackle agricultural challenges such as insect management and nutrient delivery. This research centers on the ethical issues surrounding the use of AuNPs in soil and plant systems and reveals the many facets of their significance in the fields of socioeconomics, the environment, and human health (Fig. 1).

Recognizing that technology breakthroughs frequently come with ethical responsibilities is crucial in the quest for greater agricultural sustainability and efficiency. The use of AuNPs brings a complex interplay of ethical concerns, necessitating a thorough examination to guide the correct integration of this nano-technological marvel into the agricultural landscape [2].

2 Importance and Advantages of Integrating AuNPS

Modern agriculture could benefit greatly from the incorporation of AuNPs into plant and soil systems in a number of ways [3]. Because of their distinct qualities, AuNPs are effective instruments for tackling many of the major issues the agriculture industry faces [4]. They might potentially optimize nutrient intake and increase agricultural yields by acting as carriers of nutrients, enabling precise and controlled administration to plants [5]. In addition, AuNPs promote a more productive and long-lasting soil environment by affecting microbial activity and nutrient cycling. AuNPs' antimicrobial characteristics open the door to effective pest management and disease resistance, lowering dependency on traditional chemical pesticides and boosting ecologically benign farming practices [6]. AuNPs can also improve the efficiency of plant water usage, which is especially helpful in areas where water scarcity is a problem [7]. Plants treated with AuNPs also show greater resistance to environmental challenges including drought and salinity, which contributes to better crop output [8]. Beyond this, AuNPs help detoxify soil that has been affected by contaminants or heavy metals, which contributes to environmental disposal [5]. Their meticulous application in agriculture allows for the targeted distribution of inputs, optimizing resource utilization and lowering environmental impact. Potential advantages of AuNP integration include improved crop breeding, increased food security, and the development of resilient and environmentally friendly farming systems [9]. These advantages are consistent with the tenets of sustainable agriculture. While these benefits are encouraging, it is critical to approach AuNP integration with caution by addressing ethical concerns and potential environmental repercussions for responsible and sustainable use in agriculture.

3 Assessment of Environmental Impact

The assessment of the ecological effects of AuNPs in soil is a vital component of comprehending the environmental impact of nanotechnology in agriculture. The possible impact of AuNPs on the delicate balance of ecosystems is a concern as they are introduced into soil systems. To identify any negative ecological repercussions, researchers examined existing literature, particularly emphasizing the interactions between AuNPs and non-target animals ranging from microbes to higher trophic levels. These assessments dive into the dynamics of soil microbial communities, looking for potential disruptions or modifications caused by the introduction of AuNPs. As part of the assessment, biodiversity in the soil ecosystem should also be examined with the goal of figuring out whether AuNPs have any impact on the variety and number of creatures that live in the soil. In order to comprehensively grasp the ecological impact, scientists must take into account the intricate network of interconnections among soil organisms, as well as the potential effects on nutrient cycling and decomposition mechanisms of the ecosystem.

When AuNPs are incorporated into soil systems with the intention of enhancing productivity, it is imperative to thoroughly contemplate the potential impacts on biodiversity, non-target organisms, and overall ecological well-being [10]. As these nanoparticles interact with the soil environment, concerns about potential consequences for creatures other than the intended crops develop.

4 Human Health and Safety

The investigation of potential dangers associated with nanoparticle exposure during the manufacturing of AuNP is an important aspect of maintaining occupational safety and environmental responsibility in the nanotechnology industry [11].

As a result of the possible negative consequences of exposure—which include ingestion, skin contact, and inhalation—on human health, researchers and workers may be subjected to AuNPs during the synthesis and production processes [12]. Investigators should look for and measure any possibility of toxicological effects while assessing the dangers of AuNP exposure to the pulmonary and integumentary systems. In order to fully understand the biological interactions of the produced AuNPs, the enquiry should aim to know their physicochemical characteristics, including surface modifications, size, and shape. To further lessen exposure hazards during production, efforts should prioritize to develop stringent safety regulations and control mechanisms. Setting occupational safety guidelines and reducing potential health risks for workers involved in AuNP production processes necessitate a thorough comprehension of the risks associated with nanoparticle exposure.

In light of the possible dangers of AuNP exposure during production and research operations, it is imperative to implement strong safety protocols to protect participants' health. A variety of procedures are included in safety protocols, one of which is the reduction of direct contact and inhalation of nanoparticles through the use of personal protective equipment (PPE) such as lab coats, masks, and gloves. For the purpose of reducing exposure to airborne nanoparticles, suitable ventilation systems and engineering controls are frequently in use. Regular training programmes are needed to make the employees and the participants about the possible risks associated with handling AuNPs as well as how to properly use safety precautions. For the purpose of identifying any early indicators of detrimental health consequences, health surveillance programmes and routine workplace monitoring may be implemented. The evaluation of safety procedures is a continuous process and that should be changed in response to new scientific discoveries and technical developments.

Concerns about possible negative health impacts on consumers emerge as crops treated with AuNPs find their way into the food chain [13]. The absorption capacity of consumable plant parts by AuNPs and the subsequent accumulation in human tissues following ingestion are subjects of investigation by Scientists [14]. Regular extensive testing is carried out to determine any potential toxicological effects, allergic reactions, or other health concerns that may occur from the consumption of AuNP-treated crops. Size, shape, and surface modifications are examples of the physicochemical characteristics of AuNPs that are important in influencing their bioavailability and interactions with the human body.

5 Social and Economic Consequences

In order to guarantee that the advantages of this innovative approach to agriculture are broadly spread and do not worsen the available existing societal knowledge and databases, it is essential to investigate accessibility and affordability issues pertaining to AuNP technology. Concerns about AuNP technology's affordability and suitability for use in a variety of agricultural areas grow as the technology develops [15]. In light of the financial viability for farmers with different resources, researchers and policymakers should look at the economic effects of introducing AuNPs into agricultural practices. The investigation should also include a scalability analysis of production techniques to address the needs of various farming environments. In order to find ways to improve accessibility and cost, entails a thorough examination of the possible financial obstacles that would prevent AuNP technology from being widely used.

6 Distribution of Benefits and Hazards Across Various Social Groups

When deploying AuNPs in plant and soil systems, it is important to assess how benefits and hazards should be distributed across various social groups. It's critical to closely examine how the advantages and possible risks of this technology are allocated among various societal groups as it develops as most of the nations are in the developing stage and not in a position to afford any social costs. The benefits of AuNP integration, such as higher crop yields or better farming techniques, need to be thoroughly examined in order to determine whether or not they are available to and advantageous to different socioeconomic classes. Simultaneously, possible threats, whether environmental, health-related, or economic in nature, must be evaluated for their uneven impact on various groups. It is imperative for researchers and policymakers to investigate approaches that foster equity and inclusiveness, thereby preventing the concentration of benefits associated with AuNP technology among particular factions while subjecting others to an unjust proportion of the corresponding hazards.

The examination of potential socioeconomic inequities resulting from the use of AuNP is critical to understanding the broader implications of nanotechnology in agriculture. It is important to consider carefully how these developments may impact various socioeconomic groups as AuNP technology is incorporated into farming practices. The adoption and accessibility of AuNP technology may give rise to potential disparities, whereby larger-scale or more affluent farms might experience greater advantages compared to smaller or resource-constrained farms. The expenses linked to the application of AuNP technology, such as the procurement of nano-particles and the utilization of specialized equipment, may present obstacles for farmers, hence intensifying pre-existing socio-economic disparities within the agricultural industry. Disparities may also reach the market, impacting the cost and availability of crops treated with AuNP. Understanding and managing these potential socioeconomic differences is important in ensuring that the benefits of AuNP adoption are distributed equally across farmers of various sizes and economic capacities.

7 Ethics and Nanotechnology Frameworks

When it comes to developing and implementing new technologies, especially in the field of nanotechnology, ethical issues are crucial. Transparency, accountability, fairness, and sustainability are common values included in these frameworks. Researchers and policymakers can identify gaps, obstacles, and accomplishments in prior uses of nanotechnology ethics by examining current ethical principles. This technique contributes to the creation of context-specific ethical frameworks that are adapted to the unique difficulties and opportunities given by AuNPs in agriculture. By ensuring that ethical issues are integrated into the very fabric of scientific and technical advancement, this kind of evaluation guides the ethical and responsible integration of AuNP technology for the good of society and the environment.

A critical component of improving the ethical framework controlling the use of AuNPs in agriculture is identifying the issues and difficulties in the current ethical concerns. Researchers and policymakers should examine existing ethical principles to identify instances where present frameworks may fall short or lack specifics regarding AuNP technology. Gaps could appear as imprecise risk assessment criteria, possible neglect of socioeconomic inequality, or inadequate measures to guarantee openness in the study and execution phases. Conciliating conflicting ethical positions, adjusting frameworks to fast-expanding technology settings, and adopting globally applicable standards may all present difficulties. The recognition of these deficiencies and obstacles provides a fundamental basis for the continuous improvement of ethical frameworks, promoting a more all-encompassing and flexible collection of principles.

The experimental design, methods, and findings of these investigations must all be thoroughly examined. In addition to examining the effects of AuNP integration on environmental factors and crop yield, the objective is to comprehend the practical ramifications of this process. Researchers can develop a comprehensive understanding of the usefulness and potential drawbacks of AuNPs by looking at a variety of case studies, which provide insights into the many circumstances in which AuNPs have been used.

8 Conclusion

The present study reaffirms that the potential advantages of technological innovation, such as higher crop yields and enhanced sustainability, must be realized while avoiding unfavorable effects on the environment and the socioeconomic system. The section reaffirms that ethical issues are essential to the viability and effectiveness of AuNP technology in agriculture, not only a supporting role. A responsible innovation that puts the welfare of workers, consumers, and ecosystems first can be facilitated by stakeholders including ethical issues in research, development, and implementation.

The examination of ethical considerations in practical applications of AuNPs transcends the technical perspective. This entails evaluating how the case studies' planning and evaluating phases all took ethical standards into account. Ethical concerns can include things like environmental impact evaluations, consumer and worker safety measures, and initiatives to mitigate any socioeconomic imbalances. An important way to improve ethical frameworks for future implementations is to assess how these issues were incorporated into the practical applications.

Drawing from case studies to provide valuable insights, it is imperative to determine the most ethical ways to apply AuNPs in agriculture. Compiling data on effective methods, moral dilemmas faced, and creative solutions to such dilemmas is part of this process. In order to promote ethical innovation in the agriculture industry, the scientific community should help produce guidelines and suggestions for the ethical integration of AuNPs by recording the best practices learned. These best practises could include open communication, involving the community, and ongoing observation of the effects on the environment and the economy.

Based on the examination of current rules, scholars should prioritize developing suggestions for improving or creating regulatory frameworks that are especially adapted to the special difficulties presented by AuNP technology. This may entail recommending changes to existing legislation to remedy gaps in ethical considerations, risk evaluations, or safety standards. New laws that explicitly control the manufacture, use, and removal of AuNPs in agriculture may also be included in the recommendations. Maintaining a steadfast commitment to ethical considerations, the objective is to guarantee that regulatory frameworks possess the qualities of agility, adaptability, and the capacity to confront the ever-changing domain of nanotechnology in agriculture.

Fostering international cooperation to exchange best practises, harmonize legal frameworks, and jointly address moral issues related to AuNP integration are a few possible recommendations. Fostering uniformity in moral guidelines, enabling information sharing, and guaranteeing that the conscientious application of AuNPs in farming is international undertakings are all dependent on international collaboration.

When advancing scientific knowledge, particularly in the context of AuNP technology, scientists should consider the possible conflicts that could occur. Their research should focus on the possible hazards and advantages of innovation, ethical monitoring in R&D, and the ethical implications of deploying cutting-edge technologies in real-world circumstances. Fostering a deeper awareness of how technology advancement can be in line with ethical responsibility is the goal of striking this careful balance.

In order to guarantee that ethical issues are central to the innovation process, the focus point should identify procedures that may be applied at different phases of research, development, and process. Implementing interdisciplinary collaboration to integrate multiple perspectives, providing forums for ethical conversation within the scientific community, and integrating ethical effect evaluations into research protocols are a few viable strategies. The investigation goes further and includes creating frameworks and rules to prioritize ethical ideals alongside technological advancement for researchers, politicians, and industry stakeholders.

9 Future Directions in Ethics and Research

Anticipating future developments in the field of agro nanotechnology, the Conclusion section delineates ethical guidelines and research directions. It may advocate for continuous interdisciplinary collaboration, encouraging scholars to collaborate across disciplines such as nanotechnology, agriculture, ethics, and social sciences. Taking into account emergent scientific knowledge and evolving ethical standards, future research may emphasize for continuous monitoring and evaluation of the long-term effects of AuNP integration. In order to enable consistent and ethical practises in the many applications of AuNPs in agriculture, it might promote the creation of standardized ethical protocols and guidelines. Researchers, legislators, and industry stakeholders can use the future directions provided in the conclusion as a guide to properly and ethically traverse the rapidly changing field of agricultural nanotechnology.

References

1. Mittal D, Kaur G, Singh P, Yadav K, Ali SA (2020) Nanoparticle-based sustainable agriculture and food science: recent advances and future outlook. Front Nanotechnol 2:579954
2. Xu L, Xu M, Sun X, Feliu N, Feng L, Parak WJ, Liu S (2023) Quantitative comparison of gold nanoparticle delivery via the enhanced permeation and retention (EPR) effect and mesenchymal stem cell (MSC)-based targeting. ACS Nano 17:2039–2052
3. Duhan JS, Kumar R, Kumar N, Kaur P, Nehra K, Duhan S (2017) Nanotechnology: the new perspective in precision agriculture. Biotechnol Rep 15:11–23
4. Baker S, Volova T, Prudnikova SV, Satish S, Prasad N (2017) Nanoagroparticles emerging trends and future prospect in modern agriculture system. Environ Toxicol Pharmacol 53:10–17
5. Das S, Chakraborty J, Chatterjee S, Kumar H (2018) Prospects of biosynthesized nanomaterials for the remediation of organic and inorganic environmental contaminants. Environ Sci Nano 5:2784–2808
6. Kaur S, Gill R (2022) Recent updates in plant disease management. In: Roy S, Mathur P, Chakraborty AP, Saha SP (eds) Plant stress: challenges and management in the new decade. Springer International Publishing, Cham, pp 183–198
7. Zulfiqar F, Navarro M, Ashraf M, Akram NA, Munné-Bosch S (2019) Nanofertilizer use for sustainable agriculture: advantages and limitations. Plant Sci 289:110270
8. Rajput VD, Minkina T, Kumari A, Harish Singh VK, Verma KK et al (2021) Coping with the challenges of abiotic stress in plants: new dimensions in the field application of nanoparticles. Plants 10:1221
9. Singh A, Singh NB, Hussain I, Singh H, Singh SC (2015) Plant-nanoparticle interaction: an approach to improve agricultural practices and plant productivity. Int J Pharm Sci Invent 4:25–40
10. Parray JA, Mir MY, Shameem N (2019) Sustainable agriculture: biotechniques in plant biology. Springer, Berlin
11. De Berardis B, Marchetti M, Risuglia A, Ietto F, Fanizza C, Superti F (2020) Exposure to airborne gold nanoparticles: a review of current toxicological data on the respiratory tract. J Nanopart Res 22:1–41
12. Jamuna BA, Ravishankar RV (2014) Environmental risk, human health, and toxic effects of nanoparticles. Nanomater Environ Prot 523–535

13. Maharramov AM, Hasanova UA, Suleymanova IA, Osmanova GE, Hajiyeva NE (2019) The engineered nanoparticles in food chain: potential toxicity and effects. SN Appl Sci 1:1–25
14. Surendranath A, Mohanan PV (2020) Impact of nanoparticles in balancing the ecosystem. Biointerface Res Appl Chem 11:10461–10481
15. Perumal J, Wang Y, Attia ABE, Dinish US, Olivo M (2021) Towards a point-of-care SERS sensor for biomedical and agri-food analysis applications: a review of recent advancements. Nanoscale 13:553–580

SPRINGER NATURE

GPSR Compliance

The European Union's (EU) General Product Safety Regulation (GPSR) is a set of rules that requires consumer products to be safe and our obligations to ensure this.

If you have any concerns about our products, you can contact us on ProductSafety@springernature.com

In case Publisher is established outside the EU, the EU authorized representative is:

Springer Nature Customer Service Center GmbH
Europaplatz 3
69115 Heidelberg, Germany

The manufacturer's authorised representative in the EU is Springer Nature Customer Service Centre GmbH, Europaplatz 3, 69115 Heidelberg, Germany. If you have any concerns regarding our products, please contact ProductSafety@springernature.com

Printed and bound by CPI Group (UK) Ltd, Croydon, CR0 4YY
01/10/2026
02225858-0013